Closer and Closer
Introducing Real Analysis

Carol Schumacher
Kenyon College

JONES AND BARTLETT PUBLISHERS
Sudbury, Massachusetts
BOSTON   TORONTO   LONDON   SINGAPORE

World Headquarters
Jones and Bartlett Publishers
40 Tall Pine Drive
Sudbury, MA 01776
978-443-5000
info@jbpub.com
www.jbpub.com

Jones and Bartlett Publishers
Canada
6339 Ormindale Way
Mississauga, Ontario L5V 1J2
CANADA

Jones and Bartlett Publishers
International
Barb House, Barb Mews
London W6 7PA
UK

Jones and Bartlett's books and products are available through most bookstores and online booksellers. To contact Jones and Bartlett Publishers directly, call 800-832-0034, fax 978-443-8000, or visit our website, www.jbpub.com.

> Substantial discounts on bulk quantities of Jones and Bartlett's publications are available to corporations, professional associations, and other qualified organizations. For details and specific discount information, contact the special sales department at Jones and Bartlett via the above contact information or send an email to specialsales@jbpub.com.

Copyright © 2008 by Jones and Bartlett Publishers, Inc.

All rights reserved. No part of the material protected by this copyright may be reproduced or utilized in any form, electronic or mechanical, including photocopying, recording, or by any information storage and retrieval system, without written permission from the copyright owner.

Production Credits
Acquisitions Editor: Timothy Anderson
Production Director: Amy Rose
Marketing Manager: Andrea DeFronzo
Editorial Assistant: Laura Pagluica
Production Assistant: Sarah Bayle
Composition: Northeast Compositors, Inc.
Cover Design: Kristin E. Ohlin
Cover Image: Magnifying glass: © Denis Selivanov/ShutterStock, Inc.
Background texture: © Douglas R. Hess/ShutterStock, Inc.
Printing and Binding: Malloy, Inc.
Cover Printing: John Pow Company

Library of Congress Cataloging-in-Publication Data
Schumacher, Carol.
  Closer and closer: introducing real analysis / Carol Schumacher.
    p. cm.
  ISBN-13: 978-0-7637-3593-7
  ISBN-10: 0-7637-3593-0
  1. Functions. 2. Mathematical analysis. I. Title.
  QA331.S38 2007
  515--dc22
                              2006100898

6048

Printed in the United States of America
11 10 09 08 07   10 9 8 7 6 5 4 3 2 1

To Ben, Sarah, and Glynis
who fill my life with love and happiness
and who make math a lot more fun.

About the Title

In a beginning calculus class, we might say that

> *the limit of the function f at the point a has the value L if as x gets* closer and closer *to a, the values f(x) get* closer and closer *to L.*

Notice the use of the phrase "closer and closer." Any informal description of a limiting process is certain to use this or a similar phrase. Analysis is the branch of mathematics that underlies all limiting processes; it is in analysis that we rigorously explain what is meant when we say "closer and closer." This phrase is, thus, the central focus of the book.

Contents

Part I   1

    Preliminary Remarks   3
        What Is Analysis 3
        The Role of Abstraction 5
        A Thought Experiment 6

0   **Basic Building Blocks**   9
    0.1   Sets and Set Notation 9
             Totally Ordered Sets 12
             Collections of Sets: Indexing Sets 12
             Set Operations 14
             Problems 17
    0.2   Functions 18
             Inverse Functions 20
             Images and Inverse Images 21
             Problems 24
    0.3   The Natural Numbers, the Integers, and Their Properties .. 27
             Mathematical Induction 28
    0.4   Sequences 29
             Subsequences 32
             Problems 34

1   **The Real Numbers**   37
    1.1   Constructing the Axioms 37
    1.2   Arithmetic 38
             The "Wish List" 40
             Problems 42
    1.3   Order 43
             Order and Arithmetic 45

		Absolute Values	45
		Back to the "Wish List"	46
		Problems	50
	1.4	The Least Upper Bound Axiom	52
		Problems	60

2 Measuring Distances — 63

- 2.1 Metric Spaces 63
- 2.2 The Euclidean Metric on $\mathbb{R}^n$ 65
  - The Cauchy–Schwarz Inequality 67
  - Problems 68

3 Sets and Limits — 73

- 3.1 Open Sets 73
  - Boundedness in Metric Spaces 76
  - Problems 77
- 3.2 Convergence of Sequences: Thinking Intuitively 81
- 3.3 Convergence of Sequences 82
  - Problems 85
- 3.4 Sequences in $\mathbb{R}$ 87
  - Sequence Convergence and Order 87
  - Sequence Convergence and Arithmetic 89
  - Problems 91
- 3.5 Limit Points 93
  - Problems 95
- 3.6 Closed Sets 96
  - Problems 97
- 3.7 Open Sets, Closed Sets, and the Closure of a Set 98
  - Problems 102

4 Continuity — 105

- 4.1 Thinking Intuitively 105
- 4.2 Limit of a Function at a Point 106
  - Problems 109
- 4.3 Continuous Functions 110
  - Problems 112
- 4.4 Uniform Continuity 115
  - Problems 116

5 Real-Valued Functions — 119

- 5.1 Limits, Continuity, and Order 120
  - Some Useful Special Cases 122

		Problems 124
	5.2	One-Sided Limits 124
		Problems 126
	5.3	Limits, Continuity, and Arithmetic 127
		Problems 128

6 Completeness — **131**
- 6.1 Cauchy Sequences 131
  - Problems 132
- 6.2 Complete Metric Spaces 133
  - Problems 134

7 Compactness — **135**
- 7.1 Compact Sets 136
  - Problems 140
- 7.2 Continuity and Compactness 143
  - Problems 144
- 7.3 Compactness in $\mathbb{R}^n$ 145
  - Problems 149

8 Connectedness — **151**
- 8.1 The Intermediate Value Theorem 151
  - Problems 153
- 8.2 Connected Sets 153
  - Problems 156

9 Differentiation of Functions of One Real Variable — **159**
- 9.1 Regarding Domains 159
- 9.2 The Derivative 161
  - Problems 166
- 9.3 What Does the Derivative Tell Us about the Function? . . . 170
- 9.4 Proving the Mean Value Theorem 171
  - Problems 174
- 9.5 Monotonicity and the Mean Value Theorem 176
  - Problems 179
- 9.6 Inverse Functions 180
  - Problems 182
- 9.7 Polynomial Approximation and Taylor's Theorem 182
  - Problems 186

10 Iteration and the Contraction Mapping Theorem — 189

- 10.1 Iteration and Fixed Points 189
  - Attractors and Repellors 195
  - Problems 196
- 10.2 The Contraction Mapping Theorem 198
  - Why You Should Care About Fixed Points 201
  - Problems 202
- 10.3 More on Finding Attracting Fixed Points 205
  - Problems 207

11 The Riemann Integral — 209

- 11.1 What Is Area? 209
- 11.2 The Riemann Integral 210
  - Problems 216
- 11.3 Arithmetic, Order, and the Integral 217
  - Problems 218
- 11.4 Families of Riemann Sums 219
  - The Riemann "Envelope": Upper and Lower Sums 220
  - Refinements 221
  - Cauchy Criteria for the Existence of the Integral 222
  - Problems 227
- 11.5 Existence of the Integral 228
  - Problems 231
- 11.6 The Fundamental Theorem of Calculus 234
  - Problems 235

12 Sequences of Functions — 237

- 12.1 Pointwise Convergence 237
- 12.2 Uniform Convergence 240
  - Problems 243
- 12.3 Series of Functions 246
  - Problems 248
- 12.4 Interchange of Limit Operations 248
  - Problems 254

13 Differentiating $f : \mathbb{R}^n \to \mathbb{R}^m$ — 257

- 13.1 What Are We Studying? 257
  - Problems 259
- 13.2 Thinking Intuitively 259
  - Tangent Planes 260

	13.3	Analysis in Linear Spaces 262
		Linear Transformations 265
		Linear Algebra and Analysis 267
		Problems 270
	13.4	Local Linear Approximation for Functions of Several Variables 272
		Connections—Total and Partial Derivatives 274
		Problems 282
	13.5	The Mean Value Theorem for Functions of Several Variables 287
		Problems 290

Part II   291

A Truth and Provability   293

B Number Properties   295

C Exponents   298

	C.1	Integer and Rational Powers 298
		Positive Integer Powers 298
	C.2	Irrational Powers 303

D Sequences in $\mathbb{R}$ and $\mathbb{R}^n$ 307

	D.1	Sequence Convergence in $\mathbb{R}$ and $\mathbb{R}^n$ 307
	D.2	Epsilonics: Playing the Game 311
		Voodoo Mathematics? 311
		Scratch Work—Devising a Strategy 312
		Problems 314
	D.3	Infinite Limits 315
	D.4	Some Important Special Sequences 316

E Limits of Functions from $\mathbb{R}$ to $\mathbb{R}$ 319

	E.1	Example Proofs 319
	E.2	Epsilonics: Some General Principles 321
		Problems 322

F Doubly Indexed Sequences   325

	F.1	Double Sequences and Convergence 326

G Subsequences and Convergence   330

	G.1	Subsequential Limits 330
	G.2	Limits Superior and Inferior 331

H Series of Real Numbers — 334
- H.1 Definition and Basic Properties 334
  - Geometric Series 335
  - Cauchy Criterion for Series Convergence 336
  - N^{th} Term Test 337
  - Absolute vs. Conditional Convergence 338
  - Problems 340
- H.2 Comparing Series 342
  - Problems 343
- H.3 Relatives of the Geometric Series 344
  - Comparing the Root and the Ratio Tests 346
  - Problems 347
- H.4 Rearranging the Terms of a Series 348
  - Problems 353
- H.5 Multiplying Series 354

I Probing the Definition of the Riemann Integral — 358
- I.1 Regular Riemann Sums 358
- I.2 Why the Generality? 360
  - Problems 361

J Power Series — 363
- J.1 Definitions and Convergence of Power Series 364
  - Problems 367
- J.2 Integration and Differentiation of Power Series 368
  - Problems 369
- J.3 Taylor Series 370
  - Problems 372

K Everywhere Continuous, Nowhere Differentiable — 374
- K.1 Introduction 374
- K.2 Constructing the Function 375

L Newton's Method — 380
- L.1 Setting the Stage 380
- L.2 Iterating the Newton Function 383
  - Problems 384
- L.3 Experimenting with Newton's Method 385
- L.4 On Choosing x_0 386
- L.5 Convergence Rate 387
  - Problems 389

M	**The Implicit Function Theorem**		**390**
	M.1	Solving Systems of Equations	390
	M.2	The Implicit Function Theorem	393
		What on Earth?!?	395
		Properties of the Solution Function	398
		Problems	399
	M.3	Connections: Quasi-Newton's Methods	402
	M.4	The Inverse Function Theorem	406
		Problems	407
N	**Spaces of Continuous Functions**		**408**
	N.1	The Metric Space $C(K)$	410
	N.2	Compactness in $C(K)$	412
	N.3	The Stone–Weierstrass Theorem	414
		Problems	418
O	**Solutions to Differential Equations**		**421**
	O.1	Definitions and Motivation	421
		Why Existence? Why Uniqueness?	423
	O.2	Picard Iteration Route to Existence and Uniqueness	424
		Problems	426
	O.3	Systems of Equations	427
		Problems	430

Bibliography — 431

Index — 433

Preface

Closer and Closer is a first introduction to real analysis for upper-level undergraduate mathematics majors. The book assumes that the students who use it have already had a fairly thorough calculus course, as well as some introduction to abstract mathematics and proof. It emphasizes the real numbers, n-dimensional Euclidean spaces, real-valued functions, and the close cousins of all of these; but the material is discussed in the context of general metric spaces. The book fosters active learning, drawing the student into the development of the mathematics with a conversational writing style, punctuated by exercises that help to understand the ideas. Furthermore, the student is a key player in the development of the subject, since many of the main results are left for the reader to prove.

Structure of the Book
Closer and Closer is divided into two parts: the "Core" and the "Excursions." I decided to structure the book in this way because of something I observed in other books. Analysis books seem to come in two basic types: there are thin books that take the shortest path through the theory, giving the reader a very good sense for how the essential ideas fit together; unfortunately, these fail to show where analysis leads or what it is for. This leaves many students wondering why anyone would care about it. On the other hand, there are large books that intertwine their development of central themes with discussions of special cases and applications. These books fill out the subject, but no clear distinction is made between the "big ideas," the theoretical tools of analysis, and their uses. A "middle-sized" approach is worse; such a compromise tends to have the advantages of neither and the disadvantages of both of the other types of books. I believe that the physical separation of the Core from the Excursions embodies the strengths of both models.

The Core Chapters begin with the most important tools used in analysis: the axioms of the real number system, distance, open and closed sets, sequential limits, continuity, completeness, compactness, and connectedness. The Core culminates with discussions of differentiation and integration of one variable real functions, iteration, sequences and series of functions, and differentiation of functions of several variables. These topics are arranged more or less in a linear sequence, with most chapters depending heavily on previous chapters.

The Excursions should not be thought of as "Part II" of the book, nor should they be considered less important than the chapters of the Core. As the title suggests, the Excursions are ways to depart from the central linear track of the Core. The Excursions complement the Core chapters in different ways. Here are some examples:

- Commentary:
  - Truth and Provability
  - Probing the Definition of the Riemann Integral

- Important applications:
  - Series of Real Numbers
  - Power Series
  - Newton's Method
  - The Implicit Function Theorem
  - Picard iteration and solutions to differential equations

- Special cases and examples:
  - Sequence convergence in $\mathbb{R}$ and $\mathbb{R}^n$
  - Limits of Functions from $\mathbb{R}$ to $\mathbb{R}$.
  - Everywhere Continuous, Nowhere Differentiable Functions

- Useful Side Trips:
  - Number Properties
  - Exponentiation of Real Numbers
  - Doubly Indexed Sequences

- Deepening the theory:
  - Subsequences and Convergence
  - $C(K)$ Spaces.

As the Excursions have various goals, they are written in different styles. Some are written in much the same style as the Core chapters—discussion followed by problems. Others are explorations for the student, sometimes no more than a long series of interconnected exercises. As often as possible, I leave the important results for the students to prove. But I sometimes provide a proof outline or sketch, leaving the details to the reader. Some excursions are just readings, discussion that helps shed light on some part of the theory discussed in the Core.

Flexibility of Use

The two-part structure of *Closer and Closer* makes it very flexible for classroom use. Each chapter of the Core is heavily dependent on previous chapters, but none of the Core chapters depend on material developed in the Excursions. Some Excursions depend on material in earlier Excursions, but most require only information from Core chapters. Moreover, most Excursions depend only on some parts of the Core, so there is a choice about when to cover them. Thus instructors that use the book have a great deal of flexibility about what to include in their syllabus. Each Excursion begins with a list of 'pre-requisite' information. Many core sections also end by suggesting Excursions for further study.

The book can be used in either a one-semester or a two-semester course. Excursions that are not included in the course syllabus can be used to enrich the experience of exceptionally talented or ambitious students. Some work well as student projects.

Why a Metric Space Approach?

There are several benefits, but mathematical clarity is the most important. Mathematical reasoning is a tool for understanding connections. Chains of reasoning tell us which mathematical ideas are linked and in what ways. I believe that one goal of any introduction to a new mathematical field should be to clearly communicate the fundamental relationships that underlie it. Such clarity requires that unrelated issues be addressed separately so the fact that they are unrelated is made clear to the students.

Most recent introductory Analysis books concentrate on the real line. But the real line is, paradoxically, too rich in mathematical structure. Arithmetic and order are not really analytical ideas. Least upper bounds and greatest lower bounds are closely connected with the analysis of the real line, but their link to the order structure can be a misleading distraction. Distance is the fundamental analytical idea and concentrating on the study of distance frees the students from issues that are irrelevant or peculiar to the real line. Theorems relating limits and arithmetic or limits and order are addressed in the book. But they are addressed in such a way that it is clear when they apply and when they don't.

Other Benefits of a Metric Space Approach:
- *Beyond calculus*: In calculus, students gain an intuitive understanding of limits and continuity, differentiation and integration of real functions. An approach that stays focused on the real line may appear to do no more than "cross t's and dot i's." If the only theorems that are proved in Real

Analysis are those the students became familiar with in calculus and the truth of which they did not doubt when they started the course, it is difficult to blame a student who wonders just what was achieved.

- *Important Tools*: Completeness, connectedness, and compactness are, from my point of view, essential analytical ideas. But it is difficult, or at least unnatural, to fully address these issues in the context of the real line. Once again, it is a matter of separating out unrelated issues. Completeness in the real numbers is intimately tied to the least upper bound property. The only connected subsets of $\mathbb{R}$ are intervals. "Compact" is equivalent to "closed and bounded," even in n-dimensional Euclidean space. Thus it is nearly impossible to fully do justice to "The three C's" except in the context of general metric spaces.

- *Beyond Real Analysis*: The metric space approach provides a better vantage point from which to move on to deeper analytical study. Real Analysis long ago outgrew its original intent, which was to make rigorous the ideas of the calculus (though this is still one of its jobs). Limiting processes are important in everything from atmospheric modeling to computer graphics to statistical theory. The general metric space perspective makes it easy to build on students' earlier intuition. At the same time, it broadens their perspective and deepens their understanding of analytical ideas.

Active Learning

Analysis is not just a body of mathematical knowledge; it is the style of thinking that is required to understand and apply that knowledge. Learning to think like an analyst is thus an essential part of any beginning analysis course. Moreover, I would argue that students cannot understand (or even fully recognize!) the subtleties of real analysis without struggling with and sorting out many details for themselves. Active learning is not just a trendy pedagogical phrase. It is an essential part of any successful attempt to introduce students to analysis as a discipline.

An active learning approach for the students does not mean that the teacher can take a passive role, as some mistakenly think. A teacher that supports the students' exploration by helping them make connections and see important relationships is vital to the success of active learning. My goal in writing *Closer and Closer* was to create a textbook that assists the teacher in this task.

How Does Closer and Closer Do This?

- *Student involvement*: The best textbook for an active learning approach is one in which the students don't so much read the book as interact with it. Thus, as far as rigor allows, I have adopted a conversational writing tone that is meant to draw the students into a dialogue with the book.

I often ask the reader to stop and work out a simple exercise or verify some fact about an example. Furthermore, I believe that students need to prove many of the important theorems themselves. If students are to engage the subject in a serious way, they must know that their work is central to its development. In particular, it must be clear that their work does not begin only after everything important has been done for them! Thus I leave the proofs of crucial results as problems. I do this frequently throughout the book, and I provide fewer proofs as the students gain in experience.

- *Support for Student Work*: My student readers are given a great deal of responsibility for their own learning, but they are also given support in their work. This support comes in the form of motivation for the ideas, help with general things like proof techniques, and help in making connections between intuitive ideas and mathematical rigor.

- *Motivation*: Students have a good intuitive understanding of limiting processes from their calculus courses. One of my goals is for them to understand how the ideas discussed in the text make these informal notions precise and generalize these informal notions. Thus I usually introduce new ideas and definitions with an informal discussion that appeals to my reader's intuition and previous knowledge.

- *Help with proof techniques*: For instance, I have included two in-depth discussions of the "mysteries" of "epsilonics" in early excursions. These help to understand both the strategy for devising a proof and the proper form for writing it down—and the connection between the two.

- *Bridges between intuition and abstraction*: I include many easy exercises and examples that help to connect the abstract concepts being discussed to the student's own understanding of more concrete ideas. Several of the Excursions are also written with this in mind. In addition, there are problems that introduce the readers to an unfamiliar setting and then ask them to prove something. These problems help the students understand the nuances of the abstract ideas that are being considered.

As a teacher, I make sure that my students are constantly thinking about the ideas in the context of the real line, of the plane, and so forth. This helps to keep the ideas grounded in the contexts that they understand best. The book does this, too. Many problems ask students to provide examples or counterexamples to illustrate important aspects of the analytical principles they are studying. Students are sometimes asked to draw a detailed diagram to illustrate the idea of a proof.

Other Features of the Book

- *Boxed Asides*: Tips on proof techniques, comments about the significance of theorems, remarks about notational conventions, and more are provided in boxes throughout the text.

- *Modern Insights*: Though iteration and fixed-point theorems have played a theoretical role in analysis for a long time, the insights of the last couple of decades (and the easy access to computer power) make iteration more relevant than ever. It is, indeed, an increasingly central tool for analysts. Thus I have included a chapter on iteration in the Core and it plays a central role in several of the excursions.

- *Commentary on the Problems*: Some problems are accompanied by explanatory notes that explain the context of the problem or invite the student to reflect on the implications of the result. This helps students to appreciate the broader significance of the ideas in the problem.

Class Testing

I started using early versions of this book in my real analysis classes here at Kenyon College about a decade ago. I have taught a two-semester sequence in real analysis from the manuscript for about half that long. I teach the course as an "inquiry-based learning" seminar (sometimes called modified Moore method). The students are given reading assignments and problems to work on. A typical class will be a mixture of student presentations and discussion. I rarely lecture.

In addition to my own class testing, the manuscript has been used successfully by various people at different sorts of institutions. I have received some extremely valuable feedback from these colleagues, and the book is considerably better thanks to their excellent advice. Earlier versions of the book were used here at Kenyon College by Professors Noah Aydin and Lewis Ludwig. Professor Ludwig is now a faculty member at Denison University and has continued to use the book there. Professor Matthias Kawski of Arizona State University used the book in one course populated mostly by math majors and in another populated mostly by prospective teachers. The book was also used in an independent study by Professor Robb Sinn at North Georgia College and State University, and later adopted for use in a class by his colleague, Professor John Cruthirds.

Ancillaries

Because *Closer and Closer* has some unusual features, instructors may be interested in a more detailed description of what I do when I use it in my classroom. The *Instructor's Resource Manual* will discuss my goals for the course, how I construct my syllabus, what I do in the classroom, and how I assess my students' work. It will also contain more detailed information about connections in the book. (For instance, I will try to give more detailed information about how, and to what extent, later sections depend on earlier sections. Furthermore, some problems from earlier sections are referred to in later sections. I will also detail these connections.) In addition, a *Solution's Manual* will be made available only to instructors, by means of a password protected website. These ancillaries will be available from Jones and Bartlett Publishers.

Acknowledgments
Though only one name goes on the cover, a book is the work of many hands and minds. My sincere thanks to the wonderful team who took my rough manuscript and helped me whip it into shape. I am especially grateful for the sharp eyes and excellent word of copyeditor Jill Hobbs and proofreader Jenny Bagdigian. My diagrams now look much better thanks to artist George Nichols. Fortunately, my "camera ready copy" didn't go directly to the camera. It is much sleeker and more elegant thanks to the efforts of compositors Mike and Sigrid Wile at Northeast Compositors. The finishing touch is a fabulous cover designed by Kristin Ohlin. And, of course, none of this could have come about without the hard work of the team at Jones and Bartlett Publishers: my most sincere thanks to senior acquisitions editor Tim Anderson and his editorial assistant Laura Pagluica, to production director Amy Rose and her production assistants Mike Boblitt and Sarah Bayle, and to marketing manager Andrea DeFronzo. They have been helpful and cheerful and patient throughout this process, and I am truly grateful.

All this came at the end of a very long process, of course. I wish to thank reviewers George Androulakis (University of South Carolina), Jiu Ding (University of Southern Mississippi), Ho Kuen Ng (San Jose State University), and Stephen Schecter (North Carolina State University), whose comments on the manuscript helped me to improve many sections and to sharpen the overall focus of the book. I am also grateful for the comments and suggestions from colleagues who class tested the materials. Significant improvements in the book are due to the detailed comments and suggestions of Noah Aydin (Kenyon College) and Matthias Kawski (Arizona State University). I would like also to thank my wonderful colleagues here at Kenyon College, most especially in this context Judy Holdener and Steve Slack for much support and many good comments over the years. I have learned so much from them.

Most of all, however, I would like to thank my students, who have patiently and graciously put up with my typographical (and worse) errors over the years. They have helped me to figure out when things didn't work well by running into trouble and then kindly forgiving me for putting them through the ordeal. Their encouraging comments have buoyed me up when the writing got tough. The book got done because of them and for them. If I could, I would thank, by name, every student who worked with a rough preliminary version of the book. I would like to acknowledge especially Tsvetan Asamov, Adam Atwell, Jim Bell, Robin Blume-Kohout, Christine Breiner, Matt Buckley, Teena Carroll, Dave

Carroll, Matt Chesnes, Laura Czarnecki, Marian Frazier, Mike Furr, Kjersten Hild, Jesse Horowitz, Irina Ivan, Llewellyn Jones, Brian Karrer, Lee Kennard, Steve Klise, Joe Kloc, Mary Kloc, Adam Knapp, Jeff Lanz, Max Lavrentovich, Mike Lewandowski, Jun Ma, Joel McCance, Agnese Melbarde, Gary Mitchell, Joey Neilsen, Chad Rothschild, Will Stanton, Heather Van Ligten, Atul Varma, and Matt Zaremsky for all the many suggestions they've made and for the typos they've helped to correct. The book is infinitely better because of them all.

Last of all, I would like to thank my family: my parents Jayne and LeGrand Smith who filled me with a love of learning and of truth; and my husband Ben and my daughters Sarah and Glynis who feed my love of mathematics with their own.

Carol S. Schumacher
SchumacherC@kenyon.edu
Gambier, Ohio

A Note to the Student—What This Book Expects from You.

As a mathematician, I know that learning mathematics requires an actively engaged mind. As an author, my job is to help you engage yours. My writing style is deliberately conversational because that is what I am aiming for: a conversation with you. And a conversation requires two participants. Throughout the text you will find "Exercises." These are your contributions to the reading. They are designed to help deepen your understanding of the ideas being discussed. When you read, read with paper and pencil in hand. When you come to an exercise, stop right then and do what it tells you to do. Joel, who is finishing his second semester studying out of a pre-print of *Closer and Closer*, passes on "the best advice I can give a reader of your books."

> It's easy to read the questions and exercises without really thinking about them. Don't do this! Those questions aren't just there for rhetorical effect; you'll come away with a much better understanding of the material if you think them through carefully right when you read them. It may take more time then, but it will make the later work go much more smoothly.

Irina, who took Real Analysis with me last year, also urges you to engage the ideas actively.

> Taking Real Analysis is very much like real life: it can get tough sometimes but you keep going and keep flipping the pages. For Real Analysis though, it is also essential to turn the pages the other direction, to flip to and pay constant attention to the previous chapters, lemmas, and theorems just as much as to the current ones. This 'looking in the past' approach will shed much needed light on a problem that is difficult to understand in the present.

This book is your guide on a journey. But it is you who will do the actual traveling. The journey will take you to some important mathematical destinations. Jim (a third student) advises you to have fun along the way. He says, "The writing is easy to read, and fun to read, as well. It is 'human' and not a heap of information as some textbooks are. [My friend] Elliot and I always compared the book to a children's puzzle book for big kids." So be sure to take in the scenery on your journey. Enjoy the side trips. Ask yourself about the significance of the things you are seeing. Joel, Irina, and Jim not only arrived at their destination—a mastery of the fundamentals of Real Analysis—they came to understand what learning mathematics is all about. If you, likewise, become a savvy and self-reliant mathematical traveler, this book will have done its job. And so will you.

Carol S. Schumacher
SchumacherC@kenyon.edu
Gambier, Ohio

PART I
Central Ideas

Preliminary Remarks

What Is Analysis?

This book is a first introduction to real analysis. "Real" refers to the real numbers, denoted by $\mathbb{R}$. Much of our work will revolve around the real number system. This does not mean that the only objects we consider will be real numbers, nor that all the sets we study will be sets of real numbers. Instead, this statement should be interpreted in a broader sense. For instance, many of the functions that we study will take on real values—that is, each object in our domain will be associated in some way with a real number. For instance, if we have a herd of milk cows, we might associate with each cow the average number of liters of milk it produces each day. To give a more mathematical example, we might talk very generally about measuring distances. We can talk about measuring the distance between all sorts of mathematical objects: between two points in the plane, between two spheres in three dimensions, or between two functions. However, the distance itself will always be a real number.

"Analysis" is one of the principal branches of mathematics, though saying this certainly does not answer the question "what is analysis?" In fact, this question does not have a simple answer. One often hears that analysis is the theoretical underpinnings of the calculus, but it is an answer that misleads by oversimplification. Certainly analysis had its inception in the attempt to give a careful, mathematically sound explanation of the ideas of the calculus.[1] Over the last century, analysis has grown out of its original packaging and is

---

1. *The birth of analysis*: Scientists and mathematicians were enthusiastically using calculus to solve problems for centuries before anyone thought there was any need to examine the details very carefully. But in the early nineteenth century some troubling theoretical difficulties arose, prompting a century-long debate about how and why it all worked!

now much more than simply the theory of the calculus. It is useful to think of analysis as the "mathematics of closeness."

Both derivatives and integrals are the end results of taking a limit. Continuity can be described in terms of limits. There are other examples, but these should make it clear that a thorough mathematical understanding of limiting processes necessarily lies at the heart of analytical thinking. This was true historically, and it is still true today. In a beginning calculus class, we might say that

> The limit of the function f at the point a has the value L if as x gets *closer and closer* to a, the values $f(x)$ get *closer and closer* to the value L.

Notice the use of the phrase "closer and closer." An informal description of any limiting process is almost certain to use this phrase or something similar. The main challenge facing the earliest analysts was to find a way to replace this phrase with a mathematically well-defined (but equivalent) formulation. Essentially, they had to answer the question, "What is closeness?"

Historically, due to the difficulties of manual calculation, mathematicians expended a great deal of effort finding techniques that would allow them to approximate things like square roots, values of the trigonometric functions, and so forth, by using only addition, subtraction, multiplication, and division. But then mathematicians began to ask how good these approximations were: "How close are we?" For iterative processes, they asked, "How fast does this converge? How many iterations do we need to be sure we are 'close enough'? How can we be sure it converges at all?" These approximating techniques remain important today partially because that's how our calculators or computers get such values, but the wide availability of machines make their usefulness seem remote to us. In recent decades, however, we have come to the stark realization that many natural processes that surround us and affect our daily lives are governed by equations that cannot (even in principle) be solved exactly. Such nonlinear processes describe many complex systems—weather patterns, population questions, economic fluctuations, the behavior of one fluid flowing through another. Aside from qualitative techniques, iteration and approximation are our only options. Questions of closeness and convergence remain at the center of attention

This does not tell the whole story of analysis. Many of my fellow analysts will surely say, "She threw out one oversimplification and replaced it with another." Or "How can she fail to talk about ... ?!" I would probably agree with most of their objections. However, my goal here is to help you to see that analysis goes well beyond proving *beyond the shadow of a doubt* that the derivative of $f(x) = x^2$ is $f'(x) = 2x$. It is even bigger than establishing mathematical conditions under which the fundamental theorem of calculus is guaranteed to work the way we think it should (although analysis *will* show us these things). The methodologies and ideas of analysis remain important tools for telling us things that we don't already know.

The Role of Abstraction

In Chapter 2, we will study a structure called a metric space. *Metric space* is the technical phrase we use to describe any set in which we can measure distances. For instance, we can measure the distance between any two points in the Cartesian plane; thus the plane $\mathbb{R}^2$ is a metric space. A metric space is one instance of an "abstraction." An abstraction is a mathematical structure that arises from the observation that several important mathematical examples, though differing in other ways, share some common property or properties (such as our ability to measure distance in all of them). Focusing on this commonality moves us away from the concrete examples that motivated us and focuses us on the desired properties in the abstract.

Abstraction gives us the "less is more" approach—fewer assumptions leave us in touch with more mathematical objects (and thus the theorems we prove apply in more instances). More importantly, however, it gives us the "less is clearer" approach. Because we haven't made a myriad of assumptions, our mathematical choices are more clearly defined, our mathematical arguments are cleaner, and the conclusions we draw are less likely to be cluttered up by irrelevant issues. We can, therefore, sort out the essential mathematical ingredients that support a particular fact. For instance, we will be able to see that certain properties of the real numbers depend only on our ability to measure distance there and not at all on the fact that $\mathbb{R}$ is, say, an ordered set. Moreover, the same observation shows us that these properties are common to the real line and the real plane.

As we look at a set of ideas in the abstract, we are likely to find a variety of examples that have the desired properties. Some of them may be quite unlike the natural "specimens" that inspired our definitions. Some might even seem a bit bizarre, at first. But considering an assortment of examples is important. Examples help us to understand the nuances of new definitions and unfamiliar theorems. The most "pathological" examples help us test the limitations of our theory. (How bad can things get and still be this good?) Surprisingly, some examples that seem strange at first turn out to be very useful—even natural—in a larger mathematical context. In these cases, having already proved a number of theorems about them, we can begin an in-depth consideration of these spaces from a position of strength. At the same time, it is good to remain cautious. Paying too much attention to the "mathematical zoo" can distract us from our central purpose—namely, to better understand the mathematics of limiting processes.

Thus, as we consider various notions in the abstract, it is important to keep ourselves grounded by frequently shifting our focus from the abstract to the concrete. The real numbers, real functions, and their close cousins are always at the center of attention. We are interested in what the theory tells us about limiting processes in these familiar objects. For their part, the well-known characters give us insight into the abstract ideas we are studying. We learn more about concrete situations by thinking abstractly:

- What does this theorem say about sequences in $\mathbb{R}^2$?

- What does that theorem tell me about continuity in real-valued functions?

- Under what circumstances can I closely approximate a given function by using a polynomial? What might be a reasonable measure for a "good" approximation?

We also develop our intuition about the abstractions by thinking about concrete examples:

- Can I picture how this definition works in the three-dimensional space $\mathbb{R}^3$?

- How would I prove this theorem for a function from $\mathbb{R}$ to $\mathbb{R}$?

- What is this proof saying? Can I draw a helpful picture in the plane?

Abstract thinking is very powerful. It is the tool of choice for establishing long chains of reasoning that link mathematical ideas together. It is a route to clarity, in that it distinguishes between essential mathematical connections and those that are incidental or contextual. Nevertheless, the *purpose* of abstraction is to tell us things about the mathematical objects and ideas that we care about. Thus our work will be an interplay between abstract thinking and a detailed consideration of specific mathematical objects.

A Thought Experiment

Over the course of this book, we will consider questions about the real numbers, about limiting processes, about the nature of real-valued functions. You do not come to these topics completely unprepared. Indeed, you should have a fair store of ideas and intuition built up about them. You are beginning this book because most of your knowledge is incomplete and, from a mathematical standpoint, fairly imprecise. In addition, you probably have some misconceptions. As you proceed, you will be integrating your previous knowledge with the ideas discussed in the book and updating it to make it more precise, more correct. But you should constantly have that previous knowledge on the table.

Following is a list of statements like those we will consider in the book. Using your previous knowledge and intuition about these things, try to determine whether each statement is true or false. (In some cases, this will be easy;

in other cases, it will be harder for you. And sometimes you may not be able to decide.)

For each statement, if you believe it is false, give a counterexample. If you believe it is true, give a proof or at least make the best argument you can in support of your idea.

> All the notation and language used in the questions will be carefully defined later on. But it is likely that you are already somewhat familiar with most of the ideas. If you don't recognize some bit of notation, ask your teacher or a fellow student. If you still can't make anything of it, ignore that statement and move on to others. The key idea here is for you to begin to think about your own experience with and intuition about the sorts of questions we will consider carefully as we proceed through the book.

1. If x and y are real numbers such that $x \cdot y = 0$, then $x = 0$ or $y = 0$.

2. Let (x,y) and (v,w) be points in the plane $\mathbb{R}^2$. Then the distance from (x,y) to (v,w) is the same as the distance from (v,w) to (x,y).

3. Given any open interval (a,b) in $\mathbb{R}$ there is a closed interval $[c,d]$ such that $[c,d] \subseteq (a,b)$.

4. Let $f : \mathbb{R} \to \mathbb{R}$ be a function. If K is a bounded subset of $\mathbb{R}$, then $f(K)$ is also bounded.

5. Let $f : \mathbb{R} \to \mathbb{R}$ be a one-to-one function. Then f is either strictly increasing or strictly decreasing.

6. If (a_n) and (b_n) are sequences of real numbers converging to a and b, respectively, then
$$\lim_{n \to \infty} (a_n, b_n) = (a, b).$$

7. Every bounded sequence of real numbers has a convergent subsequence.

8. Let $f : [a,b] \to \mathbb{R}$ be a continuous function. Then f achieves an absolute maximum and an absolute minimum value on $[a,b]$.

9. Let $f : \mathbb{R} \to \mathbb{R}$ be a differentiable function. Suppose that $f'(x) = 0$ for all x. Then f is constant.

10. Let $f : \mathbb{R} \to \mathbb{R}$ be a continuous function. Then f is a differentiable function.

11. Let $f : \mathbb{R} \to \mathbb{R}$ be a differentiable function. Then f is a continuous function.

12. Let $f : \mathbb{R} \to \mathbb{R}$ be a differentiable function. Then f' is a continuous function.

13. Let $f : \mathbb{R} \to \mathbb{R}$ and $g : \mathbb{R} \to \mathbb{R}$ be functions that are integrable on $[a, b]$. Then $f + g$ is integrable on $[a, b]$, and

$$\int_a^b (f + g)(x)\, dx = \int_a^b f(x)\, dx + \int_a^b g(x)\, dx.$$

14. Let $f : \mathbb{R} \to \mathbb{R}$ and $g : \mathbb{R} \to \mathbb{R}$ be functions that are integrable on $[a, b]$. Then $f \cdot g$ is integrable on $[a, b]$, and

$$\int_a^b (f \cdot g)(x)\, dx = \left(\int_a^b f(x)\, dx \right) \cdot \left(\int_a^b g(x)\, dx \right).$$

Chapter 0
Basic Building Blocks

0.1 Sets and Set Notation

The basic building blocks of mathematics are **sets**, which you should think of as collections of objects. The objects that appear in a set are called **elements**. If A is a set and z is an element of A, we denote this relationship symbolically by $z \in A$. A set is determined entirely by its contents. First, the order in which the elements appear doesn't matter; for instance, $\{1, 2, 3\}$, $\{2, 1, 3\}$, and $\{3, 1, 2\}$ are all the same set. Second, a given object can appear at most once in a set; that is, the number 1 may or may not be an element of the set A but if it's there, A cannot contain two "copies" of it.

Sets may be denoted in several ways. If the set has only a few elements, we can list its elements, separated by commas, inside a pair of curly brackets, as shown in the preceding paragraph. Sometimes we stretch this notation a bit, when we can use a pattern to make clear exactly what is intended. For instance, we might write $\{1, 2, 3, \ldots\}$ for the set of **natural numbers**, or $\{1, 2, 3, 4, \ldots, 20\}$ instead of writing out all of the first 20 natural numbers. This notation is somewhat ambiguous, of course, because you cannot be sure that my intention was not to skip every other block of four:

$$\{1, 2, 3, 4, 9, 10, 11, 12, 17, 18, 19, 20\}$$

This "suggestive" notation is useful in practice, so we do use it. We employ it sparingly and sensibly, however, so as to avoid confusing the reader. Something like $\{3, \frac{4}{3}, -22, \pi, \ldots\}$ does not really give us much of a clue as to what the set holds and should be avoided.

There are some sets that we talk about so much that we have special symbols to denote them. The natural numbers are denoted by the symbol $\mathbb{N}$. We denote the **integers**, $\{0, \pm 1, \pm 2, \ldots\}$, by $\mathbb{Z}$, and the **rational numbers**, which are ratios of integers, by $\mathbb{Q}$. Those numbers that cannot be expressed as ratios of integers are called **irrational numbers**. We denote the set of **real numbers**, comprising both rational and irrational numbers, by $\mathbb{R}$. Finally, the multidimensional spaces that we get by taking Cartesian products of $\mathbb{R}$ with itself we denote by $\mathbb{R}^2$, $\mathbb{R}^3$, $\mathbb{R}^4$, ...[1] The elements of $\mathbb{R}^n$ are n-tuples of real numbers. For instance, $(2, -\frac{1}{2})$ is an element of the Cartesian plane $\mathbb{R}^2$ and $(0, 1, 0, 0)$ is an element of the four-dimensional space $\mathbb{R}^4$.

Most often, we give a "rule" to tell us which elements are in the set and which are not. Here are some examples:

- $\{x \in \mathbb{R} : x \text{ is an integral multiple of } 3\}$.

This is read aloud as "the set of all x in $\mathbb{R}$ such that x is an integral multiple of 3."

- $\{(a, b) \in \mathbb{R}^2 : 3a - 2b = 4\}$.

Notice the format: The opening curly brace is followed by a statement that tells us what set we are taking our elements from. It is followed by a colon, which is interpreted as "such that." Finally, we have the rule that tells us exactly which elements are in the set and which are not. We close with an ending brace.

The purpose of including the larger set from which we draw our elements is precision. The sets

$$\{x \in \mathbb{Z} : \text{ there exists } y \in \mathbb{Z} \text{ for which } xy = 1\}$$

and

$$\{x \in \mathbb{Q} : \text{ there exists } y \in \mathbb{Z} \text{ for which } xy = 1\}$$

are very different sets. Sometimes, when the context makes it clear from whence we draw our elements, we can be more informal. Suppose we are working in the plane and have a fixed point **p**. If we say

$$\{\mathbf{x} : \text{ the distance from } \mathbf{x} \text{ to } \mathbf{p} \text{ is } 3\},$$

the context makes it clear that we intend our set to be the circle of radius 3 in the plane centered at **p**.

---

1. In some sense, any discussion of the real numbers in this chapter is premature, because we will carefully introduce the set of real numbers and discuss its properties only in Chapter 1. But it seems unnecessarily draconian to pretend, at this point, that you do not have any previous experience with them. In this chapter we will introduce some useful notation and define some useful (and straightforward) ideas—no more. For now, you should take discussion of ideas relating to $\mathbb{R}$, $\mathbb{R}^2$, $\mathbb{R}^3$, ... as intuitive, and not strictly in the chain of reasoning that the book will develop. Once you get to the point where you really need them, all of the ideas about the real numbers that we discuss in this chapter will be easily justifiable in retrospect.

Other common sets also have compact notation of their own. You are probably familiar, for instance, with interval notation for intervals in the real line. Let a and b be real numbers.

$[a, b] = \{x \in \mathbb{R} : a \leq x \leq b\}$ is the **closed interval** between a and b.

$(a, b) = \{x \in \mathbb{R} : a < x < b\}$ is the **open interval** between a and b.

The **half-open intervals** between a and b are

$$(a, b] = \{x \in \mathbb{R} : a < x \leq b\} \text{ and } [a, b) = \{x \in \mathbb{R} : a \leq x < b\}.$$

We end this subsection with a few elementary ideas from set theory.

Definition 0.1.1 The set with no elements is called the **empty set**. It is denoted by the symbol $\emptyset$.

Some students wonder why we need an empty set at all. Just as it is possible to do a lot of arithmetic without the number zero (in fact, people got along just fine without it for thousands of years), so it is possible to manage set theory without considering the empty set. But it is much more convenient to have it. For instance, we will be discussing the idea of the intersection of two sets. Unless we have the empty set, some set intersections are undefined.

Definition 0.1.2 Let A and S be sets. We say that S is a **subset** of A, provided that every element of S is also an element of A. We denote this relationship symbolically by $S \subseteq A$.[2]

Exercise 0.1.3 Here are a couple of easily proved facts about sets:

1. Let A be a set. Prove that if A is non-empty, then A has at least two subsets. Specifically, $\emptyset \subseteq A$ and $A \subseteq A$.

2. Let A, B, and C be sets. Prove that if $A \subseteq B$ and $B \subseteq C$, then $A \subseteq C$.

■

Definition 0.1.4 If B is a subset of a set X, and $B \neq X$, then we say that B is a **proper subset** of X.

---

2. The statement "every element of S is also an element of A" in the definition of a subset can be recast as the implication "If $x \in S$, then $x \in A$." So we prove that one set is a subset of another by proving this implication as we would any other. We assume $x \in S$, and then we work to prove that $x \in A$. This proof technique is called an **element argument**.

Definition 0.1.5 Two sets A and B are said to be **equal** provided that each is a subset of the other.

Totally Ordered Sets

Sometimes we add structures onto our sets. For instance, some sets (such as $\mathbb{N}$, $\mathbb{Z}$, and $\mathbb{R}$) are totally ordered.

Definition 0.1.6 Let A be a set. We say that A is a **totally ordered** set, provided that there is a relation[3] $\leq$ on A that satisfies the following properties:

i.: For every $a \in A$, $a \leq a$.

ii.: If $a, b \in A$, then either $a \leq b$ or $b \leq a$.

iii.: If $a, b \in A$ with $a \leq b$ and $b \leq a$, then $a = b$.

iv.: If $a, b, c \in A$ with $a \leq b$ and $b \leq c$, then $a \leq c$.

In this context, the symbols $\geq$, $<$, and $>$ have their usual interpretation. For instance, $a \geq b$ means $b \leq a$, and $a < b$ means $a \leq b$ and $a \neq b$.

Collections of Sets: Indexing Sets

We sometimes need to consider a whole class of sets at once. For instance, we might want to talk about the set of closed intervals that join adjacent integers in $\mathbb{R}$:

$$\ldots [-4, -3], [-3, -2], [-2, -1], [-1, 0], [0, 1], [1, 2], [2, 3], [3, 4], \ldots$$

The ideal way to describe such a collection is to think of some way of organizing the sets so that we can refer to them easily.

In this case, it is clear that there is one set for each integer. So we can let $I_n = [n, n+1]$. This lets everyone know that when you write I_3, you mean the interval $[3, 4]$. The **index** 3 is used as a referent for "calling up" the interval $[3, 4]$. Then we can denote the entire collection of intervals by $\{I_n\}_{n \in \mathbb{Z}}$. Individual integers $0, \pm 1, \pm 2, \ldots$ serve as our **indices** for the set of intervals. The set $\mathbb{Z} = \{0, \pm 1, \pm 2, \ldots\}$ is called the **indexing set** for the collection of intervals.

---

3. Formally, a relation is just a set of ordered pairs. We won't need to make explicit use of this word except in the context of orderings.

> When considering indexed collections, some students confuse the collection of sets, the indices, and the indexing set. But the distinctions are not really very difficult. To help you along, consider the Library of Congress system for indexing library collections. The symbols QA248.K36 1977 will help you to find the book *Set Theory and Metric Spaces* by Irving Kaplansky on your library shelves. (Something which I highly recommend doing, because it is a lovely book!) Obviously, QA248.K36 1977 is not Kaplansky's book—it is just a way of looking up the book. The whole set of Library of Congress designations is the indexing set. There are a whole bunch of them, but none of them are books, either. Keep this parallel set of distinctions in mind when you are thinking about indexed sets, their indices, and the corresponding indexing set.

Exercise 0.1.7 Here are some more examples of indexed collections of sets. For each example, describe the collection of sets, say what the indexing set is, and explain how a given index "calls up" its assigned set.

1. Let $C_r = \{(x,y) \in \mathbb{R}^2 : x^2 + y^2 = r^2\}$. Let P be the positive real numbers. Our collection is $\{C_r\}_{r \in P}$.

2. Let $L_t = \{(x,y) \in \mathbb{R}^2 : y = 3x + t\}$. Our collection is $\{L_t\}_{t \in \mathbb{R}}$.

■

Exercise 0.1.8 Find a way of indexing the collection

$$(-1, 1), \left(-\frac{1}{2}, \frac{1}{2}\right), \left(-\frac{1}{3}, \frac{1}{3}\right), \dots$$

of open intervals. Give both the indexing and the indexing set explicitly, as shown in Exercise 0.1.7. ■

Remark: Any set can, in principle, be an indexing set. If the collection of sets is finite, however, we usually use the set $\{1, 2, \dots, n\}$ to index it.

$$\{A_1, A_2, A_3, \dots, A_n\} = \{A_i\}_{i \in \{1, 2, \dots, n\}}$$

In this case, we often use the more compact notation

$$\{A_i\}_{i=1}^n.$$

Similarly, if the set is indexed by $\mathbb{N}$, we often use the notation $\{A_i\}_{i=1}^\infty$ as an alternative for $\{A_i\}_{i \in \mathbb{N}}$.

Arbitrary or General Indexing Sets

Sometimes we don't have an explicit collection of sets, just a "generic" collection of sets that we need to deal with. In such a situation, we will always assume, for ease of notation, that some indexing scheme is possible. That is, we assume the existence of an **arbitrary indexing set**,[4] and a fixed indexing scheme for our collection of sets. For example:

Let λ be an arbitrary indexing set, and let $\{B_\alpha\}_{\alpha \in \Lambda}$ be a collection of sets indexed by Λ.

It is traditional to denote arbitrary indexing sets and the corresponding indices by Greek letters. In this case, the capital letter lambda (Λ) signifies the indexing set and the lowercase alpha (α) stands for the indices.

Language:

- We read $\{B_\alpha\}_{\alpha \in \Lambda}$ as "the set of B-alpha's over alpha in lambda."

- If the element x is in at least one of the B_α's, we say that there exists $\alpha \in \Lambda$ such that $x \in B_\alpha$.

- If the element x is in every one of the B_α's, we say that, for all $\alpha \in \Lambda$, $x \in B_\alpha$.[5]

Set Operations

If we have more than one set, we sometimes want to combine them in various ways. For example, if we have two or more sets, we can take their union or intersection.

Definition 0.1.9 Let A and B be sets. Then the **union** of the two sets A and B is the set
$$A \cup B = \{x : x \in A \text{ or } x \in B\}.$$

More generally, we can define the union of an arbitrary collection of sets.

Definition 0.1.10 Let Λ be an arbitrary indexing set and let $\{B_\alpha\}_{\alpha \in \Lambda}$ be a collection of sets indexed by Λ. Then the **union over $\alpha \in \Lambda$ of the B_α's** is the set
$$\bigcup_{\alpha \in \Lambda} B_\alpha = \{x : \text{ there exists } \alpha \in \Lambda \text{ such that } x \in B_\alpha\}.$$

We can also take intersections.

---

4. Also called a **general indexing set**.

5. Our usual language patterns make it more natural to turn this around and say "$x \in B_\alpha$ for all $\alpha \in \Lambda$", but putting the quantifier after the statement can sometimes be ambiguous. You can use the natural phrasing, but beware of imprecision in the mathematical meaning.

Definition 0.1.11 Let A and B be sets. Then the **intersection** of the two sets A and B is the set
$$A \cap B = \{x : x \in A \text{ and } x \in B\}.$$

Definition 0.1.12 Let Λ be an arbitrary indexing set and let $\{B_\alpha\}_{\alpha \in \Lambda}$ be a collection of sets indexed by Λ. Then the **intersection over $\alpha \in \Lambda$ of the B_α's** is the set
$$\bigcap_{\alpha \in \Lambda} B_\alpha = \{x : \text{ for all } \alpha \in \Lambda, x \in B_\alpha\}.$$

Intersection distributes over union, and union distributes over intersection.

Theorem 0.1.13 Let Λ be an arbitrary indexing set and let $\{B_\alpha\}_{\alpha \in \Lambda}$ be a collection of sets indexed by Λ. Let C be any set. Then the following identities hold:

1. $C \cap \bigcup_{\alpha \in \Lambda} B_\alpha = \bigcup_{\alpha \in \Lambda} (C \cap B_\alpha).$

2. $C \cup \bigcap_{\alpha \in \Lambda} B_\alpha = \bigcap_{\alpha \in \Lambda} (C \cup B_\alpha).$

Proof: The proof is Problem 3 at the end of this section. $\square$

The operation of complementation is also extremely important.

Definition 0.1.14 Let U be a set and let $B \subseteq U$. The set of all elements of U that are not in B is called the **complement of B in U**. It is denoted by
$$B_U^{\mathcal{C}} = \{x \in U : x \notin B\}.$$

Remark: Complements must always be taken relative to a larger set.[6] If the set U is understood from the context, however, we can sometimes just write $B^{\mathcal{C}}$ and think of the complement of B as the set of elements that are not in B.

We end this section with some useful set algebraic identities that deal with the connection between complementation and union or intersection. These identities are called the DeMorgan laws.

---

6. Otherwise, we end up dealing with Russell's paradox, which is a paradox of set theory. It is a story for another time, but you should look it up. It's kind of cool and weird.

Theorem 0.1.15 [DeMorgan Laws] Let U be a set. Let Λ be an arbitrary indexing set for a collection $\{B_\alpha\}_{\alpha \in \Lambda}$ of subsets of U. Then the following identities hold (all complements are taken relative to U):

1. $\left(\bigcup_{\alpha \in \Lambda} B_\alpha \right)^c = \bigcap_{\alpha \in \Lambda} B_\alpha^c.$

2. $\left(\bigcap_{\alpha \in \Lambda} B_\alpha \right)^c = \bigcup_{\alpha \in \Lambda} B_\alpha^c.$

Proof: The proof is Problem 4 at the end of this section. $\square$

We need one last set theoretic operation.

Definition 0.1.16 Let A and B be sets. We define the **set difference** of A and B to be the set of all elements of A that are not also in B.

$$A \setminus B = \{x \in A : x \notin B\}.$$

Exercise 0.1.17 Consider the sets A and B given in each of the items that follow. What is $A \setminus B$?

1. Let A and B be the sets shown in the **Venn diagrams**. (You can think of mentally "shading in" the desired region.)

    (a) "Generic" intersection:

    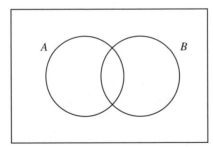

    (b) Disjoint sets:

    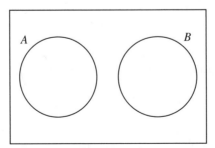

2. Let $A = [-2, 2] \subseteq \mathbb{R}$ and $B = \{1\}$.

3. Let $A = \{(x, y) \in \mathbb{R}^2 : x^2 + y^2 \leq 3\}$.

   (a) $B = \{(x, y) \in \mathbb{R}^2 : x^2 + y^2 > 1\}$,
   (b) $B = \{(x, y) \in \mathbb{R}^2 : x^2 + y^2 < 1\}$,
   (c) $B = \{(x, y) \in \mathbb{R}^2 : x^2 + y^2 \leq 5\}$,
   (d) $B = \{(-2, -1)\}$ (Here $(-2, -1)$ is the ordered pair, not the interval.)

   ∎

Problems 0.1

1. For each $n \in \mathbb{N}$, let $I_n = \left[0, \dfrac{1}{n}\right]$.

   (a) Find $\bigcup_{n \in \mathbb{N}} I_n$.
   (b) Find $\bigcap_{n \in \mathbb{N}} I_n$.

2. For each $r \in \mathbb{Q}$, let $K_r = \{r\}_\mathbb{R}^\complement$.

   (a) Find $\bigcup_{r \in \mathbb{Q}} K_r$.
   (b) Find $\bigcap_{r \in \mathbb{Q}} K_r$.

3. Prove that intersection distributes over union, and vice versa (Theorem 0.1.13).

4. Prove the DeMorgan laws (Theorem 0.1.15).

5. Natural questions can be asked about the interaction of set difference and the other set theoretic operations. For instance, does intersection distribute over set difference? That is, is it true that for all sets A, B, and C,
$$A \cap (B \setminus C) = (A \cap B) \setminus (A \cap C)$$

   (a) Answer this question with a proof or a counterexample.
   (b) Formulate two other such "natural" equations. In each case, prove that the equation is correct or give a counterexample to show that it is not.

0.2 Functions

Functions are, of course, a fundamental tool of mathematics. Dynamically, we think of a function as a sort of transformation that takes elements in one set and associates each with one and only one element in another set. Thus a function has three ingredients: a set of inputs, a set of outputs, and a "procedure" by which we associate elements in the first set with elements in the second set. We use the vague term "procedure" because it can take many forms. In high school, you learned the familiar notation $f(x) = x^2 + 4$, in which a function is given by an equation. This equation gives us our procedure: We take a real number as an input, square that number, and add 4; the result will be a real number, which is our output. For other functions, we might read values off a table or a diagram—and many other possibilities exist. Basically, for each input value, we must have a rule that allows us to associate with it exactly one element in the set of outputs.

Before we make this idea more precise, we need some language to help us talk about the familiar concept of an **ordered pair**. Suppose that A and B are sets. Let $a \in A$ and $b \in B$. Then (a, b) is called the ordered pair with **first coordinate** a and **second coordinate** b. Although (a, b) can be defined precisely using sets and set theory, it is enough for us to think of it using only our intuitive notion of "first" and "second." Two ordered pairs (a, b) and (c, d) are equal if and only if $a = c$ and $b = d$. The set of all ordered pairs for which the first coordinate is in A and the second coordinate is in B is called the **Cartesian product of A and B** and is denoted by $A \times B$. The most familiar example of a Cartesian product is, of course, the real plane: $\mathbb{R} \times \mathbb{R} = \mathbb{R}^2$.

Definition 0.2.1 Let A and B be sets. A **function f** from A to B—denoted symbolically by $f : A \to B$—is a set of ordered pairs in which each first coordinate is an element of A and each second coordinate is an element of B. [In keeping with the usual function notation, we denote $(a, b) \in f$ by $f(a) = b$.] In addition, f must satisfy the following property:

> For each $a \in A$, there exists one and only one $b \in B$, such that $b = f(a)$.

In this case, b is called the **function value** that f associates with a, or we might say that f **maps** a to b.

The set A is called the **domain** of f. The set B is called the **codomain** of f. The set $\mathcal{R}an(f) = \{b \in B : \text{there exists } a \in A \text{ such that} f(a) = b\}$ of function values of f is called the **range** of f.

> **Equality of Functions**
>
> Note that $f : A \to B$ is equal to $g : C \to D$ if and only if $A = C$, $B = D$, and for all x in their (common) domain,
> $$f(x) = g(x).$$
> In other words, to show that two functions are equal, we need to verify that they have the same domain and the same codomain (this is usually just a trivial matter of checking) and that they give the same function value for the same input.

Definition 0.2.2 Let $f : A \to B$ be a function. We say that f is **one-to-one** if
$$f(a_1) = f(a_2) \implies a_1 = a_2.$$
In other words, if each element of B has *at most* one element of A mapping to it.

We say that f is **onto** provided that each element of B has *at least* one element of A mapping to it.

A function that is both one-to-one and onto is sometimes called a **one-to-one correspondence** or a **bijection**.

Remark: The definition of function requires every element of A be "used" by f, but some elements in B may not actually be function values. The words "codomain" and "range" allow us to distinguish between the "potential" outputs and the "actual" outputs. The notion of an "onto" function tells us that the codomain and the range are equal. Every "potential" output is realized as an "actual" output.

Definition 0.2.3 Let A, B, and C be sets. Let $f : A \to B$ and $g : B \to C$ be functions. We define a new function $g \circ h : A \to C$ by
$$g \circ f(x) = g(f(x)).$$
The function $g \circ f$ is called the **composition of f and g**. The expression $g \circ f$ is read as "g composed with f."

Theorem 0.2.4 Let A, B, and C be sets. Let $f : A \to B$ and $g : B \to C$ be functions.

1. If f and g are both one-to-one, then $g \circ f$ is one-to-one.

2. If f and g are both onto, then $g \circ f$ is onto.

Proof: The proof is Problem 2 at the end of this section. $\square$

Theorem 0.2.5 [Composition of Functions Is Associative] Let A, B, C, and D be sets. Let $f : A \to B$, $g : B \to C$, and $h : C \to D$ be functions. Then

$$h \circ (g \circ f) = (h \circ g) \circ f.$$

Proof: Problem 3 asks you to critique several "proofs" of Theorem 0.2.5. □

When considering a function, it is frequently useful to restrict our attention to a (proper) subset of the domain.

Definition 0.2.6 Let A and B be sets, and let $X \subseteq A$. Let $f : A \to B$ be a function. Then the function

$$f_{|X} : X \to B \text{ given by } f_{|X}(x) = f(x)$$

is called the **restriction** of f to X.

Although the restriction of a function f is, in fact, a different function, it is usually acceptable to continue to call it f unless doing so will cause confusion (for instance, if we need to distinguish f from its restriction).

Inverse Functions

Theorem 0.2.7 Let A and B be sets. Let $f : A \to B$ be a function. The following statements are equivalent:

1. f is both one-to-one and onto.

2. There exists a function $g : B \to A$, such that

$$g \circ f(x) = x \text{ for all } x \in A, \text{ and } f \circ g(x) = x \text{ for all } x \in B.$$

Furthermore, the function g, when it exists, is unique.

Proof: The proof is Problem 4 at the end of this section. □

Definition 0.2.8 Let A and B be sets, and let $f : A \to B$ be a function. We say that f is **invertible** if it is one-to-one and onto. The function g that is given by Theorem 0.2.7 is called the **inverse** of f. To emphasize its connection to f, we usually denote the function g by the symbol f^{-1}.

Note that the notation in Definition 0.2.8 and Theorem 0.2.7 tell us that if $f : A \to B$ is one-to-one and onto, then

$$f^{-1} \circ f(x) = x \text{ for all } x \in A, \text{ and } f \circ f^{-1}(x) = x \text{ for all } x \in B.$$

Furthermore, the uniqueness clause in Theorem 0.2.7 tells us that f^{-1} is the *only* function that satisfies this condition.

Exercise 0.2.9 Let A and B be sets, and let $f : A \to B$ be a function. Consider for a moment that the function f is one-to-one and onto and, therefore, that f^{-1} exists as asserted in Theorem 0.2.7. Use the property in provision 2 of the theorem to explain why the function f^{-1} must consist precisely of the set of pairs

$$\{(f(a), a) : a \in A\}.$$

That is, f^{-1} "reverses" the inputs and the outputs of f. ∎

Images and Inverse Images

Theorem 0.2.7 tells us that unless a function is one-to-one and onto, there is no meaningful way to define an inverse for the function. Nevertheless, it is often useful to think about starting in the codomain and "pulling" back to the domain using the function as an engine. In this case, we can make use of the idea of the inverse image of a set under a function.

Definition 0.2.10 Let A and B be sets, and let $f : A \to B$ be a function. Let $S \subseteq B$. Then the set

$$f^{-1}(S) = \{x \in A : f(x) \in S\}$$

is called the **inverse image of the set S under the function f**.

Exercise 0.2.11 Let A and B be sets, and let $f : A \to B$ be a function.

1. Suppose that A, B, S, and f are as shown in the figure. What is $f^{-1}(S)$?

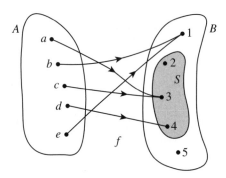

2. Let $S \subset B$. Complete the following sentence: $x \in f^{-1}(S)$ if and only if _____.

This language will be useful when you are trying to prove theorems about inverse images.

■

It is useful to see how the inverse image operation interacts with other set operations. Here is a theorem that tells us how it works with unions and intersections.

Theorem 0.2.12 Let A and B be sets, and let $f : A \to B$ be a function. Let Λ be an arbitrary indexing set and let $\{S_\alpha\}_{\alpha \in \Lambda}$ be a collection of subsets of B indexed by Λ. Then

1. $f^{-1}\left(\bigcup_{\alpha \in \Lambda} S_\alpha\right) = \bigcup_{\alpha \in \Lambda} f^{-1}(S_\alpha)$.

2. $f^{-1}\left(\bigcap_{\alpha \in \Lambda} S_\alpha\right) = \bigcap_{\alpha \in \Lambda} f^{-1}(S_\alpha)$.

Proof: The proof is Problem 7 at the end of this section. □

Here is a theorem that tells us how this concept works with complementation.

Theorem 0.2.13 Let A and B be sets, and let $f : A \to B$ be a function. Let S be any subset of B. Then

$$f^{-1}\left(S_B^{\mathcal{C}}\right) = \left(f^{-1}(S)\right)_A^{\mathcal{C}}.$$

Proof: The proof is Problem 8 at the end of this section. □

Parallel to the idea of the inverse image of a set under a function is the image of a set under a function.

Definition 0.2.14 Let A and B be sets, and let $f : A \to B$ be a function. Let $T \subseteq A$. Then the set

$$f(T) = \{x \in B : x = f(t) \text{ for some } t \in T\}$$

is called the **image of the set T under the function f**.

Exercise 0.2.15 Let A and B be sets, and let $f : A \to B$ be a function.

1. Suppose that A, B, T, and f are as shown in the figure. What is $f(T)$?

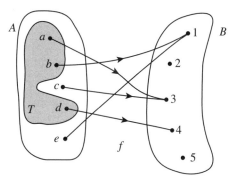

2. Let $T \subset A$. Complete the following sentence: $x \in f(T)$ if and only if _____.

■

Once again, we want to see how images interact with the algebra of sets.

Theorem 0.2.16 Let A and B be sets, and let $f : A \to B$ be a function. Let Λ be an arbitrary indexing set, and let $\{T_\alpha\}_{\alpha \in \Lambda}$ be a collection of subsets of A indexed by Λ. Then

1. $f\left(\bigcup_{\alpha \in \Lambda} T_\alpha\right) = \bigcup_{\alpha \in \Lambda} f(T_\alpha).$

2. $f\left(\bigcap_{\alpha \in \Lambda} T_\alpha\right) \subseteq \bigcap_{\alpha \in \Lambda} f(T_\alpha).$

Proof: The proof is Problem 9 at the end of this section. □

In Theorem 0.2.13, you proved that inverse images "distribute" over complements. The same is not true for images. In fact, by looking at a few examples, you can easily verify that *nothing* can be said in general about the relationship between $f(T)^C$ and $f(T^C)$.

Problems 0.2

1. Prove that the composition of two one-to-one functions is one-to-one and that the composition of two onto functions is onto (Theorem 0.2.4).

2. This problem explores a sort of converse to Theorem 0.2.4. Let A, B, and C be sets. Let $f : A \to B$ and $g : B \to C$ be functions. For each of the following statements, determine whether it is true or false. If it is true, prove it. If it is false, give a counterexample, and determine what hypothesis needs to be added to make the conclusion true. Then prove the revised statement.

    (a) If $g \circ f$ is one-to-one, then f is one-to-one.
    (b) If $g \circ f$ is one-to-one, then g is one-to-one.
    (c) If $g \circ f$ is onto, then f is one-to-one.
    (d) If $g \circ f$ is onto, then g is one-to-one.

3. Theorem 0.2.5 asserts that the two functions $h \circ (g \circ f)$ and $(h \circ g) \circ f$ are equal. Notice that each of these functions has A as its domain and D as its codomain.

   Consider the following "proofs" of the statement that the functions give the same function value for each input. (Refer to the box on page 19.) Evaluate each argument. Decide which one is correct and to what extent each of the others is incorrect. The arguments may fall short in a variety of ways: They may be ill conceived, starting and ending in the wrong places; they may contain lapses in logic; or they may apply the notion of function composition incorrectly. Each varies in how far it "strays."

Argument 1	Argument 2
$(h \circ (g \circ f))(a) = ((h \circ g) \circ f)(a)$ $h \circ (g(f(a))) = (h \circ g) \circ f(a)$ $h(g(f(a))) = h(g(f(a)))$	$(h \circ (g \circ f))(a) = h((g \circ f)(a))$ $= h(g(f(a)))$ $= (h \circ g)(f(a))$ $= ((h \circ g) \circ f)(a)$

Argument 3	Argument 4
$(h \circ (g \circ f))(a) = h((g \circ f)(a))$ $= h(g(f(a)))$ and $((h \circ g) \circ f)(a) = h \circ (g(f(a)))$ $= h(g(f(a)))$	$(h \circ (g \circ f))(a) = ((h \circ g) \circ f)(a)$ $h((g \circ f)(a)) = (h \circ g)(f(a))$ $h(g(f(a))) = h(g(f(a)))$

4. Prove Theorem 0.2.7. (*Hint*: If you are having trouble with the existence part of $1 \implies 2$, look at Exercise 0.2.9.)

5. Let A and B be sets. Let $f : A \to B$ be an invertible function. Show that f^{-1} is also invertible and that $(f^{-1})^{-1} = f$.

6. Let A and B be sets. Let $f : A \to B$ be an invertible function. Suppose that $J \subseteq A$. Show that $f|_J : J \to f(J)$ is invertible. What is the relationship between $(f|_J)^{-1}$ and $(f^{-1})|_{f(J)}$?

7. Prove that the inverse image distributes over union and intersection (Theorem 0.2.12).

8. Prove that the inverse image of the complement is the complement of the inverse image (Theorem 0.2.13).

9. Prove the theorem that relates the image of a union or intersection of sets with the union or intersection of the images of those sets (Theorem 0.2.16).

10. This problem examines the fact that

$$f\left(\bigcap_{\alpha \in \Lambda} T_\alpha\right) \subseteq \bigcap_{\alpha \in \Lambda} f(T_\alpha)$$

but that the reverse inclusion does not hold in general.

   (a) Give an example of a function $f : A \to B$ and two subsets X and Y of A such that $f(X \cap Y) \neq f(X) \cap f(Y)$.

   Let A and B be sets, and let $f : A \to B$ be a function.

   (b) Prove that $f\left(\bigcap_{\alpha \in \Lambda} T_\alpha\right) = \bigcap_{\alpha \in \Lambda} f(T_\alpha)$ for all choices of $\{T_\alpha\}_{\alpha \in \Lambda}$ if and only if f is one-to-one. (*Hint*: Prove $\Longrightarrow$ by contrapositive. Be careful not to ignore the quantifiers. They are absolutely crucial!)

11. Let A and B be sets, and let $f : A \to B$ be a function. Suppose that K and J are subsets of A and that $W \subseteq B$.

   (a) Prove that $f(K) \subseteq W$ if and only if $K \subseteq f^{-1}(W)$.

   (b) Prove that if $K \subseteq J$, then $f(K) \subseteq f(J)$.

   (c) Prove that the converse of part (b) is not, in general, true. What added condition on f will ensure that the conclusion holds? Prove your conjecture.

12. Suppose that $f : X \to Y$ is a function. Let $U \subseteq Y$. Show that if $A = U \cap f(X)$, then $f^{-1}(A) = f^{-1}(U)$.

13. Let A and B be sets, and let $f : A \to B$ be a function. In this problem, you will consider sets of the form $f(f^{-1}(S))$. Is this set equal to the set S? If not, what condition on f will ensure that the sets *are* equal? Prove your assertions.

14. Let A and B be sets, and let $f : A \to B$ be a function. In this problem, you will consider sets of the form $f^{-1}(f(T))$. Is this set equal to the set T? If not, what condition on f will ensure that the sets *are* equal? Prove your assertions.

0.3 The Natural Numbers, the Integers, and Their Properties

We will take for granted our knowledge of the set of natural numbers, $\mathbb{N}$, and the set of integers, $\mathbb{Z}$. This section makes precise which ideas and properties we will take for granted. The assumptions that are detailed here are almost certainly familiar to you, so it should be easy to skim the section and use it subsequently as a reference. The assumptions are listed as "bullets" to highlight them.

To make use of set theory, we begin with the following assumption:

- $\mathbb{N}$ and $\mathbb{Z}$ are non-empty sets, and $\mathbb{N} \subseteq \mathbb{Z}$ in the familiar way.

In fact, we will assume, without elaborating further, that sets $\mathbb{N}$ and $\mathbb{Z}$ can be built from the axioms of set theory.[7] Next we add arithmetic.

Definition 0.3.1 Let S be any set. A function $*$ from $S \times S$ to S is called a **binary operation on S**. For elements x and $y \in S$, we denote the image of (x, y) under $*$ by $x * y$.

> The definition of binary operation may seem strange and abstract. In fact, it just says that a binary operation is an operation that takes two elements of one set and produces another. The most familiar examples of binary operations are addition and multiplication of real numbers.

Definition 0.3.2 Let $*$ be a binary operation on a set S. Then $*$ is said to be **commutative** if for all x and $y \in S$, $x * y = y * x$.

The binary operation $*$ is said to be **associative** provided that for all x, y, and $z \in S$, $x * (y * z) = (x * y) * z$, where the parentheses, indicate which operations are performed first.

- $\mathbb{N}$ and $\mathbb{Z}$ are endowed with two commutative and associative binary operations called addition $(+)$ and multiplication $(\cdot)$.

- Multiplication distributes over addition. That is, for all integers x, y, and z,
$$x(y + z) = xy + xz$$
where, in the absence of parentheses, multiplications are performed before additions.

---
7. Making this claim explicit would take us away from analysis and into axiomatic set theory, and would not really add to our understanding of $\mathbb{N}$ and $\mathbb{Z}$. If you are interested in how this is done—or even in what it *means*—this information can be found in any book on axiomatic set theory. See, for instance, [H&J]. The process is also spelled out in an appendix in [SCH].

> I have suppressed the symbol · in the expressions $x \cdot (y+z)$ and $x \cdot y + x \cdot z$ and represented multiplication in the conventional way by concatenating the elements we wish to multiply. I will continue to observe this and similar algebraic conventions in cases where no confusion can arise.

- There are two "special" integers, 0 and 1. When 0 is added to any integer n, the result is n. When 1 is multiplied by any integer n, the result is n. That is, 0 is an identity for addition and 1 is an identity for multiplication. Furthermore, these numbers are unique in this respect. (Note that $1 \in \mathbb{N}$ and, therefore, acts as a multiplicative identity for that set as well. $\mathbb{N}$ does not have an additive identity.)

Next, we add ordering.

- $\mathbb{N}$ and $\mathbb{Z}$ are totally ordered under their "usual" ordering $\leq$.

- Let $x \in \mathbb{Z}$. Then there is no integer between x and $x+1$, nor is there an integer between x and $x-1$. (Integers have **immediate successors** and **immediate predecessors** under $\leq$.)

- We assume the usual notions of counting associated with the natural numbers. Specifically, we can associate with any finite string or set, a natural number that denotes the number of elements in that string or set.

Mathematical Induction

Finally, we presume that the axiom of induction holds.

The Axiom of Induction: Let S be a subset of $\mathbb{N}$. Suppose that the following conditions hold:

1. $1 \in S$.

2. If $k \in S$, then $k+1 \in S$.

Then $S = \mathbb{N}$.

You can now easily prove that the principle of mathematical induction follows immediately.

Theorem 0.3.3 [Principle of Mathematical Induction] Let (P_n) be a sequence of statements. Suppose that the following conditions hold:

1. P_1 is true.

2. If P_k is true, then P_{k+1} is true.

It follows that P_n is true for all $n \in \mathbb{N}$.

The principle of mathematical induction is a powerful tool. It allows us to prove linked sequences of statements using a sort of "domino effect." Theorem 0.3.4 about the natural numbers follows from assuming that mathematical induction holds. The proof is left to you as an exercise.

Theorem 0.3.4 Every non-empty subset of $\mathbb{N}$ has a least element. Every non-empty subset of $\mathbb{Z}$ that is bounded below has a least element.

Hint: Prove this theorem by contrapositive. Assume that $K \subseteq \mathbb{N}$ has no least element and use induction to show that its complement is all of $\mathbb{N}$. (That is, $K = \emptyset$.)

If you would like more detailed information about mathematical induction, you can find it in many places, such as Chapter 3 in [SCH].

0.4 Sequences

Sequences are very important in mathematics and play an especially important role in analysis. Loosely speaking, a sequence in a set A is an infinite "string" of elements of A, with one element being designated as the first term of the sequence, another element (not necessarily distinct from the first) as the second term of the sequence, and so on. The following are sequences of real numbers:

- $1, 0, 1, 0, 0, 1, 0, 0, 0, 1, \ldots$

- $3, 6, 2, 9, 14, 16, 10, 100, 99, 23, 15, 1003, \ldots$

- $1, \frac{1}{2}, \frac{1}{3}, \frac{1}{4}, \frac{1}{5}, \ldots$

The last of these examples has 1 as its 1st term, $\frac{1}{2}$ as its 2nd term, and so on. In general, $\frac{1}{n}$ is called the nth term of the sequence. With each natural number n we are associating a real number, $\frac{1}{n}$. This discussion should remind you of the concept of a function. We can now give a mathematical definition for a sequence.

Definition 0.4.1 Let A be a non-empty set. Any function $s : \mathbb{N} \to A$ is called a **sequence in A**.

We do not use the standard function notation $s(1)$, $s(2)$, ... to refer to the terms of the sequence.

Sequence Notation: If s is a sequence in A, we refer to the 1st term in the sequence as s_1, the 2nd term in the sequence as s_2, the 45th term in the sequence as s_{45}, and so forth. The notation s_i encodes two pieces of information:

- s_i is an element of the set A and indicates the *value* of the term.

- i is a natural number and indicates the *position* of the term in the sequence.

Our shorthand for the sequence as a whole is (s_n).

Exercise 0.4.2 Although we may (cheerfully) confuse them, it is useful—often crucial—to distinguish between a sequence (a_n) in a set S and the subset of S given by $\{a_n \in S : n \in \mathbb{N}\}$. This set is called the **range of the sequence** and is denoted by $\mathcal{R}an(a_n)$. What exactly is the distinction between the two? (*Hint*: Consider the sequence $0, 1, 0, 1, 0, \ldots$ of real numbers.) ■

Definition 0.4.3 Let A be a set. Let (s_i) be a sequence in A.

1. (s_i) is a **sequence of distinct terms** if $s_i \neq s_j$ for distinct $i, j \in \mathbb{N}$.

2. (s_i) is a **constant sequence** if there exists $a \in A$ such that $s_i = a$ for all $i \in \mathbb{N}$.

Sequences in a totally ordered set sometimes have special properties that relate to the way in which the set is ordered. (We will most often apply these ideas in the context of sequences of real numbers.)

Definition 0.4.4 Let (s_i) be a sequence in a totally ordered set $(A, \leq)$.

1. (s_i) is said to be **increasing** if whenever $n \leq m$, $s_n \leq s_m$. (s_i) is **strictly increasing** if whenever $n < m$, $s_n < s_m$.

2. (s_i) is said to be **decreasing** if whenever $n \leq m$, $s_n \geq s_m$. (s_i) is **strictly decreasing** if whenever $n < m$, $s_n > s_m$.

3. (s_i) is said to be **monotonic** if it is either increasing or decreasing.

4. (s_i) is said to be **bounded from below** if there exists $b \in \mathbb{R}$ such that for all $i \in \mathbb{N}$, $b \leq s_i$. In this case, b is called a **lower bound** for (s_i).

5. If there exists $c \in \mathbb{R}$ such that for all $i \in \mathbb{N}$, $s_i \leq c$, then we say that (s_i) is **bounded from above**. In this case, c is called an **upper bound** for (s_i).

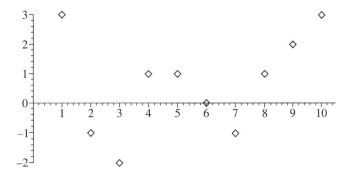

Figure 0.1 Here is a useful way to picture sequences of real numbers. The sequence illustrated here is $3, -1, -2, 1, \ldots$. (What are the next few terms?)

6. If (s_i) is bounded from above and from below, we say that it is **bounded**.

Figure 0.1 illustrates sequences of real numbers.

Exercise 0.4.5 Construct the following examples of sequences of real numbers.

1. A strictly increasing sequence
2. A decreasing sequence that is neither strictly decreasing nor constant
3. A sequence of distinct terms that is not monotonic
4. A sequence that is not increasing and is bounded from below, but not from above
5. A sequence that is bounded neither from above nor from below
6. A bounded sequence of distinct terms

∎

The following is a simple, but amazingly useful theorem.

Theorem 0.4.6 Let (n_i) be a strictly increasing sequence of natural numbers. Then $i \leq n_i$, for all $i \in \mathbb{N}$.

Proof: The proof is Problem 1 at the end of this section. □

Exercise 0.4.7 What does Theorem 0.4.6 say about the relationship between the position and the value of a term in a strictly increasing sequence of natural numbers? ∎

Subsequences

Sometimes when we have a sequence, we wish to construct a new sequence containing only some of the original terms.

Example 0.4.8 Suppose that we start with the sequence of even natural numbers:
$$(s_i) = 2, 4, 6, 8, 10, 12, 14, 16, 18, 20, \ldots.$$

We might want to extract from (s_i) the sequence consisting only of the terms of (s_i) which are also multiples of 3:
$$6, 12, 18, 24, \ldots.$$

Notice that the terms of this new sequence are
$$\underbrace{s_3}_{1^{\text{st}}}, \underbrace{s_6}_{2^{\text{nd}}}, \underbrace{s_9}_{3^{\text{rd}}}, \underbrace{s_{12}}_{4^{\text{th}}}, \underbrace{s_{15}}_{5^{\text{th}}}, \ldots.$$

That is, the ith term of this sequence is the same as the $3i^{\text{th}}$ term of the original sequence. The subscripts on the new sequence form a strictly increasing sequence of natural numbers: $3, 6, 9, 12, 15, \ldots$. This is an example of a "subsequence."

To talk about this idea properly, we need a definition.

Definition 0.4.9 Let A be a set. Let (s_i) be a sequence in A. If (n_i) is a strictly increasing sequence in $\mathbb{N}$, then the sequence
$$s_{n_1}, s_{n_2}, s_{n_3}, \ldots$$
is a **subsequence** of (s_i). We denote this subsequence by (s_{n_i}).

When thinking about subsequences, it is useful to keep in mind that encoded in the notation s_{n_i} are *three* pieces of information:

- s_{n_i} is an element of A and gives the *value* of the term.

- n_i is a natural number and indicates the *position* of the term in the original sequence.

- i is also a natural number and indicates the *position* of the term in the subsequence.

Exercise 0.4.10 Consider the following sequence (s_i) of letters:
T,H,I,[S], I,S, T,[H],[E], [S],O,[N],G, T,H,A,T, D,[O],E,S,N,[T], E,N,D,
[Y], [E],S, I,[T], G,O,E,S, O,N, A,N,[D], [O],[N], M,Y, F,R,I,[E],N,D, …
The terms that are boxed represent a subsequence (s_{n_i}) of (s_i).

1. List $s_1, s_2, s_3, \ldots, s_{10}$.

2. List $s_{n_1}, s_{n_2}, s_{n_3}, \ldots, s_{n_{10}}$.

3. What is s_3?

4. What is s_{n_3}?

5. What is n_7?

∎

I said before that Theorem 0.4.6 is "amazingly useful." It is especially useful when dealing with subsequences. The following exercise suggests why. You would do well to keep this idea in mind when dealing with subsequences in real analysis.

Exercise 0.4.11 Explain what Theorem 0.4.6 says about the relationship between the position of a term in a subsequence and its position in the original sequence. (This is meant to be an informal explanation. Phrases like "appears earlier" or "appears later" might be helpful.) ∎

"Passing to a subsequence" preserves many important properties of the original sequence. The following easy exercise will help you familiarize yourself with subsequence notation. (Don't just think the exercise through. Take the time to write the answer down so you will have to use the notation properly.) The problems at the end of the section will ask you to prove some slightly more difficult facts about sequences and their subsequences.

Exercise 0.4.12 Let A be a set, and let (s_i) be a sequence in A. Prove the following:

1. If (s_i) is constant, every subsequence of (s_i) is constant.

2. If (s_i) has distinct terms, every subsequence of (s_i) has distinct terms.

∎

It often happens in analysis that we need to define a sequence with some special characteristic or that we need to find a subsequence of an existing sequence that satisfies a desired property. Frequently, the only way to do so is to choose the successive terms one by one. Of course, a sequence contains infinitely many terms, so we cannot actually accomplish this goal. Instead, we provide a "recipe" for choosing each successive term. Informally, we can say something like "This is how I choose my first term, this is how I choose my second term, this is how I choose my third term, ..." If a pattern clearly emerges from this litany, we sometimes leave it at that. From a mathematical standpoint, however, this is not very satisfactory. The difficulty is this: How do you know for sure that when you get to the 10th or the 144th term, you won't be "stuck" and unable to continue? To make the process more rigorous, we use mathematical induction.

Constructing Sequences by Induction—General Structure: The proof comes in four parts.

i.: **The base case**: You say how the first term is chosen.

ii.: **The induction hypothesis**: You assume the first k terms have been chosen and say what properties they possess.

iii.: **The induction step (part I)**: You indicate how the $(k+1)^{\text{st}}$ term is to be chosen.

iv.: **The induction step (part II)**: You show that if you add the new term to the list of already chosen terms, the updated list will also satisfy the conditions laid out in the induction hypothesis.

This is really just a way of formally writing down the pattern that emerged in the "This is how I choose my first term, this is how I choose my second term, this is how I choose my third term, ..." process. The added bit of rigor, however, comes in Step iv, where you show that the process you describe is sustainable, as it can be repeated for term after term.

Problems 0.4

1. Prove that the value of a term in an increasing sequence of natural numbers is larger than or equal to the value of its position in the sequence (Theorem 0.4.6).

2. Let A be a totally ordered set, and let (s_i) be a sequence in A. Prove that if (s_i) is bounded above, every subsequence of (s_i) is bounded above.

3. Let A be a totally ordered set, and let (s_i) be a sequence in A. Prove that if (s_i) is increasing, every subsequence of (s_i) is increasing.

> The statement in Problem 2 is still true if the phrase "bounded above" is replaced by "bounded below." Likewise, the statement in Problem 3 is still true if the word "increasing" is replaced by "decreasing." If you try to write down one of these alternate proofs, you will see that the proofs are very similar. Basically, every $\geq$ in the proof becomes a $\leq$, or vice versa. It is common practice in mathematics to refer to such pairs of arguments as "parallel arguments."

4. Let (s_i) be a sequence in a set A. Part (a) will serve as a lemma for proving the more interesting and useful statement in part (b).

    (a) Prove that (s_i) has a constant subsequence if and only if there exists $a \in A$ such that for all $n \in \mathbb{N}$ there exists $j \geq n$ such that $s_j = a$.

    (b) Prove that either (s_i) has a constant subsequence or (s_i) has a subsequence of distinct terms. (*Hint*: You will need to be able to say what it means for a sequence *not* to have a constant subsequence. In preparation for that, negate the statement in part (a).)

5. Let (s_i) be a sequence in a totally ordered set A. Part (a) will serve as a lemma for proving the more interesting and useful statement in part (b).

    (a) Prove that any sequence in a totally ordered set that does not have a decreasing (alternatively, increasing) subsequence must have a smallest (alternatively, largest) term.

    (b) Use part (a) to prove that every sequence in a totally ordered set has a monotonic subsequence. (In particular, once we have $\mathbb{R}$ in hand, this will tell us that every sequence of real numbers has a monotonic subsequence.)

Chapter 1
The Real Numbers

1.1 Constructing the Axioms

The real numbers $\mathbb{R}$ are, without a doubt, the fundamental mathematical structure. Our purpose in this chapter is to introduce the axioms for the real number system. We will discuss the desirability and necessity of the axioms that we choose:

- *Desirability*: Our axioms should describe familiar properties of the real number system.

- *Necessity*: The real number system has many properties—it is possible to list hundreds of "facts" about the real numbers—but, of course, it is not necessary to accept them all as axioms. Many of these facts can be proved from other facts. We will want to choose for our axioms a few crucial statements from which other facts about the real number system can be proved.

We will start with a set $\mathbb{R}$, whose elements we will call *real numbers*. Our study of $\mathbb{R}$ depends, first and foremost, on set theory. Clearly, however, $\mathbb{R}$ has a structure that goes beyond a simple description of its contents. The properties of arithmetic, order, and so forth, must be specified by means of axioms.

1.2 Arithmetic

Small children begin their study of mathematics with the binary operations of addition, subtraction, multiplication, and division. We want to have these operations at our disposal.

Given two real numbers, a and b, we can think of $a - b$ as $a + -b$ and of $\frac{a}{b}$ as $a \cdot \frac{1}{b}$; thus our axioms need only discuss addition and multiplication. These axioms must, however, tell us what is meant by the additive inverse $-b$ and the multiplicative inverse $\frac{1}{b}$. A discussion of inverses presupposes the existence of additive and multiplicative identities. As a consequence, our axioms will discuss addition and multiplication, incorporating identities and inverses for each.

It is also important to say something about the way that addition and multiplication interact with each other, so our axioms must also say something about this relationship.

Axiom I [Field Axioms] Let $\mathbb{R}$ be a set. We assume the existence of two binary operations on $\mathbb{R}$, which we will call $+$ (addition) and $\cdot$ (multiplication). We assume that these binary operations have the following properties:

- $+$ and $\cdot$ are commutative operations.

- $+$ and $\cdot$ are associative operations.

- **Additive and multiplicative identities.** There exists a real number, which we will call 0 (zero), such that for all real numbers x, $x + 0 = x$. There exists a real number distinct from zero, which we will call 1 (one), such that for all real numbers x, $x \cdot 1 = x$.

- **Additive and multiplicative inverses.** For each real number x, there exists a real number, which we will call $-x$ (minus-x), such that $x + (-x) = 0$. For each non-zero real number x, there exists a real number, which we will call $\frac{1}{x}$ (one-over-x), such that $x \cdot \frac{1}{x} = 1$. (The real number $\frac{1}{x}$ is sometimes denoted by x^{-1}.)

- **Distributive property.** Multiplication distributes over addition. For all real numbers x, y, and z,

$$x(y+z) = xy + xz.$$

In effect, $\mathbb{R}$ is an algebraic structure called a **field**. Although we will not dwell on this terminology, a field is any algebraic structure that satisfies the field axioms that were just given.

We now *define* subtraction and division.

Definition 1.2.1 [Subtraction and Division] Let x and y be real numbers. We define binary operations $-$ and $\div$ on $\mathbb{R}$ as follows:

$$x - y = x + (-y) \text{ and } x \div y = x \cdot \frac{1}{y}.$$

Following the usual convention, we often write $\frac{x}{y}$ instead of $x \div y$.[1]

Theorem 1.2.2 says that additive and multiplicative inverses are unique.

Theorem 1.2.2 [Uniqueness of Inverses] Let x be a real number. Prove that there exists only one real number y such that $x + y = 0$.

Similarly, if $x \neq 0$, prove that there exists only one real number y such that $x \cdot y = 1$.

Proof: We begin by assuming that y and z are real numbers such that

$$x + y = x + z = 0.$$

Then by the well definedness[2] of the binary operation $+$, we know that $-x + (x + y) = -x + (x + z)$. Using the fact that $+$ is associative, we deduce that $(-x + x) + y = (-x + x) + z$. By the definition of $-x$, we see that this implies $0 + y = 0 + z$. Finally, the definition of 0 allows us to conclude that $y = z$.

□

Exercise 1.2.3 The proof that multiplicative inverses are unique is similar to the proof that additive inverses are unique. This proof is left to the reader. ■

---

1. If you were paying close attention here, you probably caught a little lie. Strictly speaking, $\div$ is not a binary operation on $\mathbb{R}$ because it is not defined on all pairs of real numbers. Zero has no multiplicative inverse, so $x \div 0$ does not make any sense—this fits in with our usual notion that we cannot divide by zero. Division, however, allows us to take two real numbers and obtain from them a third real number—so it *is* a lie, but still a useful shorthand.

2. All functions are well defined. This means that every possible input value is associated with an output value. It also means that there is *only* one output value. If Kristin evaluates the function at a specific value and then Paul evaluates the same function at the same point, he will get the *same* answer that Kristin did—*even if he computes the answer in a different way!* Make sure you understand why well definedness is the issue here!

Exercise 1.2.4 Use Theorem 1.2.2 to prove the following facts:

1. $-0 = 0$.

2. If x is any real number, $-(-x) = x$.

∎

The "Wish List"

Consider the following list of statements about the real numbers. Our previous experience tells us that they are true, but they are not explicitly assumed in the field axioms:

I. For all real numbers x, $x \cdot 0 = 0$. (Zero has multiplicative properties as well as additive properties.)

II. For all real numbers x, $-x = (-1) \cdot x$. (Think carefully about this. Many students think that this is the *definition* of $-x$, but the description of $-x$ in the field axioms never mentions multiplication at all!)

III. For all non-zero real numbers a, $a \neq -a$.

IV. $\mathbb{R}$ is an infinite set.

V. For all positive real numbers a, there exists a real number x such that $x^2 = a$.

We certainly want a set of axioms that will allow us to prove all of these things. So we can think of this set of statements as a (partial) "wish list" and use it as a measure of how well our axioms are doing their job. We can get the first statement right away.

Theorem 1.2.5 For all real numbers x, $x \cdot 0 = 0$.

Proof: Given that Theorem 1.2.5 discusses multiplication, and that the only thing we know about 0 is that it behaves in a certain way with respect to addition, we will clearly have to use the only axiom that talks about the connection between addition and multiplication.
$$x \cdot 0 = x \cdot (0 + 0) = x \cdot 0 + x \cdot 0$$
thus
$$-(x \cdot 0) + (x \cdot 0) = -(x \cdot 0) + (x \cdot 0 + x \cdot 0),$$
and therefore
$$0 = (-(x \cdot 0) + x \cdot 0) + x \cdot 0$$
$$= 0 + x \cdot 0 = x \cdot 0.$$

□

Exercise 1.2.6 The proof of Theorem 1.2.5 hinges on a series of algebraic computations. Justify each step by referring explicitly to a relevant axiom or to the fact that the arithmetic operations are well defined. ∎

Exercise 1.2.7 Use Theorem 1.2.2 and Theorem 1.2.5 to prove the following facts:

1. If x and $y \in \mathbb{R}$, $(-x)y = -(xy) = x(-y)$.

2. If x is any non-zero real number, then $x^{-1} \neq 0$ and $(x^{-1})^{-1} = x$.

3. If x and y are non-zero real numbers, then $\dfrac{x}{y} \neq 0$ and $\dfrac{1}{\frac{x}{y}} = \dfrac{y}{x}$.

∎

Theorem 1.2.8 For all real numbers x, $-x = (-1) \cdot x$.

Proof: The proof is Problem 3 at the end of this section. □

Exercise 1.2.9 will tell you that there are mathematical objects that are *not* $\mathbb{R}$, but that nevertheless satisfy the field axioms.

Exercise 1.2.9 Let A be a set containing two elements: the set of even integers and the set of odd integers. (We will call the set of even integers E and the set of odd integers O.) We define two binary operations $\oplus$ and $\otimes$ on A in the following intuitive way. Because the sum of two even integers is even, we define $E \oplus E$ to be E. Similarly, because the product of an even integer and an odd integer is even, we define $O \otimes E$ to be E. The remaining sums and products are defined analogously and are summarized in the following tables.

$\oplus$	E	O
E	E	O
O	O	E

and

$\otimes$	E	O
E	E	E
O	E	O

In the problems you will be asked to show that this algebraic structure satisfies the field axioms. For now, however, you may assume this fact. Your job in this exercise is to identify "0","1", and "-1". ∎

You just saw a mathematical structure composed of a set and two binary operations that satisfies all of the axioms we have adopted to date. This means that, assuming only the field axioms, any theorem that can be proved about the real numbers will also be true of the example given in Exercise 1.2.9.[3] In

---

3. For a detailed explanation of why this is true, see Excursion A.

particular, since A is finite, we will need something more than the field axioms to prove statement IV on our wish list. Furthermore, given that $1 = -1$ in the example, we have no hope of proving statement III, either.

What about statement V? It talks about our ability to take square roots of positive numbers. Of course, we understand intuitively what is meant by "positive numbers," but as yet we have no mathematical definition for them. We will need a way of getting at the positive numbers; statement V does not even make sense without it.

How might we define the positive numbers? Let's see—the positive numbers are those that are greater than zero. Unfortunately, this definition requires an order on $\mathbb{R}$, something *else* that we do not yet have. On the other hand, if we know how to recognize positive numbers, we can use this knowledge to define an order on $\mathbb{R}$: $x < y$ if $y - x$ is positive. Knowing when something is positive gives us an order on $\mathbb{R}$, and having an order on $\mathbb{R}$ tells us which real numbers are positive. The two concepts are mathematically equivalent.

You are now ready for Excursion A.

Excursion A. *Truth and Provability*

Synopsis: This excursion gives a brief discussion of the distinction between truth and provability. It works well as a reading supplement for Sections 1.2 and 1.3.

Problems 1.2

In the following problems, assume nothing beyond the field axioms and the results that have already been proved using them.

1. Let x, y, and z be real numbers. Prove that $(y + z)x = yx + zx$.

2. Let x be a non-zero real number and let a be any real number. Prove that if $x = ax$, then $a = 1$.

3. Prove that for all real numbers x, $(-1) \cdot x = -x$ (Theorem 1.2.8).

4. Let a, b, and c be real numbers.

    (a) Suppose that $a + b = a + c$. Prove that $b = c$.
    (b) Suppose that $ab = ac$. Is it true that $b = c$? Prove or disprove.

5. Prove that if x and y are in $\mathbb{R}$ and $x \cdot y = 0$, then $x = 0$ or $y = 0$.

6. Let a and b be real numbers. Prove that $(a+b)^2 = a^2 + 2ab + b^2$ and that $(a-b)^2 = a^2 - 2ab + b^2$ (where 2 is defined to be $1+1$ and we follow the convention that for all real numbers x, x^2 means $x \cdot x$).

7. If $x \neq 0$ and $y \neq 0$, show that $(xy)^{-1}$ exists and that $(xy)^{-1} = x^{-1}y^{-1}$.

8. Prove that $A = \{E, O\}$ with the defined operations of $\oplus$ and $\otimes$ described in Exercise 1.2.9 satisfies the field axioms.

1.3 Order

As indicated at the end of Section 1.2, statements III and V both touch on the question of positivity in their own way. The following axiom gives us the positive numbers.

Axiom II [Order Axiom] There exists a subset $\mathbb{R}^+$ of $\mathbb{R}$, called the **set of positive numbers**. This set satisfies two conditions:

1. The positive numbers are closed under addition and multiplication. That is, for all $x, y \in \mathbb{R}^+$,
$$x + y \in \mathbb{R}^+ \text{ and } x \cdot y \in \mathbb{R}^+.$$

2. Given any real number a, one and only one of the following is true:
   - a is positive.
   - a is zero.
   - $-a$ is positive.

We call the elements of $\mathbb{R}^+$ **positive numbers**. We call the complement of $\mathbb{R}^+ \cup \{0\}$ in $\mathbb{R}$ the **set of negative numbers** and denote it by $\mathbb{R}^-$.

Exercise 1.3.1 Now that we have the order axiom, the following facts can be easily deduced. Verify each one.

1. Let a be a real number. If a is positive, then $-a$ is negative. Conversely, if a is negative, then $-a$ is positive.

2. For all real numbers a, one and only one of the following holds:
   - a is a positive number.
   - a is zero.
   - a is a negative number.

3. The real number 1 is a positive number. (You will need both the order axiom and the field axioms to get this one!)

4. There exists a negative number.

5. For all non-zero real numbers a, $a \neq -a$.

∎

Now that we have the concept of the positive numbers, we can define an order on $\mathbb{R}$.

Definition 1.3.2 We define a relation $\leq$ on $\mathbb{R}$ as follows. For all real numbers x and y, we say that $x \leq y$ if $y - x \in \mathbb{R}^+ \cup \{0\}$.

We interpret the symbols $<$, $\geq$, and $>$ in the usual ways.

Exercise 1.3.3 Show that the positive numbers are precisely those numbers that are greater than zero and the negative numbers are precisely those numbers that are less than zero. ∎

Here are some important theorems about the order relation on $\mathbb{R}$. They are followed by some theorems about the way that $\leq$ and the arithmetic operations interact. You will be asked to provide their proofs in the problems at the end of this section.

Theorem 1.3.4 Show that $\leq$ on $\mathbb{R}$ satisfies the following three properties:

Reflexivity: For all $x \in \mathbb{R}$, $x \leq x$.

Antisymmetry: For all $x, y \in \mathbb{R}$, if $x \leq y$ and $y \leq x$, then $x = y$.

Transitivity: For all x, y, and $z \in \mathbb{R}$, if $x \leq y$ and $y \leq z$, then $x \leq z$.

Although we will not use the language, standard mathematical parlance would now say that $(\mathbb{R}, \leq)$ is a **partially ordered set**.

Proof: The proof is Problem 6 at the end of this section. □

Theorem 1.3.5 [Law of Trichotomy] Show that for all real numbers a and b, one and only one of the following is true:

- $a < b$.
- $a = b$.
- $a > b$.

Theorem 1.3.5, together with Theorem 1.3.4, tells us that $(\mathbb{R}, \leq)$ is a **totally ordered set**.

Proof: The proof is Problem 7 at the end of this section. □

Order and Arithmetic

Theorem 1.3.6 [Order and Arithmetic] Let a, b, c, and $d \in \mathbb{R}$.

1. If $a > b$ and $c \geq d$, then $a + c > b + d$.
2. If $a > b > 0$, and $c \geq d > 0$, then $ac > bd$.
3. If $a > b$, and $c < 0$, then $ac < bc$.

Proof: The proof is Problem 8 at the end of this section. □

Absolute Values

We have said that the ability to measure distances will play a crucial role in our discussion of closeness. The distance between two real numbers a and b is the absolute value of their difference. To get at this notion, we must first define absolute value.

Definition 1.3.7 Let a be a real number. The **absolute value** of a is denoted by $|a|$ and is defined to be

$$|a| = \begin{cases} a & \text{if } a \geq 0 \\ -a & \text{if } a < 0 \end{cases}.$$

Theorem 1.3.8 Let a and b be real numbers.

1. $|a| = a$ if a is positive, $|a| = -a$ if a is negative, and $|a| = 0$ if $a = 0$.
2. $|a| \geq 0$.
3. $|a| = \max\{a, -a\}$.[4]
4. $|a| = |-a|$.
5. $|a \cdot b| = |a||b|$.
6. $|a + b| \leq |a| + |b|$.
7. $|a - b| \geq |a| - |b|$, and $|a - b| \geq |b| - |a|$. Therefore, $|a - b| \geq ||a| - |b||$.
8. Let $r > 0$. Then $|a - b| < r$ if and only if $b \in (a - r, a + r)$.

Proof: The proof is Problem 10 at the end of this section. □

---

4. For two real numbers x and y, we define $\max\{x, y\}$ to be x if $x \geq y$ and y if $x < y$.

> Part 8 of Theorem 1.3.8 gives a *geometric* way of expressing the *analytical* fact that the real number b is closer than a distance of r to the real number a.
>
> > If the distance from a to b is less than r, b lies somewhere in the interval from $a - r$ to $a + r$.
>
> Many analytical ideas have parallel geometric interpretations. You will find that an ability to easily move back and forth between the analytical and geometric viewpoints will be very useful as you try to prove theorems in analysis.

Back to the "Wish List"

Where do we stand? Exercise 1.3.1 took care of statement III. We have also addressed the question of positivity, which puts us a little farther down the road toward discussing square roots. What about the infinitude of the set of real numbers?

To get at this question, we will need to be able to count. The natural numbers, which we use for counting, can be constructed using only set theory,[5] and given that we are accepting axiomatic set theory as part of our axiomatic structure, theorems about natural numbers (such as counting and induction results) are available for our use. However, there is a subtlety here. The set-theoretic "natural numbers" are not necessarily connected with the set $\mathbb{R}$ that we are studying in this chapter. We cannot *assume* that one is a subset of the other. We won't start out by doing so. Nevertheless, we shall soon identify a subset of $\mathbb{R}$ that "looks just like" the natural numbers constructed from set theory.

Let a be a positive real number. Let $k \in \mathbb{N}$. We define

$$ka = \underbrace{a + a + a + \cdots + a}_{k \text{ times}}$$

---

5. This is *not* obvious and does require proof. Assume that it has already been established. The proof can be found in any good book on axiomatic set theory. One of my favorites is *Introduction to Set Theory*, third edition, by Karel Hrbacek and Thomas Jech (Marcel Dekker, 1999).

> Note that ka is *not* assumed to be k multiplied by a. In fact, it would not make any sense to interpret it this way, because k is not an element of $\mathbb{R}$. The natural number k serves only as a counter: We are adding a to itself this number of times. Moreover, in the course of proving Lemma 1.3.9, we will refer to two natural numbers n and m, and we will assert that $n > m$. This, too, should be understood in the context of counting. It really means that na is adding up more a's than ma. Be sure to keep these things in mind as you read the proof of the lemma.

Lemma 1.3.9 Let a be a positive real number. Then for all distinct pairs of natural numbers n and m,

$$m a \neq n a.$$

Proof: We proceed by contradiction. Assume that n > m and that ma = na. Because + is well defined, we can add $-ma$ to both sides of this expression and maintain the equality. So, by using distribution of multiplication over addition, Theorem 1.2.8, and the associativity of addition (repeatedly), we see that

$$\begin{aligned}
0 = -ma + ma &= -ma + na \\
&= -1 \cdot ma + na \\
&= -1 \cdot \underbrace{(a + a + \cdots + a)}_{m \text{ times}} + \underbrace{a + a + \cdots + a}_{n \text{ times}} \\
&= \underbrace{-1 \cdot a + \cdots + -1 \cdot a}_{m \text{ times}} + \underbrace{a + \cdots + a}_{n \text{ times}} \\
&= \underbrace{-a + \cdots + -a}_{m \text{ times}} + \underbrace{a + \cdots + a}_{n \text{ times}} \\
&= \underbrace{a + a + \cdots + a}_{(n - m) \text{ times}}
\end{aligned}$$

Since n > m, we have n − m ≥ 1, the first provision of the order axiom and an easy inductive argument (see Problem 3c at the end of this section) tell us that, as a sum of positive quantities,

$$\underbrace{a + a + \cdots + a}_{n - m \text{ times}} \text{ is in } \mathbb{R}^+.$$

The second provision of the order axiom tells us that it cannot be both positive and 0. This yields the desired contradiction. □

Theorem 1.3.10 Show that $\mathbb{R}$ contains a sequence of distinct terms and is therefore infinite.

Proof: Exercise 1.3.1 tells us that $\mathbb{R}^+$ is non-empty. Let $a \in \mathbb{R}^+$. Lemma 1.3.9 tells us that

$$a, (a+a), (a+a+a), (a+a+a+a), \ldots$$

is a sequence of distinct terms. Therefore $\mathbb{R}$ is infinite. $\square$

If we take a to be 1 in Lemma 1.3.9, we obtain the following sequence:

$$1$$
$$2 = 1 + 1$$
$$3 = 1 + 1 + 1$$
$$4 = 1 + 1 + 1 + 1$$
$$\vdots$$
$$k = \underbrace{1 + 1 + \cdots + 1}_{k - \text{times}}$$
$$\vdots$$

This sequence is mathematically identical to the set-theoretic natural numbers we mentioned earlier. In particular, it satisfies all the properties that were assumed for the natural numbers in Chapter 0.[6] From now on, we will think of the natural numbers as sitting inside $\mathbb{R}$ and list them in the usual way: $\mathbb{N} = \{1, 2, 3, \ldots\}$.

> **Giving Names**
>
> Given that I have been very insistent that you not use any "facts" that you knew in a previous life without first proving them, you might be tempted to ask how I *know* that $1 + 1 = 2$. The answer is that 2 is the name that I am giving to the element $1 + 1 \in \mathbb{R}$. It is simply easier to talk about "5" than it is to talk about $1 + 1 + 1 + 1 + 1$ all the time. The only thing that is being assumed here (and it is a biggie, I'll grant you) is the stuff about this sequence being "just like" the natural numbers that are constructed from the axioms of set theory.

We obtain the integers $\mathbb{Z}$ by including 0 and the additive inverses of the natural numbers. Finally, we get the rational numbers by taking quotients of integers. Real numbers that are not rational, of course, we call "irrational." We

---

6. We would have a way to go to establish this, starting with an actual *definition* of what we mean by "the set-theoretic natural numbers"—not to mention some discussion of what is meant by "mathematically identical"!

have shown that $\mathbb{R}$ contains the natural numbers. The field axioms tell us that we can then construct integers and rational numbers. How do we know there *is* anything else? On the face of it, we do not. We have to do further work to get our hands on irrational numbers.

Theorem 1.3.11 There is no rational number whose square is 2.

Proof: The proof is Problem 13 at the end of this section. □

Question to Ponder: Theorem 1.3.11 does not, in itself, prove that there are irrational numbers. Explain why not.

Before we introduced the order axiom, we saw a mathematical structure (very different from the real numbers) that satisfied the field axioms, and we were thus able to conclude that we could not prove several important properties of the real numbers by using only the field axioms. Some of those problems have been resolved by accepting the order axiom, but not all of them, as you will see in Exercise 1.3.12. The rational numbers (considered not as a subset of $\mathbb{R}$, but as a mathematical structure in themselves) satisfy all the axioms given so far. This fact tells us that we will need another axiom to distinguish the real numbers from the rational numbers.

Exercise 1.3.12 Consider the rational numbers under ordinary addition and multiplication of real numbers. In Excursion B, you will be asked to verify that the rational numbers satisfy both the field axioms and the order axiom.

Assuming this fact, will you be able to prove that every positive real number has a square root by assuming only the field axioms and the order axiom? Explain. ■

You now have the theory necessary to prove many fundamental arithmetic and algebraic properties about the real numbers. The problems at the end of this section ask you to explore a few of these properties. Other important and useful facts can be found in Excursion B. In the interest of time, you may assume and use all of the usual arithmetic and algebraic facts about $(\mathbb{R}, +, \cdot)$ unless you are explicitly asked to prove one. However, for your own edification, I suggest that you pause from time to time when you are using something that is "common knowledge" and think about how it is connected to the theoretical structure we have built here. Can you see how you might prove it?

Problems 1.3

1. Prove that if $a \in \mathbb{R}$ and $b \in \mathbb{R}^+$, then $a < a + b$.

2. The fact that adding or multiplying two positive numbers together yields a positive number is the first provision of the order axiom. The standard "sign rules" for addition and multiplication follow. Let a and b be real numbers. Prove the following:

    (a) If $a < 0$ and $b < 0$, then $a + b < 0$.
    (b) If $a < 0$ and $b < 0$, then $a \cdot b > 0$.
    (c) If $a > 0$ and $b < 0$, then $a \cdot b < 0$.
    (d) There are analogous sign rules with $\geq$ and $\leq$. (For example, "If $a \geq 0$ and $b > 0$, then $a + b > 0$.") Formulate four such possible rules and prove them.

    (*Hint*: Keep in mind Theorem 1.2.8 as well as the provisions of the order axiom.)

3. Adding and multiplying finitely many real numbers:

    (a) Use mathematical induction to prove that for all $k \in \mathbb{N}$, the sum of k positive real numbers is a positive real number.
    (b) Use the result from part (a) to deduce directly that the sum of k negative real numbers is a negative real number.
    (c) Use mathematical induction to prove that for all $k \in \mathbb{N}$, the product of k positive real numbers is a positive real number.

4. Prove that $a \in \mathbb{R}^+$ if and only if $\frac{1}{a} \in \mathbb{R}^+$. Likewise, prove that $a \in \mathbb{R}^-$ if and only if $\frac{1}{a} \in \mathbb{R}^-$.

5. Prove that $\frac{1}{2} < 1$.

6. Prove that $\leq$ is a partial ordering on $\mathbb{R}$ (Theorem 1.3.4).

7. Prove that $(\mathbb{R}, \leq)$ satisfies the law of trichotomy and, therefore, that $\mathbb{R}$ is totally ordered under $\leq$ (Theorem 1.3.5).

8. Prove Theorem 1.3.6, which shows how order and arithmetic interact.

9. Here are a few standard arithmetic facts involving order. Try your hand at proving them using only the axioms and statements that we have proved so far.

    (a) Let a be a real number. Prove that if $a \geq 1$, then $a^2 \geq a$.

(b) Let $a \in \mathbb{R}^+$. Prove that if $a < 1$, then $a^2 < a$.

(c) Let a and $b \in \mathbb{R}^+$. Prove that if $a < b$, then $a^2 < b^2$.

(d) Let a and b be non-zero real numbers of the *same* sign (that is, both are positive or both are negative). Prove that if

$$a \leq b, \text{ then } \frac{1}{a} \geq \frac{1}{b}.$$

What can you say if a and b have different signs?

(e) Let a be a positive real number. Prove that

$$a > 1 \text{ if and only if } 0 < \frac{1}{a} < 1.$$

Likewise, prove that

$$0 < a < 1 \text{ if and only if } \frac{1}{a} > 1.$$

What can you say if a is a negative real number?

(f) Suppose that a, b, and $c \in \mathbb{R}$ and that $a \leq c$. Prove that

$$\frac{a}{b} \leq \frac{c}{b} \text{ if } b > 0,$$

and

$$\frac{a}{b} \geq \frac{c}{b} \text{ if } b < 0.$$

10. Prove Theorem 1.3.8, which establishes the properties of the absolute value.

11. Let K be a non-empty subset of $\mathbb{R}$. We say that K is **bounded** if K is bounded both above and below—that is, if there exist real numbers m and M such that for all $k \in K$, $m \leq k \leq M$. Prove that K is bounded if and only if there exists a real number T such that for all $k \in K$, $|k| \leq T$.

12. Footnote 4 on page 45 defined $\max\{x, y\}$ for two real numbers x and y.

    (a) Prove that if $x = y$, then $\max\{x, y\} = x = y$.

    The definition can be generalized to pick out the maximum of finitely many real numbers:

    $$\max\{x_1, x_2, x_3, \ldots, x_n\} = \max_{1 \leq i \leq n} \{x_i\}[7]$$

---

7. How would you do this carefully? Can you write down a rigorous mathematical definition?

(b) Let $x_1, x_2, \ldots, x_n$ and $y_1, y_2, \ldots, y_n$ be two finite collections of real numbers. Prove that

$$\max_{1 \leq i \leq n} \{x_i + y_i\} \leq \max_{1 \leq i \leq n} \{x_i\} + \max_{1 \leq i \leq n} \{y_i\}.$$

(c) Show by giving an example that the inequality derived in part (b) need not be equality.

13. Assume that $\sqrt{2}$ exists. Prove that it must be irrational (Theorem 1.3.11).

    (*Hint*: Proceed by contradiction. Assume that there is some rational number $\frac{m}{n}$ whose square is 2 and derive a contradiction.)

1.4 The Least Upper Bound Axiom

We noted in Section 1.3 that the rational numbers satisfy both the field axioms and the order axiom. However, since $\sqrt{2}$ is not a rational number, it is not true that every positive rational number has a rational square root. Thus we cannot hope to prove that every positive real number has a square root using only the field axioms and the order axiom. We need one more axiom—an axiom that will "fill the holes" left by the rational numbers. This axiom is called the least upper bound axiom. We begin by stating some important definitions.

Definition 1.4.1 Let $K \subseteq \mathbb{R}$, and let $x \in \mathbb{R}$.

1. x is said to be an **upper bound** for K if $x \geq k$ for all $k \in K$.

2. If x is an upper bound for K and $x \leq u$ for all upper bounds u of K, then x is said to be the **least upper bound** or **supremum** of K. It is denoted by $\sup K$.

3. If $x = \sup K$ and $x \in K$, then we call x the **greatest element** of K.

Exercise 1.4.2 This exercise will help you get a feeling for upper bounds and least upper bounds in $\mathbb{R}$.

1. List three upper bounds in $\mathbb{R}$ of the interval $(-3, 3]$. What is the *least* upper bound of $(-3, 3]$? Is it the greatest element?

2. Give an example of a subset of $\mathbb{R}$ that is not bounded above.

3. Give an example of a subset of $\mathbb{R}$ that has a least upper bound but no greatest element.

4. List three upper bounds for the set
$$S = \left\{\frac{1}{2}, \frac{2}{3}, \frac{3}{4}, \ldots\right\}.$$

What is $\sup S$? Is it a greatest element?

■

Exercise 1.4.3 Using Definition 1.4.1 as a model, devise definitions for lower bound, greatest lower bound or infimum, and least element. Repeat Exercise 1.4.2 replacing the phrase "upper bound" with "lower bound," "least upper bound" with "greatest lower bound," "sup" with "inf," and "greatest element" with "least element."

■

Theorem 1.4.4 Let B be a non-empty subset of $\mathbb{R}$ that is bounded above. Let $b \in \mathbb{R}$ be an upper bound for B. The following statements about b are equivalent:

1. $b = \sup B$.

2. For each positive number ϵ, there exists $x \in B$ such that $|x - b| < \epsilon$.

3. For each positive number ϵ, there exists $x \in B$ such that $x \in (b - \epsilon, b]$.

An analogous statement holds for non-empty subsets of $\mathbb{R}$ that are bounded below and greatest lower bounds.

Proof
($1 \implies 2$) We proceed by contraposition. Suppose there exists $\epsilon > 0$ such that for all $x \in B$, $|x - b| \geq \epsilon$. Then, by part 8 of Theorem 1.3.8, $(b - \epsilon, b + \epsilon) \cap B = \emptyset$. But b is an upper bound for B, so every element of B must be smaller than or equal to $b - \epsilon$. That is, $b - \epsilon$ is an upper bound for B that is strictly smaller than b; b cannot be the least upper bound of B.

($2 \implies 3$) Let $\epsilon > 0$. Choose $x \in B$ such that $|x - b| < \epsilon$. Then, by part 8 of Theorem 1.3.8, $x \in (b - \epsilon, b + \epsilon)$. But we know that b is an upper bound for B, so no element of B can be *larger* than b. Thus x must be in $(b - \epsilon, b]$.

($3 \implies 1$) We proceed by contraposition. Suppose that $b \neq \sup B$. Then there exists $u < b$ such that u is an upper bound for B. That is, there are no elements of B that are larger than u. Then if $\epsilon = b - u$, $(b - \epsilon, b] \cap B = \emptyset$. In other words, there exists a positive number ϵ such that no element of B is in the interval $(b - \epsilon, b]$.

□

> Theorem 1.4.4 does two things for us. First, it gives us several ways of looking at the concept of the least upper bound in $\mathbb{R}$. Second, and more importantly, the two new ways of thinking about least upper bounds are easier to handle analytically. When you are proving a theorem in which the least upper bound plays a role, these statements are concrete and easy to work with.
>
> Once again, we see the idea from both the analytical point of view, which is phrased in terms of distances, and the geometric point of view, which is phrased in terms of sets.

We can now state the final axiom for the real numbers.

Axiom III [Least Upper Bound Axiom] Every non-empty subset of $\mathbb{R}$ that is bounded above has a least upper bound. (As a shorthand, we say that $\mathbb{R}$ has the **least upper bound property**.)

> The least upper bound axiom is quite deep. To underscore this fact, it is worth making a special note of what is being assumed. There are subsets of $\mathbb{R}$ that are bounded below but that do not have a least element; indeed, you saw an example of such a set in Exercise 1.4.3. The least upper bound axiom says that any non-empty collection of real numbers that also happens to be the set of upper bounds of a non-empty subset of $\mathbb{R}$ *must* have a least element.

Theorem 1.4.5 Every non-empty subset of $\mathbb{R}$ that is bounded below has a greatest lower bound. (We might call this condition the *greatest lower bound property.*)

Proof: Let K be a non-empty subset of $\mathbb{R}$ that is bounded below. Then L, the set of lower bounds of K, is non-empty.

It is easy to see that every element of K is an upper bound for L. Because K is non-empty, L is bounded above.

Because L is non-empty and bounded above, the least upper bound axiom guarantees that L has a least upper bound. We shall call it b. Our claim is that b is also the greatest lower bound of K. To establish this, we must show two things:

- b is a lower bound for K.

- b is greater than or equal to every lower bound for K.

The second of these conditions follows automatically from the fact that b is an upper bound for L. Let's go back to the first condition. Suppose that b is not a lower bound for K. Then there exists $a \in K$ such that $a < b$. But, as an element of K, a is an upper bound for L. a cannot be an upper bound for L that is strictly smaller than the least upper bound. This contradiction shows that b is, indeed, a lower bound for K. $\square$

The least upper bound property of $\mathbb{R}$ has some very useful consequences. The first tells us that there are arbitrarily large natural numbers. Second, there are multiplicative inverses of natural numbers arbitrarily close to zero.

Theorem 1.4.6 [Archimedean Property of $\mathbb{R}$] For every real number x, there is a natural number n such that $n > x$.

Proof: We proceed by contradiction. Suppose that there exists a real number x that is larger than or equal to every natural number. Then $\mathbb{N}$ is bounded above and the least upper bound axiom tells us that it must have a least upper bound—say, R.

Now Theorem 1.4.4 tells us that there is a natural number n such that $|R - n| < \frac{1}{2}$. In turn, part 8 of Theorem 1.3.8 tells us that $R \in (n - \frac{1}{2}, n + \frac{1}{2})$ and, therefore, $n + 1 > R$. But this cannot be: $\mathbb{N}$ is closed under addition, so $n + 1$ is a natural number. At the same time, $n + 1$ is bigger than R, which is meant to be an upper bound for $\mathbb{N}$. This contradiction establishes our theorem. $\square$

Theorem 1.4.7 is an easy consequence of the Archimedean property. It is an extremely useful and important result in its own right.

Theorem 1.4.7 For every positive number ϵ, there exists a positive integer n such that $1/n < \epsilon$.

Proof: The proof is Problem 1 at the end of this section. $\square$

We are finally in a position to prove that every positive real number has a square root. The theorem is not trivial, as you will see.

Theorem 1.4.8 For every positive real number a, there exists a real number x such that $x^2 = a$.

Proof: This is an existence proof; the first thing we have to do is to produce a candidate for x. We know that we will have to use the least upper bound property to do this (*Why?*), so we need a non-empty subset of $\mathbb{R}$ that is bounded above. Define

$$S = \{y \in \mathbb{R} : y^2 < a\}.$$

S is clearly non-empty, since $0^2 = 0 < a$. To see that S is bounded above, we must consider two cases.

Case i.: $a < 1$. Fix $y \in S$. We have $y^2 < a < 1$, so the contrapositive of Problem 9a on page 50 tells us that $y < 1$, also. This shows that 1 is an upper bound for the set S.

Case ii.: $a \geq 1$. Fix $y \in S$. If $y < 1$, then $y < a$. If $y \geq 1$, then $y \leq y^2 < a$ (again, Problem 9a on page 50). Hence all elements of S are smaller than a, and a is an upper bound for S.

Let $x = \sup S$. Note that $x > 0$. (*Why?*) The claim is that $x^2 = a$. To establish this, we will show that $x^2 \not> a$ and that $x^2 \not< a$, then appeal to the law of trichotomy (Theorem 1.3.5).

Suppose that $x^2 < a$. Let $r = a - x^2$. Now choose a real number $\epsilon > 0$ that satisfies the following properties:

- $\epsilon < 1$.
- $\epsilon < \dfrac{r}{2x+1}$.

> **Notational Tip**
>
> We could state these two conditions more compactly by specifying that
> $$\epsilon < \min\left(1, \frac{r}{2x+1}\right).$$

Now consider $(x+\epsilon)^2$:

$$(x+\epsilon)^2 = x^2 + 2x\epsilon + \epsilon^2$$
$$< x^2 + 2x\epsilon + \epsilon$$
$$= x^2 + \epsilon(2x+1)$$
$$< x^2 + r = a.$$

Thus $x + \epsilon$ is larger than x and an element of S, contradicting the fact that x is an upper bound for S. We conclude that $x^2 \not< a$.

Suppose that $x^2 > a$. Let $r = x^2 - a$. Choose a real number $\epsilon > 0$ so that $\epsilon < \min(x, \frac{r}{2x})$. Notice that $x - \epsilon > 0$. Consider $(x - \epsilon)^2$.

$$(x-\epsilon)^2 = x^2 - 2x\epsilon + \epsilon^2$$
$$> x^2 - 2x\epsilon$$
$$> x^2 - 2x\left(\frac{r}{2x}\right)$$
$$= x^2 - r = a$$

Thus we have a positive number smaller than x whose square is larger than a. But the square function is increasing for positive numbers, so there can be no real number between $x-\epsilon$ and x whose square is less than a. By Theorem 1.4.4, this contradicts the fact that x is the least upper bound of S. We conclude that $x^2 \not> a$.

Thus x is a positive number whose square is a. □

Corollary 1.4.9 There exists an irrational number.

We now have a set of axioms that gives us properties I–V as stated in Section 1.2. In fact, we have all the axioms we need to completely describe the real number system, in a technical sense that I will not elaborate on here.[8]

Another very useful consequence of the least upper bound axiom tells us that intersections of non-empty, nested, closed intervals are always non-empty.

Theorem 1.4.10 [Nested Interval Theorem] Every nested sequence of non-empty closed intervals in $\mathbb{R}$ has a non-empty intersection.

To be more explicit, let (a_i) and (b_i) be sequences of real numbers satisfying the following properties:

- For all $i \in \mathbb{N}$, $a_i \leq b_i$.

- For all $i \in \mathbb{N}$, $a_i \leq a_{i+1}$ and $b_i \geq b_{i+1}$.

Then $\bigcap_{i=1}^{\infty} [a_i, b_i] \neq \emptyset$.

Proof: The proof is Problem 6 at the end of this section. □

Definition 1.4.11 Let A be a set. We know that A is infinite if it contains a sequence of distinct terms—that is, if we can find distinct elements $a_1, a_2, a_3, \ldots$ in A.

- An infinite set A is said to be **countable** if there is a one-to-one correspondence between $\mathbb{N}$ and A, that is, if there is a sequence of distinct terms in A that exhausts all of A:

$$A = \{a_1, a_2, a_3 \ldots\}.$$

---

8. The proof is nontrivial and requires a detour into set theory that we will not take.

- An infinite set A is said to be **uncountable**, if it is not countable. It is not too hard to show that this is equivalent to saying that no function $f : \mathbb{N} \to A$ is onto.

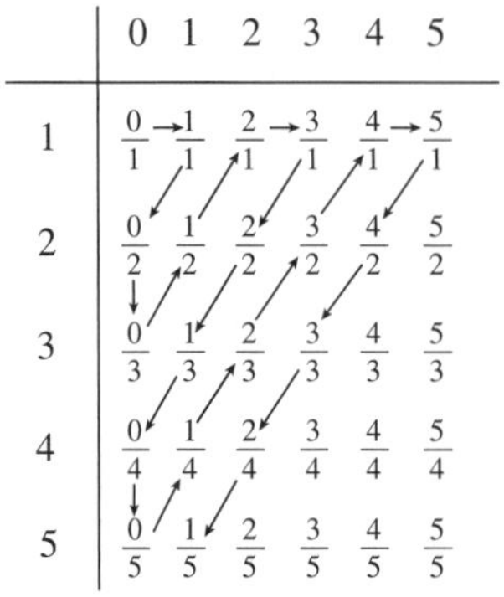

	0	1	2	3	4	5
1	$\frac{0}{1}$	$\frac{1}{1}$	$\frac{2}{1}$	$\frac{3}{1}$	$\frac{4}{1}$	$\frac{5}{1}$
2	$\frac{0}{2}$	$\frac{1}{2}$	$\frac{2}{2}$	$\frac{3}{2}$	$\frac{4}{2}$	$\frac{5}{2}$
3	$\frac{0}{3}$	$\frac{1}{3}$	$\frac{2}{3}$	$\frac{3}{3}$	$\frac{4}{3}$	$\frac{5}{3}$
4	$\frac{0}{4}$	$\frac{1}{4}$	$\frac{2}{4}$	$\frac{3}{4}$	$\frac{4}{4}$	$\frac{5}{4}$
5	$\frac{0}{5}$	$\frac{1}{5}$	$\frac{2}{5}$	$\frac{3}{5}$	$\frac{4}{5}$	$\frac{5}{5}$

Figure 1.1 The rational numbers are countable.

As indicated in Figure 1.1, $\mathbb{Q}$ is a countable set. Interestingly, the real numbers are not countable.[9] Given that the rational numbers satisfy both the field and order axioms, it follows that it must be impossible to prove the uncountability of the real numbers without using the least upper bound property. The idea is simple, but the execution requires some "delicate" handling. It can be proved, however, using the nested interval theorem.

Theorem 1.4.12 The set of real numbers is an uncountable set.

Proof: The idea of the proof is as follows. Suppose that $x_1, x_2, x_3, \ldots$ is a sequence of distinct terms in $\mathbb{R}$. The proof consists of building a sequence of nonempty, nested, closed intervals that gradually "exclude" more and more terms of (x_n). If you are careful, the intersection of the intervals will be nonempty, by the nested interval theorem, and will include none of the terms of the original sequence. That is, no sequence of real numbers can exhaust $\mathbb{R}$. The actual proof is left to you in Problem 7 at the end of this section. □

In the problems at the end of this section, you will look at how integers, rational numbers, and irrational numbers "fit together" in $\mathbb{R}$. In particular,

---

9. The best-known proof of this fact, with which you may already be familiar, is called *Cantor's diagonalization argument*, which can be found in many introductory texts (see, for instance, [SCH]). It involves writing each real number in the interval $(0, 1)$ in decimal form, assuming the existence of a function from $\mathbb{N}$ to this set, and explicitly constructing a real number that is not in the range of the function. Interestingly, this was not Cantor's first proof of this fact. His first proof more closely resembled the one we suggest here, although it was a bit more general.

you will show that any open interval in ℝ contains both rational and irrational numbers. It is somewhat amazing to contemplate how much we have gotten from so few assumptions. The set we started with was not even assumed to be infinite. We have now shown that it contains all the familiar characters—the natural numbers, the integers, the rationals and irrationals—and that these characters behave in familiar ways.

It is certain that none of the results proved in this chapter is surprising or unfamiliar to you. That's good! It means that the axioms are doing what they are supposed to be doing—describing the real numbers with which we are familiar. The real numbers hold some surprises, but they are not easy consequences of the axioms. (The easy consequences we all know about!) You will have to go further in your study of mathematics before you enter unfamiliar territory. However, we now have some evidence that the axioms are good ones. When you later learn something surprising from them, you can be confident in that result.

You are now ready for Excursions B and C (Section 1).

Excursion B. *Number Properties*

Synopsis: This excursion is a collection of useful facts about real numbers. They are left as exercises and can be used to supplement the problems given in Chapter 1.

Excursion C. *Integer and Rational Exponents*

Synopsis: Mathematics students often hear vague descriptions about the development of exponent "facts." The general hierarchy is to start with positive integer powers, then step up to the negative integer powers, then integer roots, then rational powers, and finally irrational powers. Excursion C is written as a set of interconnected exercises that take the reader through this development. Section C.1 in this excursion on integer and rational powers, requires only the background provided by Chapter 1. Section C.2 in the same excursion on irrational powers, requires information about continuity and convergence from Chapters 4 and 5 and is not accessible at this point.

Problems 1.4

1. Prove that there are multiplicative inverses of natural numbers arbitrarily close to zero (Theorem 1.4.7).

2. The following facts about the real numbers are simple to prove, but amazingly useful.

    (a) Let $r \in \mathbb{R}$, $r \geq 0$. Suppose that for all positive real numbers ϵ, $r \leq \epsilon$. Prove that $r = 0$.

    (b) Let x and $y \in \mathbb{R}$. Suppose that for all positive real numbers ϵ, $x < y + \epsilon$. Prove that $x \leq y$.

3. Let $t \in \mathbb{R}$ and let $S \subset \mathbb{R}$ that is bounded above. We can define a new set
$$tS = \{ts : s \in S\}.$$
Prove that tS is bounded above and that $\sup tS = t \sup S$.

4. Let S and T be subsets of $\mathbb{R}$ that are bounded above. Define a new set
$$S + T = \{s + t : s \in S \text{ and } t \in T\}.$$
Prove that $S + T$ is bounded above and that $\sup(S+T) = \sup S + \sup T$.

5. Let $S = \{s_\alpha\}_{\alpha \in \Lambda}$ and $T = \{t_\alpha\}_{\alpha \in \Lambda}$ be subsets of real numbers, both indexed over the same set Λ. Suppose that both S and T are bounded above. In this case, one might sensibly define $S + T$ differently than in Problem 4:[10]
$$S + T = \{s_\alpha + t_\alpha\}_{\alpha \in \Lambda}.$$

    (a) Prove that $S + T$ is bounded above and that
    $$\sup(S + T) \leq \sup S + \sup T.$$

    (b) Show by giving an example that the inequality in part (a) may be strict.

---

[10]. Note the importance of context in the two distinct ways of interpreting the symbols $S+T$!

6. Prove the nested interval theorem (Theorem 1.4.10).

7. Use the nested interval theorem to prove that $\mathbb{R}$ is uncountable (Theorem 1.4.12). *A word of caution:* As you construct your nested intervals, you will need to make sure that infinitely many of the intervals do not share an endpoint that is a term of (x_n). If they do, the intersection will be non-empty, but you will not have enough information to conclude that the intersection of the intervals contains anything other than one of the terms of (x_n).

8. Integers, rationals, reals, and how they fit together.

    (a) Show that for any $x \in \mathbb{R}$ there is an integer n such that $n-1 \leq x < n$. That is, every real number lies between two *consecutive* integers.

    (b) Show that for any $x \in \mathbb{R}$ and any positive integer N, there exists an integer n such that
    $$\frac{n-1}{N} \leq x < \frac{n}{N}.$$

    (c) Show that if $x \in \mathbb{R}$, and $\epsilon > 0$, then there exists a rational number r such that $|x - r| < \epsilon$. That is, any real number can be approximated as closely as we like by a rational number.

    (d) Let x and y be real numbers with $x < y$. Use parts (a)–(c) of this problem to show that for all $x, y \in \mathbb{R}$, there exists a rational number r satisfying $x < r < y$.

    (e) There is a complementary fact about irrational numbers. Let x and y be real numbers with $x < y$. Then there is an irrational number s satisfying $x < s < y$. (*Hint*: The easy way to get this fact is to use a cardinality argument, though it can also be done by explicit construction of a number s that satisfies the required properties.)

Chapter 2
Measuring Distances

2.1 Metric Spaces

The introduction states that analysis is the study of closeness. Without talking about distances, we cannot even begin to define closeness.

In this chapter we will consider a mathematical structure called a metric space. That's the technical phrase we use to describe a set in which we can measure distances. You already have at your fingertips several handy examples of metric spaces: The real line is a metric space; the plane and three-dimensional space are metric spaces.

Let us think about what actually happens when we determine distance. Given two objects whose distance from each other we wish to measure, we follow some procedure,[1] the end result of which is a real number that depicts the distance between the objects.

What has happened here? We have taken a pair of objects and, using some procedure, assigned a real number to that pair of objects. Casting this in mathematical language, we take our objects from a particular set. The procedure for assigning a real number to any pair of elements of that set is a *function* whose domain is the Cartesian product of the set with itself and whose range is a subset of the set of real numbers. Such a function is called a "distance function."

---

1. For instance, we might stretch a measuring tape from one object to the other and read the result or do some prespecified mathematical computation.

Definition 2.1.1 Let X be a non-empty set. Let $d : X \times X \to \mathbb{R}$ be a real-valued function with the following properties:

Positive: For all $a, b \in X$, $d(a, b) \geq 0$.
Positive Definite: For $a, b \in X$, $d(a, b) = 0$ if and only if $a = b$.
Symmetric: For all $a, b \in X$, $d(a, b) = d(b, a)$.
Triangle Inequality: For all a, b, and $c \in X$, $d(a, c) \leq d(a, b) + d(b, c)$.

The function d is called the **distance function** or **metric** on X. The set X together with d (which we denote by the ordered pair (X, d)) is called a **metric space**. The elements of a metric space are called **points**.

Exercise 2.1.2 In your own words, give a rationale for the properties that were required of the distance function. Include a rationale for the names that are associated with the various properties. ∎

Metric spaces can have a fairly complex structure (such as that of the real numbers or the Cartesian plane) or they can have a very simple structure. The example given in Exercise 2.1.3 shows that *any* set can be made into a metric space (granted, a fairly uninteresting one) by defining a rather simple metric.

Exercise 2.1.3 Let X be any non-empty set. For $a, b \in X$, define

$$d(a, b) = \begin{cases} 0 & \text{if } a = b \\ 1 & \text{if } a \neq b \end{cases}$$

For obvious reasons, this metric is often called the **trivial metric**.

1. Verify that (X, d) is a metric space.

2. Do you agree or disagree with the following statement? Why?

   When we endow an arbitrary set with the trivial metric, we really add no significant mathematical information.

∎

Exercise 2.1.4 Verify that $(\mathbb{R}, d)$ is a metric space, where $d(a, b) = |a - b|$. (*Hint*: Refer to Theorem 1.3.8.) ∎

Example 2.1.5 Let (X, d) be any metric space. Let S be any subset of X. Show that S is a metric space under the distance function obtained by restricting d to pairs of elements of S. In this case, we say that (S, d) is a **subspace** of (X, d).

Remark: If the pair (X, d) is a metric space, and the metric d is understood, it is common practice to simply say that X is a metric space. Although this terminology is technically incorrect, it rarely leads to real confusion and makes life and language a lot easier.

This is, in fact, exactly what we did when we said that $\mathbb{R}$, $\mathbb{R}^2$, and $\mathbb{R}^3$ were examples of metric spaces—we implied that everyone knew which distance function we meant. We were, of course, referring to the distance functions that are familiar to most high school students.

2.2 The Euclidean Metric on $\mathbb{R}^n$

Example 2.2.1 The distance formulas for $\mathbb{R}^2$ and $\mathbb{R}^3$ are

$$d((a_1, a_2), (b_1, b_2)) = \sqrt{(a_1 - b_1)^2 + (a_2 - b_2)^2},$$

and

$$d((a_1, a_2, a_3), (b_1, b_2, b_3)) = \sqrt{(a_1 - b_1)^2 + (a_2 - b_2)^2 + (a_3 - b_3)^2}.$$

The pattern you see here works for any $n \in \mathbb{N}$. The usual distance formula on $\mathbb{R}^n$ is

$$= \frac{d((a_1, \ldots, a_n), (b_1, \ldots, b_n))}{\sqrt{(a_1 - b_1)^2 + (a_2 - b_2)^2 + \cdots + (a_n - b_n)^2}}$$

The usual metric on $\mathbb{R}^n$ is often called the *Euclidean distance* or the *Euclidean metric*. This book follows this common practice. Whenever it refers to the metric space $\mathbb{R}^n$ without mention of a specific metric, it is understood that the metric in question is the Euclidean metric.

Remark on Notation: Suppose we have an element $(a_1, a_2, \ldots, a_n)$ in $\mathbb{R}^n$. It is not always necessary to specify the individual components; thus it is convenient to refer to elements of $\mathbb{R}^n$ in a more compact form—say, **a** for the vector $(a_1, a_2, \ldots, a_n)$ above. Note that the vector appears in boldface. Given that vectors (elements of $\mathbb{R}^n$) and scalars (elements of $\mathbb{R}$) frequently crop up in the same calculations, the bold typeface will help you keep track of which symbols represent vectors and which represent scalars. It is difficult to boldface letters when you are writing by hand, so you may prefer to write your vectors in the more standard $\vec{a}$ form with a little arrow on top.

Exercise 2.2.2 Verify that the Euclidean metric on $\mathbb{R}^n$ is positive, positive definite, and symmetric. ∎

To see that $\mathbb{R}^n$ together with the Euclidean metric is, indeed, a metric space, we need verify only that the triangle inequality holds.

As is often the case with the Euclidean distance for $\mathbb{R}^n$, practical considerations make it more convenient to perform calculations on the *square* of the distance rather than on the distance itself. That is the case here. If we can prove that

$$(d(\mathbf{a},\mathbf{c}))^2 \leq (d(\mathbf{a},\mathbf{b}) + d(\mathbf{b},\mathbf{c}))^2 = (d(\mathbf{a},\mathbf{b}))^2 + 2d(\mathbf{a},\mathbf{b})d(\mathbf{b},\mathbf{c}) + (d(\mathbf{b},\mathbf{c}))^2$$

then the fact that the square root function is increasing for positive values will allow us to take square roots of both sides of the inequality and thus to obtain the triangle inequality for $\mathbb{R}^n$. We now do some preliminary calculations for points $\mathbf{a} = (a_1, a_2, \ldots, a_n)$, $\mathbf{b} = (b_1, b_2, \ldots, b_n)$, and $\mathbf{c} = (c_1, c_2, \ldots, c_n)$ in $\mathbb{R}^n$:

$$(d(\mathbf{a},\mathbf{c}))^2 = \sum_{i=1}^{n} (a_i - c_i)^2$$

$$= \sum_{i=1}^{n} (a_i - b_i + b_i - c_i)^2$$

$$= \sum_{i=1}^{n} \left((a_i - b_i)^2 + 2(a_i - b_i)(b_i - c_i) + (b_i - c_i)^2 \right)$$

$$= \sum_{i=1}^{n} (a_i - b_i)^2 + 2\sum_{i=1}^{n} (a_i - b_i)(b_i - c_i) + \sum_{i=1}^{n} (b_i - c_i)^2$$

$$= (d(\mathbf{a},\mathbf{b}))^2 + 2\left(\sum_{i=1}^{n} (a_i - b_i)(b_i - c_i)\right) + (d(\mathbf{b},\mathbf{c}))^2.$$

Compare the following two expressions:

(We have) $\quad (d(\mathbf{a},\mathbf{c}))^2 = (d(\mathbf{a},\mathbf{b}))^2 + 2\sum_{i=1}^{n} (a_i - b_i)(b_i - c_i) + (d(\mathbf{b},\mathbf{c}))^2$

(We need) $\quad (d(\mathbf{a},\mathbf{c}))^2 \leq (d(\mathbf{a},\mathbf{b}))^2 + 2d(\mathbf{a},\mathbf{b})d(\mathbf{b},\mathbf{c}) + (d(\mathbf{b},\mathbf{c}))^2$

To complete the proof, we clearly need to prove that for all $\mathbf{a} = (a_1, a_2, \ldots, a_n)$, $\mathbf{b} = (b_1, b_2, \ldots, b_n)$, and $\mathbf{c} = (c_1, c_2, \ldots, c_n)$ in $\mathbb{R}^n$,

$$\sum_{i=1}^{n} (a_i - b_i)(b_i - c_i) \leq d(\mathbf{a},\mathbf{b})d(\mathbf{b},\mathbf{c}) = \left(\sum_{i=1}^{n} (a_i - b_i)^2\right)^{\frac{1}{2}} \left(\sum_{i=1}^{n} (b_i - c_i)^2\right)^{\frac{1}{2}}$$

In fact, we can prove a stronger relation. For all $\mathbf{a} = (a_1, a_2, \ldots, a_n)$, $\mathbf{b} = (b_1, b_2, \ldots, b_n)$, and $\mathbf{c} = (c_1, c_2, \ldots, c_n)$ in $\mathbb{R}^n$,

$$\left|\sum_{i=1}^{n} (a_i - b_i)(b_i - c_i)\right| \leq \left(\sum_{i=1}^{n} (a_i - b_i)^2\right)^{\frac{1}{2}} \left(\sum_{i=1}^{n} (b_i - c_i)^2\right)^{\frac{1}{2}}$$

This useful fact is called the Cauchy–Schwarz inequality. Before we state it formally, observe that though there appear to be three parameters in the preceding relation, there are really only two. One is the difference of a_i and b_i; the other is the difference of b_i and c_i. We can get a "cleaner" relation by setting $u_i = a_i - b_i$ and $v_i = b_i - c_i$.

The Cauchy–Schwarz Inequality

Lemma 2.2.3 [Cauchy–Schwarz Inequality] For all $\mathbf{u} = (u_1, u_2, \ldots, u_n)$, and $\mathbf{v} = (v_1, v_2, \ldots v_n)$ in $\mathbb{R}^n$,

$$\left| \sum_{i=1}^{n} u_i v_i \right| \leq \left(\sum_{i=1}^{n} (u_i)^2 \right)^{\frac{1}{2}} \left(\sum_{i=1}^{n} (v_i)^2 \right)^{\frac{1}{2}}$$

Equality holds if and only if $\mathbf{v}$ is a scalar multiple of $\mathbf{u}$. That is, equality holds if there exists $t \in \mathbb{R}$ such that $\mathbf{v} = t\mathbf{u}$.

Proof: The proof of the Cauchy–Schwarz inequality (along with accompanying commentary about the intuition behind the proof) is outlined in Problem 7 at the end of this section. □

Corollary 2.2.4 $\mathbb{R}^n$ is a metric space under the Euclidean metric.

Thinking Intuitively

Stepping outside of our strict chain of reasoning for a moment, we appeal to "previous knowledge" for intuition only. Suppose that $\mathbf{u} = (u_1, u_2)$ and $\mathbf{v} = (v_1, v_2)$ are vectors in the plane. The expression on the left of the Cauchy–Schwarz inequality is the absolute value of the dot product of $\mathbf{u}$ and $\mathbf{v}$:

$$|\mathbf{u} \cdot \mathbf{v}| = |u_1 v_1 + u_2 v_2|$$

The expression on the right is the product of their magnitudes:

$$\|\mathbf{u}\|\|\mathbf{v}\| = \sqrt{u_1^2 + u_2^2} \sqrt{v_1^2 + v_2^2}$$

As you probably know from previous experience with vectors, we can think of $\mathbf{u}$ and $\mathbf{v}$ as "arrows" with lengths $\|\mathbf{u}\|$ and $\|\mathbf{v}\|$. An alternate expression for the dot product of $\mathbf{u}$ and $\mathbf{v}$ is

$$\mathbf{u} \cdot \mathbf{v} = \|\mathbf{u}\|\|\mathbf{v}\| \cos(\theta)$$

where θ is the angle "between" $\mathbf{u}$ and $\mathbf{v}$.

> One can loosely think of the Cauchy–Schwarz inequality as a measure of how far **v** is from being a scalar multiple of **u**. If **v** is "very close" to being a scalar multiple of **u**, then $|\cos(\theta)|$ will be "very close" to 1, so the expressions on the left and right sides of the Cauchy–Schwarz inequality will be approximately equal. When the "arrows" are close to perpendicular, the left side will be considerably smaller than the right side.

Problems 2.2

1. Let (X, d) be a metric space. Let $x, y \in X$. Suppose that for all positive real numbers ϵ, $d(x, y) \leq \epsilon$. Prove that $x = y$.

   > One way to prove that two elements of a metric space are equal is by showing that the distance between them is zero. However, this is often difficult to do directly. It is often easier to show that the distance between the elements is arbitrarily small. As Problem 1 shows, the result is the same.

2. Let (X, d) be a metric space. Prove that for all $a, b, c, \in X$,
$$|d(a, b) - d(b, c)| \leq d(a, c).$$

3. Let (X, d) be a metric space. Use mathematical induction to prove the following generalization of the triangle inequality. Suppose that $a_1, a_2, a_3, \ldots, a_k$ are in X and that $k \geq 3$. Then
$$d(a_1, a_k) \leq \sum_{i=1}^{k-1} d(a_i, a_{i+1}).$$

   (*Hint:* You will understand this theorem better if you draw a picture.)

4. Suppose we want to find a way to measure the distance between two real functions. Here's a possibility: For two functions $f : \mathbb{R} \to \mathbb{R}$ and $g : \mathbb{R} \to \mathbb{R}$, define $d(f, g) = |f(0) - g(0)|$. Is this a metric? Why or why not?

5. In the text we talked about the "usual" metric on $\mathbb{R}^n$—the Euclidean metric. There are other distance functions on $\mathbb{R}^n$ as well. Show that each of the following functions is a distance function on $\mathbb{R}^n$. Let $\mathbf{a} = (a_1, \ldots, a_n)$ and $\mathbf{b} = (b_1, \ldots, b_n)$ be points in $\mathbb{R}^n$.

(a) $d(\mathbf{a}, \mathbf{b}) = \sum_{i=1}^{n} |a_i - b_i|$. This metric is usually called the ℓ_1 metric (which is read "little ell-one" metric). In $\mathbb{R}^2$ it is sometimes called the "taxicab metric." Explain why. (*Hint*: Think geometrically.)

(b) $d(\mathbf{a}, \mathbf{b}) = \max_{1 \leq i \leq n} |a_i - b_i|$.

6. Consider $\mathbb{R}^2$ under each of the metrics given in Problem 5. In each case, describe (geometrically) the set of all points that lie less than 10 units from the origin. Explain your reasoning.[2]

7. Use the following outline to prove that the Cauchy–Schwarz inequality (Lemma 2.2.3) holds.

   Step 1. Verify that the inequality holds trivially if either of the two vectors is the zero vector. Check that you get equality of the two expressions if $\mathbf{v} = t\mathbf{u}$, where t is some real number.

   > **The Main Idea**
   >
   > As suggested in the "intuitional aside" in the box on page 68, we are interested in how close $\mathbf{v}$ is to being a scalar multiple of $\mathbf{u}$.
   >
   > Let $t \in \mathbb{R}$. Note that the expression
   >
   > $$\sum_{i=1}^{n}(v_i - tu_i)^2$$
   >
   > is always positive and that it is zero if and only if $\mathbf{v} = t\mathbf{u}$. The idea of the proof is to find a value of t that will make this expression as small as possible.

   Step 2. Note that

   $$0 \leq \sum_{i=1}^{n}(v_i - tu_i)^2$$

   for all real numbers t, and assume that $\mathbf{u} \neq \mathbf{0}$. Expand the sum to get a quadratic expression in t. (The coefficients will be "relevant" portions of the Cauchy–Schwarz inequality.)

---

2. In Chapter 3, you will see that this is called "the open ball of radius 10 centered at the origin."

> **Off-Stage Whisper**
>
> Notice that the inequality derived in Step 2 is valid for all real numbers t. Use your previous knowledge of elementary calculus to find the value of t that will minimize the quadratic expression.[3]

Step 3. In your proof, let t be the value that you derived "off-stage." Substitute your value for t into the quadratic inequality you got in Step 2.

Step 4. Notice that the expression obtained in Step 3 yields the Cauchy–Schwarz inequality.

8. The metrics that were examined in Problem 5 can be viewed as special cases of a more general idea. Suppose that (X, d_X) and (Y, d_Y) are metric spaces. We can define a distance function on $X \times Y$ by "combining" the distances from the component spaces.

    (a) For (a, b) and (c, d) in $X \times Y$, define
    $$d((a,b),(c,d)) = d_X(a,c) + d_Y(b,d).$$
    Prove that d is a metric on $X \times Y$.

    (b) For (a, b) and (c, d) in $X \times Y$, define
    $$d((a,b),(c,d)) = \max(d_X(a,c), d_Y(b,d)).$$
    Prove that d is a metric on $X \times Y$.

    (c) Can the schemes in parts (a) and (b) be generalized to the situation where we have n metric spaces $(X_1, d_1), (X_2, d_2), \ldots, (X_n, d_n)$? What if we have countably many spaces (X_i, d_i) for $i = 1, 2, 3, \ldots$?

9. Consider the set S of all bounded sequences of real numbers. For each pair $\mathbf{x} = (x_1, x_2, x_3, \ldots)$ and $\mathbf{y} = (y_1, y_2, y_3, \ldots)$ of elements of S, define
   $$d(\mathbf{x}, \mathbf{y}) = \sup\{|x_i - y_i| : i \in \mathbb{N}\}.$$

    (a) Use d as defined previously to compute the distances between the following pairs of sequences:
    - $(1, 0, 0, 0, 0, \ldots)$ and $(0, 1, 0, 0, 0, 0, \ldots)$
    - $(1, 0, 1, 0, 1, 0, \ldots)$ and $(0, -1, 0, -1, 0, -1, \ldots)$

---

[3] Many proofs of the Cauchy–Schwarz inequality seem mysterious because this calculation is not part of the proof. In fact, how you get the value for t is irrelevant from a mathematical standpoint. If you prefer, you can gaze into your crystal ball or read the entrails of birds. Appealing to calculus is just more likely to get you a "judicious" choice for t.

- $(1, 2, 3, 1, 2, 3, \ldots)$ and $(-1, -2, -3, -1, -2, -3, \ldots)$
- $(\frac{1}{2}, \frac{1}{3}, \frac{1}{4}, \frac{1}{5}, \ldots)$ and $(\frac{1}{2}, \frac{2}{3}, \frac{3}{4}, \frac{4}{5}, \ldots)$

(b) Verify that d is a well-defined function. That is, show that given any two bounded sequences of real numbers, d assigns one and only one real number as their distance. (In particular, how do you know that the least upper bound exists in all cases?)

(c) Verify that d is positive and positive definite.

(d) Verify that d is symmetric.

(e) Prove that d satisfies the triangle inequality.

You have just verified that d is a distance function for S. The metric space (S, d) is ordinarily called ℓ_∞, which is pronounced "little-ell-infinity."[4]

---

[4]. ℓ_∞ and the space ℓ_1 discussed in Problem 5b are called "little" to distinguish them from two other famous spaces, $\mathcal{L}^\infty$ ("big-ell-infinity") and $\mathcal{L}^1$ ("big-ell-one").

Chapter 3

Sets and Limits

3.1 Open Sets

In our study of limiting processes we will frequently be concerned with all points that lie within a certain distance of a fixed point.

Definition 3.1.1 Let X be a metric space and let $a \in X$. Fix a real number $r > 0$. Then the set

$$B_r(a) = \{x \in X : d(a, x) < r\}$$

is called the **open ball of radius r centered at a**.

$$C_r(a) = \{x \in X : d(a, x) \leq r\}$$

is called the **closed ball of radius r centered at a**.

Exercise 3.1.2 Let X be any non-empty set. Recall the trivial metric space defined in Exercise 2.1.3. Let a be any point in the space. Describe the open and closed balls of radii $\frac{1}{2}$, 1, and 5 centered at a. ∎

Exercise 3.1.3 Describe the open and closed balls in $\mathbb{R}$, $\mathbb{R}^2$, and $\mathbb{R}^3$. ∎

Definition 3.1.4 Let (X, d) be a metric space. A subset U of X is called an **open subset of X** or simply an **open set** if it is the union of a collection of open balls. (See Figure 3.1.)

Figure 3.1 Open set in $\mathbb{R}^2$: A union of open balls.

Exercise 3.1.5 Show that open balls are open sets. ∎

Exercise 3.1.6 Verify that if X is a metric space, both X and $\emptyset$ are open subsets of X. ∎

In Theorem 3.1.7 we see, once again, the duality between a geometric viewpoint, which is given in terms of sets, and an analytic viewpoint, which is given in terms of distances. The first two provisions of the theorem are geometric descriptions of open sets; the third provision is an analytic view of open sets.

Theorem 3.1.7 Let (X, d) be a metric space. Let U be a subset of X. The following statements are then equivalent.

1. U is an open set.

2. For every $a \in U$, there exists $r > 0$ such that $B_r(a) \subseteq U$.

3. For every $a \in U$, there exists $r > 0$ such that if $d(x, a) < r$, then $x \in U$.

Proof: The proof is Problem 1 at the end of this section. □

Theorem 3.1.7 gives us different ways of looking at open sets. All of these perspectives are important to us when we are proving theorems. If we need an open set in a proof, we often build it by taking unions of open balls. If an open set is given to us, it is almost always easier to exploit this fact by making use of the second or third provision of Theorem 3.1.7, which are local and more concrete. Likewise, if we have a set and want to know whether it is open, it is usually easiest to show that the second provision holds (see Figure 3.2). Because the statements in Theorem 3.1.7 are equivalent formulations of the concept of open set, we will use them interchangeably and do so without comment.

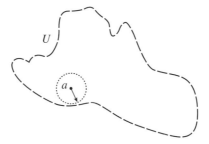

Figure 3.2 This figure illustrates Provision 2 of Theorem 3.1.7.

Theorem 3.1.8 Every union of open sets is open, and every finite intersection of open sets is open.

Proof: Each open set is a union of open balls, so a union of open sets is a union of unions of open balls, which is also a union of open balls. Thus every union of open sets is open.

To prove that every finite intersection of open sets is open, we take an arbitrary finite collection $U_1, U_2, \ldots, U_n$ of open subsets of X. If $x \in \bigcap_{i=1}^{n} U_i$, then $x \in U_i$ for each $i \leq n$. As each of these sets is open, there exist positive numbers $r_1, r_2, \ldots, r_n$ so that for each $i \leq n$, $\mathrm{B}_{r_i}(x)$ is totally contained within U_i. Let $r = \min(r_1, r_2, \ldots, r_n)$. (Note that $r > 0$.) Then for each $i \leq n$,

$$\mathrm{B}_r(x) \subseteq \mathrm{B}_{r_i}(x) \subseteq U_i.$$

Therefore $\mathrm{B}_r(x) \subseteq \bigcap_{i=1}^{n} U_i$, and this set must be open. (*Which of the formulations given in Theorem 3.1.7 is being used here?*) □

Exercise 3.1.9 It is not true that every intersection of open sets is open.

1. We start by examining a question. In the proof of Theorem 3.1.8, we proved that every finite intersection of open sets is open. We did so by identifying a radius for each set in the intersection and taking the minimum value r over all radii. It then followed that $\mathrm{B}_r(x)$ was totally contained within the intersection of the open sets. Why won't this proof work if we have an infinite intersection of open sets and replace the minimum with an infimum?

2. Using the insight you gained by answering Question 1, construct a collection of open subsets of $\mathbb{R}$ whose intersection is not open.

Boundedness in Metric Spaces

We already know that a subset of an ordered set is bounded if it is bounded both above and below. This notion does not extend to metric spaces, however, because they are not (in general) ordered. We need to expand our idea of what it means to be bounded. In $\mathbb{R}^2$, for instance, you might picture something as bounded by imagining that it cannot go out "infinitely far" in any direction. Or, if you picture yourself as "standing" at a point on the "edge" of the set, you might say that the set is bounded if you can "walk across" it. These informal ideas can be made precise using distances and open sets.

Definition 3.1.10 Let X be a metric space. Let S be a subset of X. If $S = \emptyset$, we define the diameter of S to be 0. Suppose that S is not empty. Then the **diameter** of S is said to be infinite if the set

$$\{d(a,b) : a, b \in S\}$$

is not bounded above. It is the real number

$$\mathrm{diam}(S) = \sup\{d(a,b) : a, b \in S\}$$

if the supremum exists. In this case we say that S has **finite diameter**.

Exercise 3.1.11 Some examples to consider:

1. Let X be a metric space. Let F be a finite subset of X. Prove that F has finite diameter.

2. Show that the interval (a, b) in $\mathbb{R}$ has finite diameter.

3. What is the diameter of the (closed) circle of radius 1 centered at the origin in $\mathbb{R}^2$? Give an argument, based on Definition 3.1.10, to support your answer.

4. Show that $\mathbb{N}$, as a subset of the metric space $\mathbb{R}$, has infinite diameter.

■

Theorem 3.1.12 Let X be a metric space. Let S be a subset of X. The following statements about S are equivalent:

1. S has finite diameter. (*Analytic viewpoint*)

2. There exists $a \in X$ and $r > 0$ such that $S \subseteq \mathrm{B}_r(a)$. (*Geometric viewpoint*)

3. Given any $a \in X$, there exists $r > 0$ such that $S \subseteq \mathrm{B}_r(a)$. (*Geometric viewpoint*)

Proof: The proof is Problem 10a at the end of this section. □

Definition 3.1.13 Let X be a metric space. A subset S of X is said to be **bounded** if it satisfies any one of the conditions given in Theorem 3.1.12 (and, therefore, all of them).

> When we encounter a new term, we usually give a definition first. A theorem that gives several equivalent formulations for the term sometimes follows the definition. Here we started with several statements that were equivalent and then defined a set to be bounded if it satisfies any one of them. This approach emphasizes the fact that, from a mathematical point of view, equivalent statements are simply different ways of expressing the *same* idea.
>
> In other instances, We choose one statement to be "the" definition and then show the other statements to be equivalent to it. (See, for instance, Definition 3.1.4 and Theorem 3.1.7.) Keep in mind that "the" definition was just one possible choice of phrasing. In proving theorems, we use whichever formulation is the most useful or convenient.

Let $K \subseteq \mathbb{R}$. Theorem 3.1.12 immediately shows that our notion of boundedness in the ordered set $(\mathbb{R}, \leq)$ coincides with our notion of boundedness in the metric space $(\mathbb{R}, |\cdot|)$.

Corollary 3.1.14 Let $S \subseteq \mathbb{R}$. Then S is bounded as a subset of the metric space $(\mathbb{R}, |\cdot|)$ if and only if there exists $K \in \mathbb{R}$ such that $|s| \leq K$ for all $s \in S$.

Proof: The proof is Problem 10b at the end of this section. □

Problems 3.1

1. Prove Theorem 3.1.7, which gives the various characterizations of open set. *Hint:* This is a fairly easy proof, but there is one subtlety. If $x \in \bigcup_{\alpha \in \Lambda} B_{r_\alpha}(x_\alpha)$, then $x \in B_{r_\alpha}(x_\alpha)$ for some $\alpha \in \Lambda$. It does not follow from this that $x = x_\alpha$. Thus Figure 3.3 is relevant to proving that the first statement of the theorem implies the second statement.

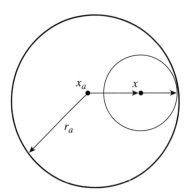

Figure 3.3 A hint for $1 \Rightarrow 2$ in the proof of Theorem 3.1.7.

2. Recall the metric space ℓ_∞ defined in Problem 9 at the end of Chapter 2. The element $\mathbf{x} = (2, -1, 0, 1, 1, 2, -1, 0, 1, 1, 2, -1, 0, 1, 1, 2, -1, 0, 1, 1, \ldots)$ is depicted as follows:.

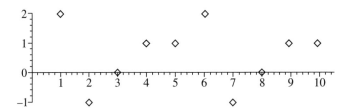

Describe the open and closed balls of radius $\frac{1}{2}$ centered at $\mathbf{x}$. (If you find it convenient to use the graphical representation to help with your description, feel free to do so.)

3. Let x be any point in a metric space X. Prove that $\{x\}^C$ is an open set.

4. Let x and y be distinct points in a metric space X. Show that there exists $r > 0$ such that $B_r(x)$ and $B_r(y)$ are disjoint.[1]

5. In Exercise 3.1.3, you were asked to describe the open balls of $\mathbb{R}$. Prove that the sets you described are open balls and that all open balls are of this form.

---

1. This theorem illustrates a more general idea. Topological spaces are generalizations of metric spaces, and topological spaces in which for every pair of points x and y there exist disjoint open sets U and V so that $x \in U$ and $y \in V$ are called *Hausdorff spaces*. In this problem you prove that metric spaces are Hausdorff.

6. Open balls in $\mathbb{R}^n$ are convex.

   Definition: A subset E of $\mathbb{R}^n$ is said to be **convex** if, given any two vectors $\mathbf{x}$ and $\mathbf{y} \in E$, the entire line segment joining $\mathbf{x}$ and $\mathbf{y}$ is also in E. More precisely, E is convex if for all real numbers $t \in (0,1)$ and for all $\mathbf{x}$ and $\mathbf{y} \in E$, $t\mathbf{x} + (1-t)\mathbf{y} \in E$.

   Let $\mathbf{x} \in \mathbb{R}^n$ and let $r > 0$. Prove that $\mathrm{B}_r(\mathbf{x})$ is convex.

7. Let (X,d) be a metric space and S be a subset of X. Example 2.1.5 tells us that if we restrict the metric d to pairs of elements of S, we get a new metric space. This problem will help you to understand how the open subsets of (S,d) are related to the open subsets of (X,d).

   (a) It will help to start by thinking about an example. Consider the case where $X = \mathbb{R}^2$ and S is the (closed) upper half-plane $S = \{(x,y) \in \mathbb{R}^2 | y \geq 0\}$. Look at the statements in parts (b), (c), and (d). Explain what each is saying in the context of S and $\mathbb{R}^2$. Be sure to consider what happens around the lower edge of the set S. Draw lots of pictures.

   For the remainder of the problem, X is an arbitrary metric space and S is an arbitrary subset of X.

   (b) Let $a \in S$ and $r > 0$. To distinguish between them, we will denote the ball of radius r about a in the metric space S by $B_r^S(a)$ and the ball of radius r about a in the metric space X by $B_r^X(a)$. Show that $B_r^S(a) = B_r^X(a) \cap S$.

   (c) Show that if U is an open subset of X, then $U \cap S$ is an open subset of S.

   (d) Show that if V is any open subset of S, then there exists an open subset U of X such that $V = U \cap S$.

8. Let X be a metric space and x be an element of X. If $\{x\}$ is an open set, then x is called an **isolated point** of X. A metric space in which every point is isolated is called a **discrete metric space**.

   (a) Suppose there exists $r > 0$ such that $\mathrm{B}_r(x)$ contains only finitely many points. Prove that x is isolated. Use this fact to conclude that every finite metric space is discrete.

   (b) Prove that every set endowed with the trivial metric (defined in Exercise 2.1.3) is a discrete metric space.

   (c) Using your intuition about the words "isolated" and "discrete," find an example of an *infinite* discrete metric space. Prove that your example is indeed discrete. No fair using the trivial metric as an example! (*Hint*: Think about subspaces of spaces you know well—say, $\mathbb{R}$ or $\mathbb{R}^2$.)

(d) Let X be a discrete metric space. Prove that every subset of X is open.

9. Let X be a metric space. Prove that every subset of X is the intersection of open sets.[2] (*Hint*: Problem 3 may be useful to you here.)

10. **Bounded sets**.

    (a) Prove the equivalence of the various formulations of the word "bounded" as given in Theorem 3.1.12.

    (b) Prove Corollary 3.1.14, which characterizes bounded subsets of $\mathbb{R}$.

11. Let $\overline{X}$ be a metric space. Prove that a union of finitely many open balls in $\overline{X}$ is a bounded subset of $\overline{X}$.

12. Let $r > 0$. Let $\mathbf{x} = (x_1, x_2, \ldots, x_n) \in \mathbb{R}^n$. The subset of $\mathbb{R}^n$ given by
$$K_r(\mathbf{x}) = \{\mathbf{z} = (z_1, z_2, \ldots, z_n) \in \mathbb{R}^n : |x_i - z_i| < r \text{ for } i = 1, 2, \ldots, n\}$$
is called the *r-cell* about $\mathbf{x}$.

> *Remark*: Part (a) asks you to draw relevant pictures in $\mathbb{R}^2$ as a prelude to proving things about cells. Picturing things in $\mathbb{R}^3$ will also be helpful.

   (a) **Preliminaries:** For a fixed $i \leq n$ and $r > 0$, consider the sets of the form
   $$K_{i,r}(\mathbf{x}) = \{\mathbf{z} = (z_1, z_2, \ldots, z_n) \in \mathbb{R}^n : |x_i - z_i| < r\}.$$

   i. For $n = 2$ and a fixed point (a, b), draw the sets $K_{1,\frac{1}{2}}(a, b)$ and $K_{2,\frac{1}{2}}(a, b)$.

   ii. What is the relationship between the sets $K_{1,r}(\mathbf{x})$, $K_{2,r}(\mathbf{x})$, $K_{3,r}(\mathbf{x})$, ..., $K_{n,r}(\mathbf{x})$ and the set $K_r(\mathbf{x})$?

   (b) Show that for each $i \leq n$ and $r > 0$, $K_{i,r}(\mathbf{x})$ is an open set.

   (c) Show that cells are open subsets of $\mathbb{R}^n$.

   (d) Let $B_r(\mathbf{a})$ be an open ball in $\mathbb{R}^n$. Prove that there exists $s > 0$ such that $K_s(\mathbf{a}) \subseteq B_r(\mathbf{a})$. Likewise, show that if $K_r(\mathbf{a})$ is a cell in $\mathbb{R}^n$, then there exists $s > 0$ such that $B_s(\mathbf{a}) \subseteq K_r(\mathbf{a})$.

---

[2]. This theorem tells us that the fact that not every intersection of open sets is open is *good news*. If this were not so, the concept of open set would be a pretty stupid one—just a fancy name for "subset."

(e) Let U be a subset of $\mathbb{R}^n$. Show that U is an open subset of $\mathbb{R}^n$ if and only if for each $\mathbf{x} \in U$, there exists $r > 0$ such that the r-cell about $\mathbf{x}$ is totally contained within U.

(f) Recall the metric
$$d(\mathbf{a}, \mathbf{b}) = \max_{1 \leq i \leq n} |a_i - b_i|$$
on $\mathbb{R}^n$ given in Problem 5b at the end of Chapter 2. Verify that the open balls in $\mathbb{R}^n$ given by this metric are precisely the cells in ($\mathbb{R}^n$, Euclidean).

(g) Use the results of parts (e) and (f) to prove that the open subsets of $\mathbb{R}^n$ under the Euclidean metric are precisely the same as the open subsets of $\mathbb{R}^n$ under the alternative metric from Problem 5b in Chapter 2.

13. You know that every open subset of a metric space is a union of open balls. In general, this union is not a disjoint union; however, something very like this occurs in the case of open subsets of $\mathbb{R}$. Let U be a nonempty open subset of $\mathbb{R}$. Prove that it can be written as the disjoint union of countably many open intervals.

3.2 Convergence of Sequences: Thinking Intuitively

Having talked about distance, we are now in a position to discuss limiting processes. In your calculus courses, you became familiar (on an informal level) with a number of limiting processes: the limit of a function at a point, the derivative as a limit, the integral as a limit, the limit of a sequence or series of real numbers.

We will ultimately consider all of these limiting processes (and others), but we start with the simplest—the limit of a sequence. Let us begin by thinking about this informally in the context of sequences of real numbers. We would like to say the following:

A sequence (a_n) of real numbers converges to a number x provided that, as we go farther and farther out in the sequence, the terms of the sequence get closer and closer to x.

For example, our intuition tells us that the sequence
$$\frac{1}{2}, \frac{2}{3}, \frac{3}{4}, \frac{4}{5}, \frac{5}{6}, \ldots, \frac{n}{n+1}, \ldots$$

converges to 1 because numbers like $\dfrac{9999}{10,000}$ and $\dfrac{99,999}{100,000}$ are very close to 1 indeed. In contrast, the sequence

$$1, 0, 0, 1, 1, 0, 0, 1, 1, 0, 0, 1, \ldots$$

does not converge at all. Its terms continually oscillate back and forth between 0 and 1 and do not approach any fixed number.

From a mathematical point of view, the key sticking points in our informal "definition" are the phrases *farther and farther out* and *closer and closer*. Any mathematically sound definition requires a rigorous understanding of what these phrases mean and how they fit together to give us the behavior that we want. To begin, we note that going "farther and farther out" refers to the fact that the terms in the sequence are numbered. The 10th term is farther out than the 3rd. Likewise, the 50th term is farther out than the 10th term, and the 1000th term is farther out than the 50th term. Roughly speaking, we are requiring that as the *index* gets larger and larger without bound, the distances between the corresponding terms and the limit get progressively smaller—eventually smaller, in fact, than any fixed positive real number.

Note that the size of the index and the distance from the corresponding term to the limit are not independent of one another. In other words, one convergent sequence may require 1000 terms to get within 0.01 units of its limit (and never, thereafter, get farther away), whereas another sequence might take 1,000,000 or even 10^{543} terms. No matter; the crucial requirement is that any fixed distance—no matter how small—*eventually* be achieved and maintained. The quantity that must be controlled is not the largeness of the index but rather the smallness of the distance. My calculus professor would have said that our worst enemy gets to specify as small a distance as he or she desires and that we are required to respond by *producing* an index large enough to meet that challenge. The point here is that given any positive distance, we must be able to find an index large enough so that all terms in the sequence beyond that point are within the specified distance of the limit.

3.3 Convergence of Sequences

Definition 3.3.1 The sequence (a_n) in a metric space (X, d) **converges** to the **limit** $x \in X$ provided that for each positive real number ϵ there exists a natural number N so that whenever $n > N$,

$$d(a_n, x) < \epsilon.$$

When (a_n) converges to x, we may write $\lim\limits_{n \to \infty} a_n = x$ or just $a_n \to x$.

Exercise 3.3.2 Let X be a metric space. Prove that all constant sequences in X converge. ∎

Remark: Our intuitive definition talked about the terms of a sequence getting "closer and closer" to the limit. This may be a bit misleading; Definition 3.3.1 actually allows for some possibilities that don't seem (quite) to go along with the informal language of convergence. For instance, Definition 3.3.1 tells us that

$$0, 0, 0, 0, 0, \ldots,$$

$$0, 1, 0, \frac{1}{2}, 0, \frac{1}{3}, 0, \frac{1}{4}, 0, \frac{1}{5}, \ldots,$$

and

$$100, 5349, -33847, 1, \frac{1}{2}, \frac{1}{3}, \frac{1}{4}, \frac{1}{5}, \ldots$$

all converge to 0 (*be sure you see why*). However, the terms of the first sequence are all 0—no term is *closer* to zero than any other. In the second and third sequences, some terms are farther away from the limit than preceding terms. Nevertheless, all terms of the sequence lying sufficiently "far out" in the sequence are closer to the limit than any distance we care to specify. The third example also illustrates the fact that one could add to or take away from the front of a sequence any number of terms without affecting the convergence of that sequence.

Because the first few (or the first billion!) terms of a sequence do not affect the convergence of the sequence in any way, we say that sequence convergence is a "tail" condition. Therefore, it is sometimes convenient to be able to talk about a particular kind of subsequence called a tail.

Definition 3.3.3 Let (a_n) be a sequence in a metric space X. Let $N \in \mathbb{N}$. Then the subsequence of (a_n) that consists of the terms

$$a_N, a_{N+1}, a_{N+2}, a_{N+3}, \ldots$$

is called the $\mathbf{N^{th}}$ **tail of the sequence** and is denoted by $(a_n)_{n \geq N}$ or $(a_n)_{n=N}^{\infty}$.

Exercise 3.3.4 Let (a_n) be a sequence in a metric space X. Let $x \in X$. Prove that (a_n) converges to x if and only if there exists $N \in \mathbb{N}$ such that the Nth tail of (a_n) converges to x. ∎

Definition 3.3.5 A sequence (a_n) is said to be **constant** if all of its terms are equal: $a, a, a, \ldots$. The sequence is said to be **eventually constant** if it has a tail that is constant.

Exercise 3.3.4 clearly implies that all eventually constant sequences converge.

Exercise 3.3.6 Below you will find several statements involving a sequence of real numbers (a_n) and a real number L. In each case, consider the statement as an alternative to the definition of $(a_n) \to L$. Think about the statement and its implications. Then provide an example of a sequence of real numbers (a_n) and a number L that satisfies the "definition" and yet $(a_n) \not\to L$. Accompany your example with a verbal explanation of the inadequacies of the "definition."

1. The sequence (a_n) converges to L if for all $\epsilon > 0$, there exists $n \in \mathbb{N}$ such that $d(a_n, L) < \epsilon$.

2. The sequence (a_n) converges to L if for all $\epsilon > 0$, there exists $N \in \mathbb{N}$ such that for some $n > N$, $d(a_n, L) < \epsilon$.

3. The sequence (a_n) converges to L if for all $N \in \mathbb{N}$, there exists $\epsilon > 0$ so that for all $n > N$, $d(a_n, L) < \epsilon$.

4. The sequence (a_n) converges to L if for all $N \in \mathbb{N}$ and all $\epsilon > 0$, there exists $n > N$ such that $d(a_n, L) < \epsilon$.

∎

The following easy theorem explicitly sets out the duality between geometric and analytic formulations for sequence convergence. It does so by connecting the ideas of open set and sequence convergence.

Theorem 3.3.7 Let (a_j) be a sequence in a metric space X. Then the following statements are equivalent:

1. The sequence (a_j) converges to a.

2. For every open set U containing a, there exists $N \in \mathbb{N}$ such that for all $n > N$, $a_n \in U$.

3. For every $\epsilon > 0$, there exists $N \in \mathbb{N}$ such that if $n > N$, then $a_n \in B_\epsilon(a)$.

Proof

$1 \Longrightarrow 2$ Let U be any open set containing a. Then there exists a positive number ϵ such that $B_\epsilon(a) \subseteq U$. The fact that (a_j) converges to a tells us that there exists $N \in \mathbb{N}$ such that for all $n > N$, $d(a_n, a) < \epsilon$. (Same ϵ!) But then, for all $n > N$, $a_n \in U$.

$2 \Longrightarrow 3$ Assume that given any open set U containing a, there exists $N \in \mathbb{N}$ such that for all $n > N$, $a_n \in U$. Fix $\epsilon > 0$. Because $B_\epsilon(a)$ is open and contains a, there exists $N \in \mathbb{N}$ such that for all $n > N$, $a_n \in B_\epsilon(a)$.

$3 \Longrightarrow 1$ Fix $\epsilon > 0$. Choose $N \in \mathbb{N}$ such that if $n > N$, then $a_n \in B_\epsilon(a)$. This means that if $n > N$, $d(a_n, a) < \epsilon$. So $a_n \to a$.

□

> **Notational Conventions**
>
> Notice the various uses of the Greek letter ϵ. By tradition, analysts use lowercase Greek letters to refer to "small" positive quantities. ϵ (epsilon), δ (delta), and η (eta) are probably the favorites—in that order. The lowercase Roman letters n, m, i, j, and k are primarily used for natural numbers (often counters of some sort such as those used in Definition 3.3.1). Both uppercase and lowercase r and the uppercase K are usually reserved for real numbers that are not necessarily "small." The letters x, y, z, and w are typically used to indicate "continuous" variables, such as those that come up in real functions.
>
> Although these are not hard-and-fast rules, they represent common usage. These notational conventions (although they are, in fact, void of any mathematical significance) are important because they give visual cues that aid in reading mathematical writing; thus it will help you to be aware of common notational conventions. Moreover, you should adopt them in your own writing so your readers have the same advantage.

You are now ready for Excursion D:

> **Excursion D.** *Sequences in $\mathbb{R}$ and $\mathbb{R}^n$*
>
> Synopsis: Whereas Sections 3.3 and 3.4 concentrate on general theoretical results about convergence of sequences, Excursion D discusses the convergence of actual numerical sequences in $\mathbb{R}$ and $\mathbb{R}^n$. Section D.1 culminates in the theorem that relates the convergence of a sequence in $\mathbb{R}^n$ to the convergence of its (real) coordinate sequences. An important special feature of this excursion is the section entitled "Epsilonics: Playing the Game." This section takes the reader "behind the scenes" with a detailed discussion of how choices are made in a convergence argument. The final sections of the excursion briefly touch on infinite limits and discuss the convergence of some especially useful special sequences.

Problems 3.3

1. Let (a_n) be a sequence in a metric space X. Let $x \in X$. Suppose that for all $n \in \mathbb{N}$, $a_n \in \mathrm{B}_{\frac{1}{n}}(x)$. Prove that $a_n \to x$.

2. Show that a convergent sequence in a metric space has a unique limit.

3. Let (X, d) be a metric space. Let (x_n) be a sequence in X and $a \in X$. Show that the following statements are equivalent

   i.: $x_n \to a$ as $n \to \infty$.

   ii.: $d(x_n, a) \to 0$ as $n \to \infty$.

4. A sequence in a metric space is said to be **bounded** if its range is bounded.

   (a) Prove that every convergent sequence is bounded.

   (b) Show by giving an example that bounded sequences are not necessarily convergent.

5. Let X be a metric space.

   (a) Let $x \in X$ be an isolated point. Prove that the only sequences in X that converge to x are the sequences that are eventually constant with tail $x, x, x, \ldots$.

   (b) Prove that the only convergent sequences in a discrete metric space are the eventually constant sequences.

   (See Problem 8 on page 79 for the definitions of "isolated" and "discrete.")

6. Let (a_n) be a sequence in a metric space X. Let $x \in X$. Show that the following statements are equivalent:

   i.: (a_n) converges to x.

   ii.: Every subsequence of (a_n) converges to x.

   iii.: Every subsequence of (a_n) has a subsequence that converges to x.

7. Let X be an unbounded metric space. Show that X contains a sequence with no convergent subsequence.

8. Let (a_i) be a sequence in a metric space X. Let $a \in X$.

   (a) Show that (a_i) has a subsequence converging to a if and only if every open ball about a contains infinitely many terms of the sequence (a_i).

   > Once again we see analytic and geometric descriptions of the same phenomenon. Do you see which is which?

   (b) Give examples of a sequence in $\mathbb{R}$ and a sequence in $\mathbb{R}^2$ that have no convergent subsequences.

3.4 Sequences in $\mathbb{R}$

Since $(\mathbb{R}, |\cdot|)$ is a metric space, all of the theorems that we have proved so far can be applied to the real numbers. But the special structures of the real numbers lend an added dimension to analysis in $\mathbb{R}$. Here, for instance, is a useful little theorem that is true in $(\mathbb{R}, |\cdot|)$, but whose statement doesn't even make sense in a general metric space.

Theorem 3.4.1 Let (a_n) be a sequence of real numbers that converges to the real number a. Then $|a_n| \to |a|$.

Proof: The proof is Problem 2 at the end of this section. $\square$

Exercise 3.4.2 Is the converse of Theorem 3.4.1 true? Prove or give a counterexample. ∎

Sequence Convergence and Order

The real numbers are not only a metric space; they are also an ordered set. A number of useful theorems connect the convergence of sequences of real numbers with the order properties of $\mathbb{R}$. Only a couple of the proofs are written out here; the rest are left to you. Use your intuition about the real numbers to construct your arguments and use the theorems from Chapter 1 to justify them.

Theorem 3.4.3 Let (a_n) be a sequence of real numbers. Suppose that k is a real number and that $a_n \geq k$ for each $n \in \mathbb{N}$. If (a_n) converges to a real number a, then $a \geq k$. Likewise, if $a_n \leq k$ for each $n \in \mathbb{N}$ and $a_n \to a$, then $a \leq k$.

Proof: Let's prove the case when $a_n \leq k$. The case when $a_n \geq k$ is Problem 3 at the end of this section.

Proceed by contradiction. Suppose that $a > k$. Thus our hypothesis says

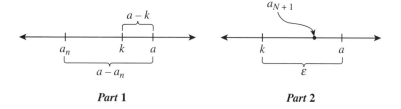

Figure 3.4 Proof of Theorem 3.4.3

that $a_n \leq k < a$ for all $n \in \mathbb{N}$. So $a - k < a - a_n$ for all $n \in \mathbb{N}$. (See Figure 3.4, Part 1.) But it is also true that, for $\epsilon = a - k$, we can choose $N \in \mathbb{N}$ such that for all $n > N$, $|a_n - a| < \epsilon$. Thus, in particular, $|a_{N+1} - a| < \epsilon$ (See Figure 3.4, Part 2.) So

$$a - k = \epsilon > |a_{N+1} - a| = a - a_{N+1}.$$

This contradicts our previous assertion. □

Exercise 3.4.4 Let (a_n) be a sequence of real numbers. Suppose that k and a are real numbers. If $k < a_n$ for each $n \in \mathbb{N}$, and $a_n \to a$, can we then assert that $k < a$? Prove or give a counterexample. ∎

Theorem 3.4.5 Let (a_n) be a sequence of real numbers that converges to the non-zero real number a. Let $k \in \mathbb{R}$ be such that $0 < k < |a|$. Then there exists $N \in \mathbb{N}$ such that for all $n > N$, $|a_n| > k$. Likewise, if $|a| < k$, then there exists $N \in \mathbb{N}$ such that for all $n > N$, $a_n \leq |a_n| < k$.

Proof: The proof is Problem 4 at the end of this section. □

Theorem 3.4.6 Let (a_n) and (b_n) be sequences of real numbers that converge to real numbers a and b, respectively. If $a_n \leq b_n$ for all $n \in \mathbb{N}$, then $a \leq b$.

Proof: The proof is Problem 5 at the end of this section. □

Corollary 3.4.7 Let (a_n) and (b_n) be sequences of real numbers that converge to real numbers a and b, respectively. If $a_n \geq b_n$ for all $n \in \mathbb{N}$, then $a \geq b$.

Exercise 3.4.8 Show by giving examples that it is possible to have two convergent real sequences (a_n) and (b_n), each consisting of distinct terms and satisfying $a_n < b_n$ for all n, but still converging to the same limit. ∎

Theorem 3.4.9 Every bounded sequence of real numbers has a convergent subsequence.

Proof: The proof is Problem 7 at the end of this section. □

Theorem 3.4.10 Let $S \subseteq \mathbb{R}$ with least upper bound ℓ. Then there exists a sequence in S that converges to ℓ. Likewise, if ℓ is the greatest lower bound of S, there exists a sequence in S that converges to ℓ.

Proof: Let's prove the case involving the least upper bound. The case involving the greatest lower bound (and some elaborations) is left to you in Problem 8 at the end of this section.

We choose the sequence inductively. Using Theorem 1.4.4 with $\epsilon = 1$, we choose $s_1 \in S$ such that
$$\ell - 1 < s_1 \leq \ell$$
(which implies that $\ell - s_1 < 1$).

Suppose we have chosen $s_1, s_2, s_3, \ldots, s_n$ in S such that
$$\ell - s_k < \frac{1}{k} \text{ for all } k \leq n$$
Theorem 1.4.4 tells us that we can find $s_{n+1} \in S$ such that
$$\ell - \frac{1}{n+1} < s_{n+1} \leq \ell$$
(which implies that $\ell - s_{n+1} < \frac{1}{n+1}$).

Because the inductive sequence satisfies
$$d(s_k, \ell) < \frac{1}{k} \text{ for all } k \in \mathbb{N}$$
$s_k \to \ell$ by Problem 1 on page 85. $\square$

Remark: I didn't really have to choose the sequence inductively in the proof of Theorem 3.4.10. (*Can you see how to rewrite the proof without using induction?*) I chose to write the proof that way because it will give you a headstart on the "elaborations" I promised; those will require that the sequence be chosen inductively.

Sequence Convergence and Arithmetic

If we have a sequence of real numbers, we can form a new sequence by taking any real number and multiplying each term of the sequence by it. Furthermore, if we have two sequences of real numbers, we can add, subtract, multiply, or divide them term by term. Theorem 3.4.11 tells us that if the original sequences converge, so do the newly created sequences; the limits of these new sequences are related in a simple way to the limits of their "parent" sequences. In other words, arithmetic and sequence convergence work nicely together.

Theorem 3.4.11 Let (a_n) and (b_n) be sequences in $\mathbb{R}$. Let k be any real number. Suppose that $a_n \to L$ and $b_n \to M$. Then

1. $(a_n + b_n)$ converges to $L + M$.
2. $(a_n - b_n)$ converges to $L - M$.

3. $(a_n b_n)$ converges to LM.

4. $(k a_n)$ converges to kL.

5. $\left(\dfrac{a_n}{b_n}\right)$ converges to $\dfrac{L}{M}$ provided that b_n is never zero and that $M \neq 0$.

Proof: Let's prove the case of the product. The rest of the cases are left to you in Problem 9 at the end of this section.

Let $\epsilon > 0$. Because (b_n) is convergent, it is bounded (Problem 4 on page 86). Thus we can find a real number $K > 0$ so that $|b_n| \leq K$ for all n. (This is Problem 1 at the end of this section.) Using the convergence of (a_n) and (b_n), we can make the following choices:

1. Choose $N_1 \in \mathbb{N}$ such that if $n > N_1$, then $|a_n - L| < \dfrac{\epsilon}{2K}$.

2. Choose $N_2 \in \mathbb{N}$ such that if $n > N_2$, then $|b_n - M| < \dfrac{\epsilon}{2(|L|+1)}$.

Now choose $N = \max(N_1, N_2)$. Suppose that $n > N$ (so n satisfies Conditions 1 and 2). Then

$$\begin{aligned} d(a_n b_n, LM) &= |a_n b_n - LM| \\ &\leq |a_n b_n - L b_n| + |L b_n - LM| \\ &= |b_n||a_n - L| + |L||b_n - M| \\ &< (K)\left(\dfrac{\epsilon}{2K}\right) + |L|\left(\dfrac{\epsilon}{2(|L|+1)}\right) \\ &= \dfrac{\epsilon}{2} + \dfrac{|L|}{|L|+1}\dfrac{\epsilon}{2} \\ &< \dfrac{\epsilon}{2} + \dfrac{\epsilon}{2} = \epsilon. \end{aligned}$$

We conclude that $(a_n b_n)$ converges to LM. $\square$

Question to Ponder In the course of the proof of the product condition, it might not have been obvious why the choice of N_2 involved $\dfrac{\epsilon}{2(|L|+1)}$ rather than the simpler $\dfrac{\epsilon}{2|L|}$. This choice was deliberate and absolutely necessary. Why? (*Hint*: It may help you to know that

$$\dfrac{\epsilon}{2|L|+1} \text{ or } \dfrac{\epsilon}{2|L|+.001}$$

would have served just as well, although the algebra would not have been as tidy.)

Another Question to Ponder Theorem 3.4.11 does more than just calculate the values of the various limits: It shows that they *exist*. (Convergence is not assumed.) Why is this significant?

Proof Strategy

In the proof of Theorem 3.4.11 (3), I had to show that the distance from $a_n b_n$ to LM is "small" when n is "large." The hypothesis told me something about $|a_n - L|$ and $|b_n - M|$, so I had to express $|a_n b_n - LM|$ in terms of these quantities. The argument used a pair of important principles that are worth keeping in mind.

> **The sum of small things is small.** You can break up an estimate into a sum with as many terms as necessary *provided* that you can make each summand as small as you like.

So I broke $|a_n b_n - LM|$ into the sum

$$|b_n||a_n - L| + |L||b_n - M|.$$

> **The product of something small and something bounded is small.** We can handle products *provided* that we can find upper bounds for all but one of the factors and that we can make the other as small as we like.

Because we could find an upper bound for $(|b_n|)$ and L was fixed, I was able to control the overall size of

$$|a_n b_n - LM|.$$

(For a more detailed discussion, read Excursion D.)

Problems 3.4

1. Let (a_n) be a sequence of real numbers. Show that (a_n) is bounded if and only if there exists $K \in \mathbb{R}^+$ such that $|a_n| \leq K$ for all $n \in \mathbb{N}$.

2. Prove that if $a_n \to a$, then $|a_n| \to |a|$ (Theorem 3.4.1).

3. Prove that $a_n \geq k$ and $a_n \to a$, then $a \geq k$. This is the second case of Theorem 3.4.3.

> This proof is very similar to the proof given earlier. It would be easy to copy the proof of the other case, changing signs here and there, but that defeats the purpose of the exercise. To get the "good" out of this problem, first work to understand the proof of the other case; then rethink this case from scratch, on your own. This process will help you with subsequent proofs.

4. Prove Theorem 3.4.5.

5. Prove Theorem 3.4.6. Explain why Corollary 3.4.7 is, indeed, an immediate corollary.

6. Let (a_n) be a sequence of real numbers that converges to the real number a. Suppose that K is a real number and that $|a_n| \leq K$ for all $n \in \mathbb{N}$. Prove that $|a| \leq K$.

7. This problem shows one possible approach for showing that every bounded sequence of real numbers has a convergent subsequence (Theorem 3.4.9).

    (a) Prove that a monotonic sequence of real numbers converges if and only if it is bounded.

    (b) Use part (a) and the fact that every sequence of real numbers has a monotonic subsequence (see Problem 5b at the end of Section 0.4) to prove Theorem 3.4.9.

8. Elaborating on Theorem 3.4.10.

    (a) Prove the greatest lower bound portion of Theorem 3.4.10. (See the box accompanying Problem 3.)

    (b) Let S and ℓ be as in Theorem 3.4.10. Prove that if the least upper bound ℓ of S is not in S, then the sequence converging to the least upper bound can be chosen to be strictly increasing. Likewise, if the greatest lower bound ℓ is not in S, the sequence converging to the greatest lower bound can be chosen to be strictly decreasing.

9. Prove the rest of Theorem 3.4.11. (*Hints*: The first three cases are much easier than the product. The quotient is similar to the product, but here is an additional hint. Show that all terms sufficiently far out in the sequence $|b_n|$ are larger than $|M|/2$.)

10. In Theorem D.4.6 at the end of Section D.4 you will be asked to prove that if $c > 1$, then $\lim_{n \to} c^{\frac{1}{n}} = 1$. Assume that result for now. Use it and Theorem 3.4.11 to prove that the same result holds if $0 < c < 1$.

11. In Section D.4 (Excursion D.4.7) you will be asked to prove that $\lim_{n\to\infty}(n+1)^{\frac{1}{n}} = 1$. Use that result to prove that $\lim_{n\to\infty}(n+1)^{-\frac{1}{n}} = 1$.

3.5 Limit Points

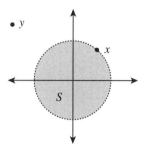

Figure 3.5 Our intuition tells us that the point x has a qualitatively different relationship to the set S than does the point y.

In Figure 3.5, x is what we technically call a *limit point* of the set S. How would we describe the difference between y and x vis-à-vis the set S? Well, we might observe that x is "right next to" the set S, whereas y is "some distance away" from S. Equivalently, we might say that points of S "cluster about" the point x, whereas they "keep their distance from" y. These notions are captured in Theorem 3.5.1.

Theorem 3.5.1 Let X be a metric space and S be a subset of X. Let $x \in X$. The following conditions on x are then equivalent:

1. There exists a sequence of points in $S \setminus \{x\}$ converging to x.

2. There exists a sequence of distinct points of S converging to x.

3. For all $r > 0$, $B_r(x)$ contains infinitely many points of S.

4. For all $r > 0$, $B_r(x)$ contains a point of $S \setminus \{x\}$.

Proof

$1 \implies 2$ Let (a_j) be a sequence in $S \setminus \{x\}$ that converges to x. Note that any sequence has either a constant subsequence or a subsequence of distinct terms.

Because any constant subsequence of a sequence converging to x would have to be the constant sequence $x, x, x, \ldots$ and (a_j) is in $S \setminus \{x\}$, (a_j) cannot have a constant subsequence. Thus (a_j) has a subsequence of distinct terms. This subsequence will be in S and will converge to x.

$2 \implies 3$ Suppose that (a_j) is a sequence of distinct points of S that converges to x. Fix $r > 0$. Then there exists $N \in \mathbb{N}$ such that for all $n > N$, $a_n \in B_r(x)$. Because the terms of $(a_j)_{j=N}^{\infty}$ are distinct, $B_r(x)$ contains infinitely many points of S.

$3 \implies 4$ Because every non-empty open ball about x contains infinitely many points in S, it must certainly contain a point of S other than x.

$4 \implies 1$ For every $n \in \mathbb{N}$, choose an element a_n in $B_{\frac{1}{n}}(x) \cap (S \setminus \{x\})$. Then $a_n \to x$ by Problem 1 of Section 3.3.

□

Definition 3.5.2 Let X be a metric space, let S be any subset of X and let $x \in X$. We say that x is a **limit point** of S if it satisfies any one (and therefore all) of the conditions in Theorem 3.5.1.

Exercise 3.5.3 Show that a finite set cannot have any limit points. ■

Exercise 3.5.4 In the definition of limit point, we talk about the set $S \setminus \{x\}$ instead of just the set S. This distinction is very important. To see why, think about the following two situations.

1. Let S be a subset of an arbitrary metric space. Show that every x in S has the property that there is a sequence of points in S converging to it.

2. Look at the "smiley face." In this figure, the subset S of $\mathbb{R}^2$ consists of the boundary of the circle, the "smile," and the two "eyes."

Which of the points in the "smiley set" satisfy the definition of limit point? Which don't? Give your intuition about why these distinctions make sense.

■

Problems 3.5

1. Let X be a metric space. Let $S \subseteq X$ and $x \in X$.

   (a) Prove that the following statements are equivalent:

   i.: There exists a sequence in S that converges to x.
   ii.: Every open ball about x contains a point of S.
   iii.: For every open set U containing x, $U \cap S \neq \emptyset$.

   (b) Show that the statement

   > Every open set U containing x contains infinitely many points of S

   is *not* equivalent to the three statements in part (a).

   (c) In the context of parts (a) and (b), say what you can about the status of the statement:

   > There exists a sequence of distinct points in S that converges to x.

2. Show that an isolated point of a metric space can never be the limit point of any set. (For the definition of isolated point, see Problem 8 in Section 3.2.)

3. Let X be a metric space.

   Definition: Fix a real number $\delta > 0$. A subset K of X is said to be **δ-separated** provided that $d(x,y) \geq \delta$ for all x, $y \in K$ such that $x \neq y$.

   (a) Find an infinite 1-separated subset of $\mathbb{R}^2$.

   (b) Let K be a δ-separated subset of X. Show that K has no limit points.

4. Let X be a metric space. Let $D \subseteq X$. We say that D is **dense** in X if every point of X is either an element of D or a limit point of D.

   (a) Prove that $\mathbb{R}$ has a countable, dense subset.

   (b) Prove that $\mathbb{R}^n$ has a countable, dense subset.

3.6 Closed Sets

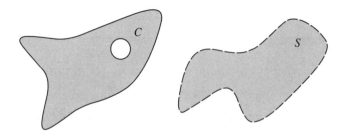

Figure 3.6 The set C is a closed set, whereas the set S is not.

Definition 3.6.1 Let X be a metric space. A subset C of X is said to be a **closed subset** of X or merely a **closed set** if C contains all of its limit points. (See Figure 3.6.)

Example 3.6.2

1. If X is any metric space, then X and $\emptyset$ are both closed subsets of X. (*You should verify this.*)

2. Every finite set is closed by Exercise 3.5.3.

3. Closed balls are closed sets. You will be asked to prove this fact in Problem 4 at the end of this section.

Theorem 3.6.3 Let X be a metric space, and let C be a subset of X. The following conditions are equivalent:

- C is closed.

- If (a_n) is a sequence in C converging to $x \in X$, then $x \in C$. (Equivalently, C contains all the limits of its convergent sequences.)

Proof: The proof is Problem 3 at the end of this section. □

Problems 3.6

1. Verify that given any two real numbers a and b with $a < b$, the half-open interval $(a, b]$ is neither open nor closed in $\mathbb{R}$.

2. Let X be a metric space, and let $x \in X$. Show that $\{x\}$ is closed.

3. Prove Theorem 3.6.3, which gives two equivalent formulations of "closed set".

4. Prove that closed balls are closed sets.

5. Let X be a metric space. Let (a_i) be a sequence in X converging to a. Show that the set consisting of all points in the range of the sequence (a_i) together with the limit a is a closed set.

6. Let X be a metric space, and let $A \subseteq X$. Prove that the set of limit points of A is a closed subset of X.

7. Let X be a metric space. A subset S of X is said to be **perfect** if S is closed and if every element of S is a limit point of S.

    (a) Prove that the empty set is perfect.

    (b) Prove that no non-empty finite set is perfect.

    (c) Give an example of an infinite closed subset of $\mathbb{R}$ that is not perfect.

    (d) Let a and b be distinct real numbers with $a < b$. Prove that $[a, b]$ is a perfect set.

8. Let X be a metric space. Let $\delta > 0$. Let K be a δ-separated subset of X. (See Problem 3 at the end of Section 3.5 for the definition of "δ-separated.") Show that K is closed.

9. After referring to Exercise 8 on page 79, show that every subset of a metric space X is both open and closed if and only if X is a discrete metric space.

10. Show that if U is a subset of $\mathbb{R}$ that is both open and closed, then $U = \mathbb{R}$ or $U = \emptyset$.

3.7 Open Sets, Closed Sets, and the Closure of a Set

Alhough their respective definitions do not seem to suggest it, you might guess from the choice of terminology (and you would be correct!) that open sets and closed sets are somehow related to each other. However, in their mathematical guises, the words "open" and "closed" are not mutually exclusive terms. Indeed, Exercise 3.1.6 and Example 3.6.2 show that there are subsets of every metric space that are both open and closed. As you saw in Problem 1 at the end of Section 3.6, it is also true that some metric spaces contain subsets that are neither open nor closed. Nevertheless, open and closed sets are "opposites" of a sort. You have already seen the major mathematical ingredients of this relationship in Problem 1a at the end of Section 3.5.

Theorem 3.7.1 Let X be a metric space, and let S be any subset of X. Then S is closed if and only if S^C is open.

Proof

$\implies$ Suppose that S is a closed set. Let $x \in S^C$. Then the definition of a closed set tells us that there cannot be a sequence in S converging to x, otherwise x would be in S. Thus, by Problem 1a at the end of Section 3.5, there exists an open ball about x that contains no points of S. But then this ball is totally contained within S^C. We conclude that S^C is open.

$\impliedby$ The proof is by contraposition. If S is not closed, then there exists a sequence (a_i) in S converging to a point $x \in S^C$. This being the case, Problem 1a at the end of Section 3.5 tells us that every ball about x must intersect S. In other words, there is no ball about x that is totally contained within S^C. Hence S^C is not open.

$\square$

This relationship makes it very easy to derive a closed set analog to Theorem 3.1.8.

Theorem 3.7.2 Every intersection of closed sets is closed, and every finite union of closed sets is closed.

Proof: Let X be any metric space and Λ be any indexing set. Suppose that $\{C_\alpha\}_{\alpha \in \Lambda}$ is a collection of closed subsets of X. We must show that $\bigcap_{\alpha \in \Lambda} C_\alpha$ is a closed set. This follows easily from the fact that the union of open sets is open (Theorem 3.1.8) and from Theorem 3.7.1.

By DeMorgan's laws,

$$\left(\bigcap_{\alpha \in \Lambda} C_\alpha\right)^\mathcal{C} = \bigcup_{\alpha \in \Lambda} (C_\alpha)^\mathcal{C}.$$

That is, the complement of $\bigcap_{\alpha \in \Lambda} C_\alpha$ is a union of open sets. (The sets $C_\alpha^\mathcal{C}$ are open because they are complements of closed sets!) We conclude that $\bigcap_{\alpha \in \Lambda} C_\alpha$ is a closed set.

By analogous reasoning, we see that if we take an arbitrary finite collection $C_1, C_2, \ldots, C_n$ of closed subsets of X, the complement of its union is the intersection of finitely many open sets—its complement is open. Therefore, any finite union of closed sets is closed. $\square$

Definition 3.7.3 Let S be any subset of a metric space X. The intersection of all closed subsets of X that contain S is called the **closure of S**. We denote this set by $\overline{S}$.

Exercise 3.7.4 establishes some important, if elementary, facts about $\overline{S}$.

Exercise 3.7.4 Let S be a subset of a metric space X. Prove the following statements:

1. The closure of S exists. (*Hint*: For the intersection of all closed sets containing S to exist, there must *be* a closed set containing S.)

2. $\overline{S}$ is a subset of every closed set that contains S. (*This is an important fact that will be used often! Make sure you understand why it is so, and keep it in mind as you go forward.*)

3. $\overline{S}$ is a closed set.

4. S is closed if and only if $S = \overline{S}$.

5. Let $x \in \overline{X}$. Prove that the following statements are equivalent:

    i. There exists a sequence in S that converges to x.
    ii. Every open ball about x contains a point of S.
    iii. For every open set U containing x, $U \cap S \neq 0$.
    iv. $x \in \overline{S}$.

    (*Hint*: Problem 1a at the end of Section 3.5 proved the equivalence of the first three. You need only make links to the fourth statement.)

Exercise 3.7.4 proves that every set S has a closure. Because that closure is a closed set, and because it is contained in every closed set that contains S, we can very reasonably think of $\overline{S}$ as the *smallest closed set containing S*. What do we have to add to S to get a closed set? It is clear that at a minimum we must "add in" any limit points for the set. Theorem 3.7.5 shows that this is always enough.

Theorem 3.7.5 Let S be a subset of a metric space X. Then

$$\overline{S} = S \cup \{x \in X : x \text{ is a limit point of } S\}.$$

Proof

$\subseteq$ The proof is by contraposition. Suppose that x is neither an element of S nor a limit point of S. Then $x \notin \overline{S}$ by Exercise 3.7.4 (part 5).

$\supseteq$ If $x \in S$, then $x \in \overline{S}$. Suppose that x is a limit point of S. Then there exists a sequence (a_n) in S converging to x. If K is any closed set containing S, then (a_n) is a sequence in K. Because K is closed, $x \in K$. That is, x is in every closed set containing S, so $x \in \overline{S}$.

□

We end this section with two useful notions: the interior of a set and the boundary of a set.

To consider what we might mean when we talk about the interior of a set, think of our use of the phrase "interior of the nation." By this phrase, we mean those portions of the country that don't lie on the edge—that is, land that is completely surrounded by more territory belonging to that same country. Intuitively, we might think of the interior of a set S as the collection of points that lie "well inside" the set.

Definition 3.7.6 Let X be a metric space and S be any subset of X. A point $x \in S$ is called an **interior point** of S provided that there is an open ball about x that is totally contained within S. The set of all interior points of S is called the **interior** of S. We denote this set by $\mathcal{I}nt(S)$.

Theorem 3.7.7 Let X be a metric space and let $S \subseteq X$. Then

1. $\mathcal{I}nt(S)$ is an open subset of X.

2. Every open subset of X that is contained in S is contained in $\mathcal{I}nt(S)$.

3. The union of all open subsets of X that are contained in S is equal to $\mathcal{I}nt(S)$.

4. S is open if and only if $S = \mathcal{I}nt(S)$.

5. $\mathcal{I}nt(S) = \left(\overline{S^C}\right)^C$.

Proof: The proof is Problem 7 at the end of this section. $\square$

Exercise 3.7.8 Is it always true that $\mathcal{I}nt(S) = \mathcal{I}nt(\overline{S})$? (Prove or give a counterexample.) ∎

Intuitively, we might think of the boundary of a set S as the collection of points that lie "right at the edge" of S or, alternatively, right at the place where S ends and S^C begins.

Definition 3.7.9 Let X be a metric space and S be a subset of X. We define the **boundary** of S to be $\overline{S} \cap \overline{(S)^C}$. We denote it by $\partial(S)$.

Theorem 3.7.10 Let X be a metric space and let $S \subseteq X$. Then

1. $\partial(S)$ is a closed set.

2. An element $x \in X$ is in $\partial(S)$ if and only if for every $r > 0$, $B_r(x) \cap S \neq \emptyset$ and $B_r(x) \cap S^C \neq \emptyset$.

3. S is closed if and only if S contains $\partial(S)$.

4. S is open if and only if S and $\partial(S)$ are disjoint.

Proof: The proof is Problem 8 at the end of this section. $\square$

You are now ready for Excursions G and H.

Excursion G. *Subsequences and Convergence*

Synopsis: Least upper bound and greatest lower bound, sequence and subsequence convergence, convergence and order, convergence and arithmetic, and limit points all come together in this excursion on the limit supremum and the limit infimum of sequence of real numbers. The excursion is written as a long set of interconnected exercises that take the reader, step-by-step, through the ideas. It can be assigned as a project or individual parts can be assigned as additional exercises.

> **Excursion H.** *Series of Real Numbers*
>
> Synopsis: The only core material needed for this excursion up through Section H.4 is in Chapter 3, but the discussion also relies on Excursion G. To be more precise, most of the excursion can be understood using only general facts about sequences of real numbers from this chapter; however, the discussion of the root and the ratio tests assumes familiarity with the limit supremum and the limit infimum, as discussed in Excursion G. Less general statements for these theorems can be given without reference to Excursion G if limits are assumed to exist, so that limsup and liminf both become ordinary limits.
>
> Section H.5, in which two series are multiplied together, also relies on Excursion F, Doubly Indexed Sequences. Excursion F technically relies only on information from this chapter, but it will be easier to understand after careful study of at least one uniformity condition such as uniform continuity (Section 4.4) or uniform convergence (Chapter 12).

Problems 3.7

1. Let U be an open subset and C be a closed subset of some metric space X. Show that $U \setminus C$ is open and that $C \setminus U$ is closed.

2. Reprove Theorem 3.7.2 using the definition of closed set directly—that is, without reference to Theorem 3.1.8.

3. Let X be a metric space. Recall that a subset S of X is perfect if S is closed and if every element of S is a limit point of S. (See Problem 7 in Section 3.6.)

    (a) Prove that any finite union of perfect sets is perfect.

    (b) Give an example of an infinite collection of perfect sets whose union is not perfect.

    (c) Let A and B be perfect subsets of X. Must $A \cap B$ be perfect? Prove or give a counterexample.

4. Let A and B be subsets of a metric space X. Prove the following:

    (a) If $A \subseteq B$, then $\overline{A} \subseteq \overline{B}$.

    (b) $\overline{A \cup B} = \overline{A} \cup \overline{B}$.

    (c) $\overline{A \cap B} \subseteq \overline{A} \cap \overline{B}$.

(d) Show by giving a counterexample that $\overline{A} \cap \overline{B}$ is not necessarily a subset of $\overline{A \cap B}$.

5. **The closure of an open ball**: Let X be a metric space, let $x \in X$, and let $r > 0$.

    (a) In Problem 4 at the end of Section 3.6, you proved that closed balls are closed sets. Use this fact to prove that $\overline{B_r(x)} \subseteq C_r(x)$.

    (b) Give an example to prove that, in general, $\overline{B_r(x)} \neq C_r(x)$.

    (c) However, the closure of the open ball and the closed ball are the same thing in $\mathbb{R}^n$. Prove that if $X = \mathbb{R}^n$, $\overline{B_r(x)} = C_r(x)$.

6. Let C be a subset of a metric space X. Prove that $\text{diam}(C) = \text{diam}(\overline{C})$.

7. Prove Theorem 3.7.7, which details some properties of the interior of a set.

8. Prove Theorem 3.7.10, which details some properties of the boundary of a set.

9. Let X be a metric space. Let $S \subseteq X$. Prove that X is the *disjoint* union of $\mathcal{I}nt(S)$, $\partial(S)$, and $\mathcal{I}nt(S^C)$.

10. Let X be a metric space.

    (a) Let $x \in X$. Prove that the following statements are equivalent:

        i. If C is a closed subset of X with $x \notin C$, then there exist disjoint open sets U and V such that $x \in V$ and $C \subseteq U$.

        ii. If U is any open subset containing x, then there exists an open subset V of U such that $x \in V$ and $\overline{V} \subseteq U$.

    (b) Prove that all metric spaces satisfy one or the other (and therefore both) of the equivalent conditions given in part (a) of this problem.

    > In general topology, spaces that satisfy the equivalent conditions given in Problem 10a are called **regular spaces**. In this problem you prove that metric spaces are regular.

11. Let X be a metric space, and let A be a closed subset of X. Prove that the following statements are equivalent:

    i. If B is a closed subset of X with $A \cap B = \emptyset$, then there exist disjoint open sets U and V of X such that $A \subseteq V$ and $B \subseteq U$.

ii. If U is any open subset containing A, then there exists an open subset V of U such that $A \subseteq V$ and $\overline{V} \subseteq U$.

> In general topology, spaces that satisfy the equivalent conditions in Problem 11 are called **normal spaces**. Problem 6 at the end of Section 4.4 indicates one way to prove that all metric spaces are normal. It is a bit trickier to do so, but it can also be proved with the information you now have. Give it a try!

Chapter 4

Continuity

4.1 Thinking Intuitively

Most high school students have an intuitive idea of what it means to say that a function from $\mathbb{R}$ to $\mathbb{R}$ is "continuous." They would tell you right away that Figure 4.1 shows the graphs of a continuous function and a discontinuous function.

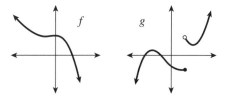

Figure 4.1 f is continuous; g is discontinuous.

Intuitively speaking, a function $f : \mathbb{R} \to \mathbb{R}$ is continuous at a particular point $x = a$ if (while drawing the graph) the pencil runs smoothly from the left, through and onto the right of the point with first coordinate a. Students of introductory calculus might see this interpreted in the following manner: For f to be continuous at $x = a$,

- The function f must be defined at a, and
- $\lim_{x \to a} f(x)$ must exist.
- Furthermore, $f(a)$ and $\lim_{x \to a} f(x)$ must be the same.

The third condition really implies the first two, because if $f(a)$ and $\lim_{x \to a} f(x)$ don't exist, they certainly cannot be equal. Thus we could simply say

$$\lim_{x \to a} f(x) = f(a)$$

Exercise 4.1.1 Following the intuitive idea that a function is continuous if its graph can be drawn "without lifting the pencil," precalculus and calculus teachers frequently lie to their students and say that functions such as $f(x) = \frac{1}{x}$ (which clearly don't meet this criterion) are discontinuous. (See Figure 4.2.) This particular fiction is useful and not too pernicious for students through the level of, say, introductory calculus, but it is nonsense from a mathematical viewpoint. Explain why. (*Hint*: Think about the various components that make up a function: domain, range, a set of ordered pairs. Where does $x = 0$ fit in with the function $f(x) = \frac{1}{x}$?)

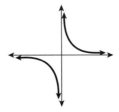

Figure 4.2 $f(x) = \dfrac{1}{x}$ is continuous.

■

4.2 Limit of a Function at a Point

Having looked in detail at one limiting process, we have established some general principles that will make others fall into place more easily. After a while, they will all begin to look a great deal alike. Nevertheless, it will be useful for us to go into some detail one more time, building on the insight we obtained last time.

Informally speaking:

> Suppose that $f : X \to Y$ is a function. Let a be a limit point of X. We say that $\lim_{x \to a} f(x) = L$ provided that as x gets *closer and closer* to a ($x \neq a$), the corresponding function values $f(x)$ get *closer and closer* to L.

By analogy with our discussion of limits of sequences, it is clear that our definition will ultimately involve *two* small positive quantities (one for each "closer and closer" in the informal definition): One of these small quantities will force values in the domain to be "close," and the other will force corresponding values in the range to be "close." By tradition, we call these small positive numbers δ and ϵ, respectively.

Once again, δ and ϵ are not independent of each other. The crucial question is this: Which depends on which? This can be a bit confusing, at first. The informal definition says that the closer x gets to a, the closer $f(x)$ gets to L. That is, closeness of the values in the domain forces the values in the range to be close. It is, therefore, easy to fall into the trap of assuming that ϵ (the small value for the range) depends on δ (the small value for the domain), but this is not so. Note that the same reasoning would have led you to assume that the value of ϵ in the definition of sequence convergence depends on N, rather than the other way around. In the case of sequences, the crucial requirement was that the terms of the sequence eventually come arbitrarily close to the limit of the sequence. For the limit of a function at a point, the crucial requirement is to assure that the function values $f(x)$ get arbitrarily close to the value L—thus ϵ must be arbitrarily small. The positive value δ will then need to be chosen so that the distances in the domain force the distances in the range to be ϵ-close.

Definition 4.2.1 Let (X, d_X) and (Y, d_Y) be metric spaces, let $K \subseteq X$, and let $f : K \to Y$ be a function. Let $a \in X$ be a limit point of K. We say that the limit of the function f as x goes to a is L, provided that for every real number $\epsilon > 0$, there exists a real number $\delta > 0$ such that if $x \in K$ and $0 < d_X(x, a) < \delta$, then $d_Y(f(x), L) < \epsilon$. We denote this by $\lim_{x \to a} f(x) = L$.

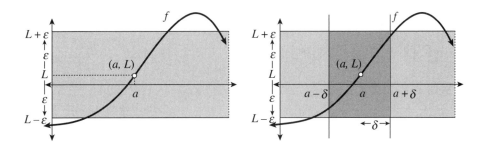

Figure 4.3 For functions from $\mathbb{R}$ to $\mathbb{R}$, Definition 4.2.1 specifies that we first fix the horizontal strip from $L - \epsilon$ and $L + \epsilon$ (shown in the diagram on the left). It says that the limit of f at a is L if we can always find δ small enough so that the part of the graph of f that lies in the vertical strip between $a - \delta$ and $a + \delta$ will also lie in the horizontal strip between $L - \epsilon$ and $L + \epsilon$ (shown in the diagram on the right).

Question to Ponder In Definition 4.2.1, we required that $0 < d_X(x,a)$; however, we did not require that $0 < d_Y(f(x), L)$. Using graphs to help you, think about the following things: Are these choices necessary? Could either of them be changed? Why or why not? What role, exactly, does each play in the definition?

Exercise 4.2.2 It is usually true that as $\epsilon \to 0$, $\delta \to 0$, too. But this is not necessarily so. Consider the graph of f shown in Figure 4.4:

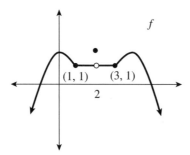

Figure 4.4 $\lim_{x \to 2} f(x) = 1$.

In a proof that $\lim_{x \to 2} f(x) = 1$, what is the smallest that δ ever need be? ∎

Theorem 4.2.3 Let X, K, Y, f, L, and a be as in Definition 4.2.1. Then the limit of f as x approaches a is unique. That is, if $\lim_{x \to a} f(x) = L$ and $\lim_{x \to a} f(x) = M$, then $L = M$.

Proof: The proof is Problem 2 at the end of this section. □

Theorem 4.2.4 Let (X, d_X) and (Y, d_Y) be metric spaces, and let $f : X \to Y$ be a function. Let $a \in X$, and let $L \in Y$. Prove that the following statements are equivalent:

1. $\lim_{x \to a} f(x) = L$.

2. For all sequences (x_n) in X of distinct points, if $x_n \to a$, then $f(x_n) \to L$.

Proof

$1 \implies 2$ Suppose that (x_n) is a sequence of distinct points in X with $x_n \to a$. We must show that $f(x_n) \to L$. To this end, let $\epsilon > 0$. Because $\lim_{x \to a} f(x) = L$, we know that we can find $\delta > 0$ so that if $0 < d_X(x, a) < \delta$, then $d_Y(f(x), f(a)) < \epsilon$. For the given value of δ, the fact that $x_n \to a$ guarantees that there exists $N_1 \in \mathbb{N}$ such that if $n > N_1$, $d_X(x_n, a) < \delta$. The fact that (x_n) is a sequence of distinct points guarantees that at most one term in the sequence is a; thus there exists $N_2 \in \mathbb{N}$ such that if $n > N_2$, $x_n \neq a$. Fix $n > \max(N_1, N_2)$. Then $0 < d_X(x_n, a) < \delta$, so $d_Y(f(x_n), L) < \epsilon$. We conclude that $f(x_n) \to L$.

$2 \Longleftarrow$ The proof is by contraposition. Suppose that $\lim_{x \to a} f(x) \neq L$. Then there exists some $\epsilon > 0$ such that for all $\delta > 0$, there exists $x \in X$ with $0 < d_X(x, a) < \delta$ yet $d_Y(f(x), L) \geq \epsilon$. For each $n \in \mathbb{N}$, choose $x_n \in B_{\frac{1}{n}}(a) \setminus \{a\}$ such that $d_Y(f(x_n), L) \geq \epsilon$. Clearly, $(f(x_n))$ is a sequence in Y that *does not* converge to L. However, (x_n) *does* converge to a. (see Problem 3 on page 86.)

$\square$

You are now ready for Excursion E.

Excursion E. *Limits of Functions from $\mathbb{R}$ to $\mathbb{R}$*

Synopsis: As always, the core chapters concentrate on general theoretical results. Excursion E looks at the specialized techniques needed to prove limit results for specific real functions. The discussion started in Excursion D, "Epsilonics: Playing the Game" (Section D.2), continues with "Epsilonics: Some General Principles." This section gives the reader some explicit tips on how to choose a δ that can do several jobs at once.

Problems 4.2

1. Negate Definition 4.2.1. Consider each of the diagrams below. For each function, using the language of Definition 4.2.1 or language like that used in Figure 4.3, explain why $\lim_{x \to a} f(x)$ does not exist. (An informal explanation is expected here, not a proof.)

    (a)   (b)

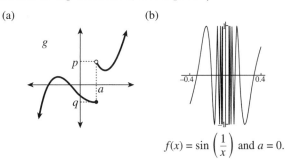

2. Prove that the limit of a function at a point (if it exists) is unique (Theorem 4.2.3).

> **Something to Ponder**
>
> The requirement in Definition 4.2.1 that a be a limit point of the domain of f is very important. For one thing, if we remove that requirement, we can no longer prove that the limit of a function at a is unique. Where in the proof you gave for Theorem 4.2.3 did you use the fact that a is a limit point of the domain of K? What would happen to your argument if it weren't?

3. Let (X, d_X) and (Y, d_Y) be metric spaces, and let $f : X \to Y$ be a function. Let $a \in X$, and let $L \in Y$. Prove that the following statements are equivalent:

   i.: For all sequences (x_n) in X of distinct terms, if $x_n \to a$, then $f(x_n) \to L$.

   ii.: For all sequences (x_n) in $X \setminus \{a\}$, if $x_n \to a$, then $f(x_n) \to L$.

4. Let X be a metric space, and let $f : X \to \mathbb{R}$ be a function. Let a be a limit point of X, and let $L \in \mathbb{R}$. Prove that the following statements are equivalent:

   i.: $\lim_{x \to a} f(x) = L$

   ii.: $\lim_{x \to a} |f(x) - L| = 0$

4.3 Continuous Functions

We now have a good mathematical understanding of $\lim_{x \to a} f(x)$. In this light, the heuristic description of our calculus days becomes a perfectly rigorous definition of continuity.

Definition 4.3.1 Let X and Y be metric spaces. Let $f : X \to Y$ be a function. Let $a \in X$. If a is a limit point of X, we say that f is **continuous** at a if

$$\lim_{x \to a} f(x) = f(a)$$

If a is not a limit point of X, then f is also defined to be continuous at a. We say that f is continuous if it is continuous at every point of its domain.

4.3 Continuous Functions

Exercise 4.3.2

1. Show that all constant functions are continuous. (A constant function takes on the same value at all points of its domain.)

2. Let X be a metric space. Let $f : X \to X$ be the identity function on X (i.e., $f(x) = x$). Prove that f is continuous on X.

■

Theorem 4.3.3 Let (X, d_X) and (Y, d_Y) be metric spaces, $f : X \to Y$ be a function, and $a \in X$. Then the following conditions are equivalent:

1. f is continuous at a.

2. For every $\epsilon > 0$, there exists a $\delta > 0$ such that if $d_X(x, a) < \delta$, then $d_Y(f(x), f(a)) < \epsilon$.

3. Given any sequence (x_n) in X converging to a, $(f(x_n))$ converges to $f(a)$.

Proof: The proof is Problem 1 at the end of this section. □

Theorem 4.3.4 Let X, Y, and Z be metric spaces. Suppose that $f : X \to Y$ and $g : Y \to Z$ are continuous functions. Show that $g \circ f$ is a continuous function.

Proof: The proof is Problem 4 at the end of this section. □

Definition 4.3.1 gives what is called a *local* definition for a continuous function. That is, it defines continuity at single points and then says that a function is continuous if it satisfies the condition at every point of its domain. The conditions given in Theorem 4.3.3 are also local conditions. *Global* conditions, conditions that discuss the function as a whole rather than point by point—can sometimes be useful as well. Theorem 4.3.5 gives a global condition for continuity. Another is described in Problem 6 at the end of this section.

Theorem 4.3.5 Let X and Y be metric spaces. A function $f : X \to Y$ is continuous if and only if the inverse image of every open subset of Y is an open subset of X.

Proof

$\implies$ Let V be an open subset of Y, and $f : X \to Y$ be a continuous function. Let $x \in f^{-1}(V)$. Then $f(x) \in V$. Because V is open, there exists $\epsilon > 0$ such that if $d_Y(z, f(x)) < \epsilon$, then $z \in V$.

Because f is continuous for (this value of!) $\epsilon > 0$, there exists $\delta > 0$ such that if $d_X(y, x) < \delta$, then $d_Y(f(y), f(x)) < \epsilon$. But then if $d_X(y, x) < \delta$, $f(z) \in V$. Finally, this means that if $d_X(y, x) < \delta$, then $y \in f^{-1}(V)$. Therefore, $f^{-1}(V)$ is open.

$\Longleftarrow$ Let $f : X \to Y$ be a function. Suppose the inverse image of every open subset V of Y is open in X.

Let $x \in X$. Set $y = f(x)$, and fix $\epsilon > 0$. Then $B_\epsilon(y)$ is an open subset of Y, so $f^{-1}(B_\epsilon(y))$ is open in X. Furthermore, $x \in f^{-1}(B_\epsilon(y))$. Hence there exists $\delta > 0$ such that $B_\delta(x)$ is totally contained in $f^{-1}(B_\epsilon(y))$. That is, every point in X that lies within a distance δ of x is sent by f to a point in Y that lies within a distance ϵ of $f(x)$. This shows that f is continuous.

$\square$

Think about Theorems 4.3.3 and 4.3.5. Which conditions do you see as geometric conditions, and which do you see as analytic conditions? (In a discussion with others, you might encounter some disagreement. As the mathematical structures become more complex, these distinctions become less clear-cut.)

Problems 4.3

1. Prove Theorem 4.3.3, which gives several equivalent conditions for continuity.

2. Let (X, d) be a metric space. Let $a \in X$. Define a function $f : X \to \mathbb{R}$ by
$$f(x) = d(a, x)$$
Show that f is a continuous function.

> This theorem says something interesting about the nature of metric spaces. When we speak intuitively about continuity, we often say, "If two points in the domain are close together, then their function values are close together." In this vein, what does the continuity of f tell you?

3. Let X, Y, and Z be metric spaces. Let $f : X \to Y$ be a function, and let $g : Y \to Z$ be a continuous function. Suppose that $\lim_{x \to a} f(x)$ exists. Prove that $\lim_{x \to a} g \circ f(x)$ exists and that
$$\lim_{x \to a} g \circ f(x) = g\left(\lim_{x \to a} f(x)\right)$$

> The fact that a limit can be taken "inside" a continuous function is used all the time. Consider, for instance, the following calculation, which might be done in a calculus class:
>
> $$\lim_{x \to 1} x^{\frac{1}{x-1}} = \lim_{x \to 1} \exp\left(\frac{1}{x-1} \ln(x)\right)$$
> $$= \exp\left(\lim_{x \to 1} \frac{1}{x-1} \ln(x)\right)$$
> $$= \exp\left(\lim_{x \to 1} \frac{\ln(x)}{x-1}\right)$$
> $$= \exp\left(\lim_{x \to 1} \frac{\frac{1}{x}}{1}\right)$$
> $$= e$$

4. Prove that the composition of continuous functions is continuous (Theorem 4.3.4).

5. Continuity and inverses: Let K be a closed, bounded subset of $\mathbb{R}$, and let $f : K \to \mathbb{R}$ be a one-to-one, continuous function. Prove that $f^{-1} : f(K) \to \mathbb{R}$ is also continuous.

   (*Hint*: Use a sequence approach and think about Problem 6 on page 86).

> This result is fairly useful. For instance, if $f : [a, b] \to \mathbb{R}$ is a one-to-one, continuous function, we can invert it and know immediately that its inverse is also continuous. For instance, it is easy to show that $f(x) = x^n$ is continuous for $n \in \mathbb{N}$. It is more challenging to show that $g(x) = \sqrt[n]{x}$ is continuous, but this theorem gives us that result for free.

6. Let X and Y be metric spaces. Show that a function $f : X \to Y$ is continuous if and only if the inverse image of every closed subset of Y is a closed subset of X. Prove this theorem in two ways:

   (a) Use Theorem 4.3.5 and complementation.

   (b) Prove the theorem directly without appealing to Theorem 4.3.5 or to the fact that the complement of a closed set is an open set. (*Hint*: In proving that if the inverse image of every closed set is closed, then the function is continuous, proceed by contraposition. Show that if the function fails to be continuous, there exists a sequence (x_n) in X converging to x and a positive real number ϵ such that

$d(f(x_n), f(x)) > \epsilon$ for all n. Use this fact to construct a closed subset of Y whose inverse image is not closed in X.)

7. Let X and Y be metric spaces. Let $f : X \to Y$ be a function. Prove that f is continuous at $a \in X$ if and only if $\lim_{x \to a} d(f(x), f(a)) = 0$. (*Hint:* You must ponder what the expression $\lim_{x \to a} d(f(x), f(a)) = 0$ means; it is the limit of a function at a point. But what function? At what point?)

8. Let (X, d) be a metric space. Show that the following statements about X are equivalent:

   i. (X, d) is discrete.

   ii. All functions defined on X are continuous. (The range can be *any* metric space.)

9. Let X and Y be metric spaces, and let D be a dense subset of X. (For the definition of "dense," see Problem 4 at the end of Section 3.5.)

   (a) Let $f : X \to Y$ and $g : X \to Y$ be continuous functions. Suppose that $f(d) = g(d)$ for all $d \in D$. Prove that f and g are the same function.

   This theorem has some bearing on the problem of extending a function on a smaller space to a function on a larger space. For instance, we might have a function whose values are defined at all the rational numbers, but we would like to have a function whose values are defined everywhere on the real numbers.

   (b) Suppose you have a function h that goes from the rational numbers to the real numbers. Discuss the following questions; illustrate with examples and/or give proofs, if appropriate.

   i. How many ways are there to extend h to a function from $\mathbb{R}$ to $\mathbb{R}$?

   ii. What is the maximum possible number of extensions of h to a *continuous* function from $\mathbb{R}$ to $\mathbb{R}$?

   iii. Can you find a function h that *cannot* be extended in a continuous way?

   iv. If h is continuous on the rational numbers, do you think it is always possible to extend it in a continuous way to a function on $\mathbb{R}$?

   We will be able to say more about the process of building extensions after Section 4.4. See Problem 7 on page 117.

4.4 Uniform Continuity

Definition 4.4.1 Let X and Y be metric spaces. Let $f : X \to Y$ be a function. If $S \subseteq X$, f is said to be **uniformly continuous on S** if for every $\epsilon > 0$, there exists $\delta > 0$ such that $d_Y(f(a), f(b)) < \epsilon$ whenever $a, b \in S$ and $d_X(a,b) < \delta$.

If f is uniformly continuous on X, we simply say that f is uniformly continuous.

Exercise 4.4.2 Carefully compare the definitions of continuity and uniform continuity. How are they different? Which is stronger?[1] Prove your answer. ∎

Definition 4.4.3 Let X and Y be metric spaces. Let $f : X \to Y$ be a function. Let k be a positive real number. If $d_Y(f(a), f(b)) \leq k\, d_X(a,b)$ for all $a, b \in X$, f is said to satisfy a **Lipschitz condition with constant k**. A function that satisfies a Lipschitz condition is called a **Lipschitz function.**

Theorem 4.4.4 If $f : X \to Y$ is a Lipschitz function, then f is uniformly continuous.

Proof: The proof is Problem 2 at the end of this section. □

You are now ready for Excursion F.

Excursion F. *Doubly Indexed Sequences*

Synopsis: This excursion considers the limits of doubly indexed sequences. The interplay between the limit of a doubly indexed sequence and the related iterated limits requires a uniform convergence condition that can be used to reinforce the uniformity condition studied in Section 4.4. (However, parts of Excursion F will be challenging at this point, as they foreshadow ideas from Chapter 12. It will be a much less challenging excursion if tackled after discussing the role that uniform convergence plays in the exchange of limit operations.) Excursion F is written as a set of interconnected exercises that can be assigned as a mini-project, or individual parts can be assigned as additional exercises.

---

1. We say that condition A is stronger than condition B if every time that condition A is satisfied, condition B is also satisfied—that is, if condition A *implies* condition B.

Problems 4.4

1. Consider the function $f : \mathbb{R} \to \mathbb{R}$ given by $f(x) = x^2$. Using only the definition of uniform continuity, prove that f is uniformly continuous on $[0, 1]$ but not on $[0, \infty)$.

2. Prove that Lipschitz functions are uniformly continuous (Theorem 4.4.4).

3. Let X and Y be metric spaces, and let $f : X \to Y$ be a function. Suppose that $S \subseteq X$.

   (a) Say what it means for f to fail to be uniformly continuous on S by negating the definition of uniform continuity.

   (b) Prove that f fails to be uniformly continuous on S if and only if there exist $\epsilon > 0$, and sequences (x_n) and (y_n) in S such that $d_X(x_n, y_n) \to 0$ and yet, for all $n \in \mathbb{N}$, $d_Y(f(x_n), f(y_n)) \geq \epsilon$.

4. Let X and Y be metric spaces.

   Definition: A function $f : X \to Y$ is said to be **bounded** if its range $f(X)$ is a bounded subset of Y. (It follows from Corollary 3.1.14 that if $Y \subseteq \mathbb{R}$, this amounts to the existence of a real number M such that $|f(x)| \leq M$ for all $x \in X$.)

   Suppose that X is a bounded metric space and Y is any metric space. Let $f : X \to Y$ be a Lipschitz function. Prove that f is a bounded function.

5. Let K be a bounded, non-empty subset of $\mathbb{R}$. Let $f : K \to \mathbb{R}$ be a uniformly continuous function. Prove that f is bounded.

6. Let X be a metric space. Let $Y \subseteq X$. For each $x \in X$, define $d(x, Y) = \inf\{d(x, y) | y \in Y\}$. Now define a function $f_Y : X \to \mathbb{R}$ by $f_Y(x) = d(x, Y)$.

   (a) Prove that $f_Y(x) = 0$ if and only if $x \in \overline{Y}$. In particular, this shows that when Y is closed, $f_Y(x) = 0$ if and only if $x \in Y$.

   (b) Prove that f_Y is a uniformly continuous function.

   (c) Let A and B be disjoint closed subsets of X. Define $f : X \to \mathbb{R}$ by $f(x) = f_A(x) - f_B(x)$. Prove that the sets $V = f^{-1}((0, \infty))$ and $U = f^{-1}((-\infty, 0))$ are disjoint open sets containing A and B, respectively.

   (d) Look back at Problem 11 in Section 3.7 for the definition of normality. Explain why the previous parts of this problem prove that metric spaces are normal.

7. Let X be a metric space. Let D be a dense subset of X. Suppose that $f : D \to \mathbb{R}$ is bounded and uniformly continuous on each bounded subset of D. In this problem, you will show that f can be extended to a continuous function on all of X.

   We will define the function $F : X \to \mathbb{R}$ as follows: Let $x \in X$. Choose any sequence (t_i) in D such that $t_i \to x$. Note that the sequence $(f(t_i))$ has a convergent subsequence. (*Why?*) Extract such a subsequence and define $F(x)$ to be its limit.

   (a) Show that the function F as defined above is well defined. That is, show that it doesn't matter which sequence (t_i) is chosen or how we extract the convergent subsequence from $(f(t_i))$; the limit is always the same.

   (b) Prove that $F|_D = f$. (F is an *extension* of f.)

   (c) Prove that the function F is uniformly continuous on each bounded subset of X and is, therefore, continuous on X.

8. This problem extends the ideas from Problem 7.

   (a) Find an *unbounded* continuous function f from the rational numbers to the real numbers that cannot be extended to a continuous function on all of $\mathbb{R}$.

   (b) Find a *bounded* continuous function f from the rational numbers to the real numbers that cannot be extended to a continuous function on all of $\mathbb{R}$.

   (c) What is the significance of parts (a) and (b)?

Chapter 5
Real-Valued Functions

A real-valued function is one whose outputs are real numbers; the inputs may lie in any set. They show up in all sorts of situations.

Example 5.0.1

1. Every function $f : \mathbb{R} \to \mathbb{R}$ is a real-valued function.

2. Real-valued functions of several real variables, functions from $\mathbb{R}^n$ to $\mathbb{R}$, arise in many contexts:

    - **Economics.** Consider a company that produces some commodity—say, bean-bag animals. The profit made by the company depends on several factors, such as how much the company spends on salaries and other overhead, how many bean-bag animals it produces, how much the materials cost, how much the company charges for each animal, and how many animals it actually sells. In this case, the profit function maps $\mathbb{R}^5$ to $\mathbb{R}$.

    - **Physics.** To each point in space, we can assign the value of the electric potential in volts. This is a function from $\mathbb{R}^3$ to $\mathbb{R}$.

3. The determinant function takes an $n \times n$ real matrix and returns a real number.

4. The distance function d for a metric space (X, d) is a real-valued function because it takes a pair of points in X and produces a real number.

5. A probability density function is a function that takes elements in some space X of possible outcomes; to each outcome, it assigns a real number. When only finitely many outcomes are possible, the assigned real number

is the probability that the outcome will occur. If a continuous range of possibilities exists, the probability that the outcome of an experiment will lie within a certain interval is the integral of the probability density function over the interval.

As you can see, real-valued functions are everywhere. They are enormously important. In this section, we examine some theorems that connect limits and continuity in real-valued functions with the order and arithmetic properties in $\mathbb{R}$.

5.1 Limits, Continuity, and Order

Let X be a metric space. Here is the sort of question that arises frequently in practice: If we have an inequality, can we take limits of both sides and maintain the sense of the inequality? Loosely speaking, does $f(x) < g(x)$ for all x "near" a imply that $\lim_{x \to a} f(x) < \lim_{x \to a} g(x)$? This section provides some partial answers to this class of questions.

Theorem 5.1.1 Let X be a metric space, let $a \in X$ be a limit point of X, and let $f : X \to \mathbb{R}$ be a function. Assume that the limit of f exists at a. Fix $t \in \mathbb{R}$.

Suppose there exists $r > 0$ such that $f(x) \geq t$ for every $x \in B_r(a) \setminus \{a\}$; then $\lim_{x \to a} f(x) \geq t$. Likewise, if there exists $r > 0$ such that $f(x) \leq t$ for all $x \in B_r(a) \setminus \{a\}$, then $\lim_{x \to a} f(x) \leq t$.

Proof: The proof is Problem 3 at the end of this section. □

Exercise 5.1.2 Let $a \in \mathbb{R}$ be a limit point of X, and let $f : \mathbb{R} \to \mathbb{R}$ be a function. Show by giving an example that, even if $f(x) > 0$ everywhere in $B_r(a) \setminus \{a\}$ for some $r > 0$, the limit of f at a may nevertheless be zero. ■

Proof Tip

Many theorems involving inequalities have two parts: one involving $>$ or $\geq$ and one involving $<$ or $\leq$. You have undoubtedly noticed that the proofs are almost always parallel, in the sense that if you can prove one clause of the theorem, the proof of the other is virtually identical; just replace each $>$ with $<$, or vice versa. Another useful trick can also help with these proofs: You can often get the second part "for free" from the first part. To begin, just multiply everything in sight by -1; this changes the sense of every inequality, putting you into the first case. Draw conclusions using the first part of the theorem and multiply the results by -1, and *presto!* You get what you want. Try it!

Exercise 5.1.3 Let f and g from $\mathbb{R}$ to $\mathbb{R}$ such that
$$f(x) < g(x) \text{ for all } x \in (-1, 0) \cup (0, 1).$$
Must it be true that $\lim_{x \to 0} f(x) \leq \lim_{x \to 0} g(x)$? (*Careful!* This is a trick question.) ∎

Theorem 5.1.4 Let X be a metric space, and let $a \in X$ be a limit point of X. Let f and g be functions from X to $\mathbb{R}$. Suppose that $\lim_{x \to a} f(x)$ and $\lim_{x \to a} g(x)$ both exist. Assume further that there exists $r > 0$ such that
$$f(x) \leq g(x) \text{ for all } x \in \mathrm{B}_r(a) \setminus \{a\}.$$
Then $\lim_{x \to a} f(x) \leq \lim_{x \to a} g(x)$.

Proof: The proof is Problem 4 at the end of this section. □

The point of Exercise 5.1.3 is that you can't take the limit of both sides of an inequality unless you know ahead of time that the limits exist! Sometimes they may not. Some theorems, however, give us the existence of a limit for free if the other hypotheses are satisfied. The "sandwich theorem" is one of these.

Theorem 5.1.5 [Sandwich Theorem] Let X be a metric space, and let $a \in X$ be a limit point of X. Let f, g, and h be functions from X to $\mathbb{R}$. Suppose that

- $\lim_{x \to a} f(x) = \lim_{x \to a} h(x) = L$.
- there exists $r > 0$ such that $f(x) \leq g(x) \leq h(x)$ for all $x \in \mathrm{B}_r(a) \setminus \{a\}$.

Then $\lim_{x \to a} g(x) = L$.

Proof: The proof is Problem 5 at the end of this section. □

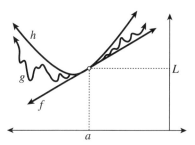

Figure 5.1 The sandwich theorem.

Theorem 5.1.5 does not assume the existence of $\lim_{x \to a} g(x)$, which is precisely what makes it useful. What usually happens in practice is that f and h are "well-behaved" functions whose limits can be deduced easily; g is more complicated. The sandwich theorem gives us a way to show that the limit of g exists and to compute it by using simple upper and lower bounds that "pinch together" at the point a. (See Figure 5.1.)

Turning things around a bit, if we know the limit of a function at a point, what does this information tell us about the values of the function close to the point? It is sometimes useful to know that a function is bounded on some region. If the limit exists at a, then it will be bounded on a region surrounding a.

Theorem 5.1.6 Let X be a metric space, let $a \in X$ be a limit point of X, and let $f : X \to \mathbb{R}$ be a function. Suppose that $\lim_{x \to a} f(x) = L$. For all $K \in \mathbb{R}$ such that $|L| < K$, there exists $\delta > 0$ such that if $x \in B_\delta(a) \setminus \{a\}$, then $|f(x)| < K$.

Proof: The proof is Problem 6 at the end of this section. $\square$

A "parallel" theorem says that if the limit of the function is not zero, then f is bounded away from zero on some region surrounding a.

Theorem 5.1.7 Let X be a metric space, let $a \in X$ be a limit point of X, and let $f : X \to \mathbb{R}$ be a function. Suppose that $\lim_{x \to a} f(x) = L \neq 0$. For all $K \in \mathbb{R}$ such that $|L| > K > 0$, there exists a positive real number δ such that if $0 < d(x, a) < \delta$, then $|f(x)| > K$.

Proof: The proof is Problem 7 at the end of this section. $\square$

Exercise 5.1.8 Actually, the conditions $L \neq 0$ and $K > 0$ can be removed from Theorem 5.1.7. The theorem remains true, but is not very interesting. Explain why. ∎

Some Useful Special Cases

Some of the theorems in this section have special cases that are especially useful in practice. We will identify those explicitly for easy reference.[1]

Corollaries 5.1.9 and **??** are special cases of Theorem 5.1.1.

Corollary 5.1.9 Suppose that X is a metric space. Let $f : X \to \mathbb{R}$ be a function, and let $a \in X$ be a limit point of X. Suppose that the limit of f

---

1. These are special cases and require no further proof, but you should check for yourself to see why this is so.

exists at a. If there exists $r > 0$ such that $f(x) \geq 0$ for every $x \in B_r(a) \setminus \{a\}$, then $\lim_{x \to a} f(x) \geq 0$.

Theorem 5.1.6 is frequently stated and used in the following form:

Corollary 5.1.10 Let X be a metric space, and let $f : X \to \mathbb{R}$ be a function. Suppose that $\lim_{x \to a} f(x) = L$. Then for all positive real numbers ϵ, there exists $\delta > 0$ such that if $0 < d(x, a) < \delta$, then $|f(x)| < |L| + \epsilon$.

In other words, not only is the function bounded on a region about a, but the bound can also be as close to $|L|$ as we like. (Of course, if ϵ is very small, we will likely pay a price in the size of δ.)

Continuous Functions: Each of the limit theorems has an immediate corollary in which the existence of the limit at a is replaced by continuity at a.

Corollary 5.1.11 Let X be a metric space, let $a \in X$ be a limit point of X, and let $f : X \to \mathbb{R}$ be a function. Assume that the f is continuous at a. Fix $t \in \mathbb{R}$.

Suppose there exists $r > 0$ such that $f(x) \geq t$ for every $x \in B_r(a) \setminus \{a\}$; then $f(a) \geq t$. Likewise, if there exists $r > 0$ such that $f(x) \leq t$ for all $x \in B_r(a) \setminus \{a\}$, then $f(a) \leq t$.

Corollary 5.1.12 Let X be a metric space, let $a \in X$ be a limit point of X, and let f and g be functions from X to $\mathbb{R}$ that are continuous at a. Assume that there exists $r > 0$ such that

$$f(x) \leq g(x) \text{ for all } x \in B_r(a) \setminus \{a\}$$

Then $f(a) \leq g(a)$.

Corollary 5.1.13 [Sandwich Theorem Revisited] Let X be a metric space, and let $a \in X$ be a limit point of X. Let f, g, and h be functions from X to $\mathbb{R}$. Suppose that the following conditions hold:

- f and h are continuous at a.

- $f(a) \leq h(a)$.

- There exists $r > 0$ such that $f(x) \leq g(x) \leq h(x)$ for all $x \in B_r(a) \setminus \{a\}$.

Then $\lim_{x \to a} g(x) = f(a) = h(a)$.

Corollary 5.1.14 Let X be a metric space, let $a \in X$, and let $f : X \to \mathbb{R}$ be a function that is continuous at a. For all $K \in \mathbb{R}$ such that $|f(a)| < K$, there exists $\delta > 0$ such that if $x \in \mathrm{B}_\delta(a)$, then $|f(x)| < K$.

Corollary 5.1.15 Let X be a metric space, let $a \in X$, and let $f : X \to \mathbb{R}$ be a function that is continuous at a. Suppose that $f(a) \neq 0$. For all $K \in \mathbb{R}$ such that $|f(a)| > K > 0$, there exists a positive real number δ such that if $d(x, a) < \delta$, then $|f(x)| > K$.

Problems 5.1

1. Let X be a metric space. Let f and g be functions from X to $\mathbb{R}$. Let $a \in X$ and let $r > 0$. Suppose that $f(x) = g(x)$ for all $x \in \mathrm{B}_r(a) \setminus \{a\}$.

    (a) Prove that it is not always the case that $\lim_{x \to a} f(x) = \lim_{x \to a} g(x)$.

    (b) Prove that if $\lim_{x \to a} g(x)$ exists, then $\lim_{x \to a} f(x)$ also exists and $\lim_{x \to a} f(x) = \lim_{x \to a} g(x)$.

2. Let X be a metric space and let $a \in X$ be a limit point of X. Suppose that $f : X \to \mathbb{R}$ is a function and that $\lim_{x \to a} f(x)$ exists. Show that $\lim_{x \to a} |f(x)|$ also exists by showing that $\lim_{x \to a} |f(x)| = |\lim_{x \to a} f(x)|$.

3. Prove Theorem 5.1.1.

4. Prove Theorem 5.1.4.

5. Prove the sandwich theorem (Theorem 5.1.5).

6. Prove Theorem 5.1.6.

7. Prove Theorem 5.1.7.

5.2 One-Sided Limits

When the *domain* of a function is a subset of $\mathbb{R}$, the fact that $\mathbb{R}$ is ordered allows us to talk about one-sided limits.[2]

---

2. Sorry! False advertising! This section isn't really about real-valued functions, because it is the domain—not the range—that must be a subset of $\mathbb{R}$ in this discussion.

5.2 One-Sided Limits

Definition 5.2.1 Let K be a subset of $\mathbb{R}$, let X be any metric space, and let $f : K \to X$ be a function. Fix $c \in \mathbb{R}$ and $\eta > 0$ such that $(c - \eta, c) \subseteq K$. We say that the **limit of f as x approaches c from the left** is equal to L if for every positive number ϵ, there exists a positive number $\delta \leq \eta$ such that whenever $x \in (c - \delta, c)$, then $d(f(x), L) < \epsilon$. We denote this symbolically by

$$\lim_{x \to c^-} f(x) = L$$

Exercise 5.2.2 Using Definition 5.2.1 as a model, define the **limit of f as x approaches c from the right**, which is denoted by

$$\lim_{x \to c^+} f(x) = L$$

Be complete and precise. ∎

Theorem 5.2.3 is the crucial link between one-sided and two-sided limits.

Theorem 5.2.3 Let K be a subset of $\mathbb{R}$, let X be any metric space, and let $f : K \to X$ be a function. Suppose that $c \in \mathbb{R}$ and that for some $\eta > 0$, $(c - \eta, c) \cup (c, c + \eta) \subseteq K$. The limit of f as x approaches c exists if and only if the limit of f as x approaches c from the right and the limit of f as x approaches c from the left both exist and are equal.

Proof: The proof is Problem 1 at the end of this section. □

Exercise 5.2.4 Let K be a subset of $\mathbb{R}$, and let $f : K \to \mathbb{R}$ be a function. Suppose that $\lim_{x \to c} f(x)$ exists, that $\lim_{x \to c^-} f(x) \geq 0$, and that $\lim_{x \to c^+} f(x) \leq 0$. Show that $\lim_{x \to c} f(x) = 0$. ∎

Left- and right-handed limits allow us to describe mathematically the distinction between different sorts of discontinuities.

Exercise 5.2.5 Look again at Problem 1 at the end of Section 4.2. In that problem, you were asked to consider two discontinuous functions; the discontinuities you saw there were different in kind. The first function is said to have a **jump discontinuity**; the second is said to have an **essential discontinuity**. Use the definitions of left- and right-handed limits to give careful definitions for jump discontinuity and essential discontinuity for functions with domains that are subsets of $\mathbb{R}$. ∎

Problems 5.2

1. Prove Theorem 5.2.3, which establishes the relationship between one- and two-sided limits.

2. Let $K \subseteq \mathbb{R}$. We say that $f : K \to \mathbb{R}$ is a **monotonic function** provided that f is either increasing or decreasing. In other words, f is monotonic provided that it satisfies one of the following two conditions:

$$\text{If } x, y \in K \text{ and } x < y, \text{ then } f(x) \leq f(y)$$

or

$$\text{If } x, y \in K \text{ and } x < y, \text{ then } f(x) \geq f(y)$$

Let $a, b \in \mathbb{R}$, and let $f : [a, b] \to \mathbb{R}$ be a monotonic function.

(a) Prove that for all $c \in (a, b)$, $\lim_{x \to c^-} f(x)$, and $\lim_{x \to c^+} f(x)$ both exist. Use this result to deduce that if f is not continuous, it can have only jump discontinuities.

(b) Prove that f can have at most countably many discontinuities.

3. Let X and Y be metric spaces. Let $f : X \to Y$ be a function.

   Definition: If f is discontinuous at a, we say that f has a **removable discontinuity** at a provided that there exists a function $g : X \to Y$ such that g is continuous at a and $g(x) = f(x)$ for all $x \neq a$.

   (a) Give an example of a function $f : \mathbb{R} \to \mathbb{R}$ with a removable discontinuity.

   (b) Prove that f has a removable discontinuity at $a \in X$ if and only if f is discontinuous at a and $\lim_{x \to a} f(x)$ exists.

4. Let $I \subseteq \mathbb{R}$ be an interval and let $f : I \to \mathbb{R}$ be a function. We say that f has the **intermediate value property** provided that for all $a, b \in I$, if γ lies between $f(a)$ and $f(b)$, there exists $c \in (a, b)$ such that $f(c) = \gamma$. Prove that if $f : I \to \mathbb{R}$ has the intermediate value property, then it cannot have any jump discontinuities.

5.3 Limits, Continuity, and Arithmetic

Real-valued functions with the same domain can be combined in the usual ways by adding, subtracting, multiplying, and dividing.[3] If the limits of both component functions exist, so does the limit of the combined function. If the component functions are continuous, the resulting functions will be as well.

Theorem 5.3.1 Let X be a metric space. Suppose that $f : X \to \mathbb{R}$ and $g : X \to \mathbb{R}$. Suppose further that $\lim_{x \to a} f(x) = L$ and $\lim_{x \to a} g(x) = M$. Then

1. $\lim_{x \to a} (f(x) + g(x)) = L + M$.

2. $\lim_{x \to a} (f(x) - g(x)) = L - M$.

3. $\lim_{x \to a} (f(x)g(x)) = LM$.

Assume further that $g(x) \neq 0$ on some interval containing a and that $M \neq 0$.

4. Then $\lim_{x \to a} \dfrac{f(x)}{g(x)} = \dfrac{L}{M}$.

Proof: The proof is Problem 2a at the end of this section. $\square$

Question to Ponder 5.3.2 Can Theorem 5.3.1 be generalized to functions whose range is not a subset of $\mathbb{R}$? Why or why not?

Theorem 5.3.3 is really a corollary to Theorem 5.3.1. I call it a theorem to emphasize its importance as a theorem in its own right.

Theorem 5.3.3 Let X be a metric space. Suppose that $f : X \to \mathbb{R}$ and $g : X \to \mathbb{R}$ are continuous functions. Then the following functions are continuous:

1. $f + g$

2. $f - g$

3. fg

4. $\dfrac{f}{g}$ (provided that 0 is not in the range of g)

---

3. Note: such combinations can be formed even if the arithmetic operations make no sense in the domain!

Proof: The proof is Problem 2b at the end of this section. □

You are now ready for the last part of Excursion C.

Excursion C: Section 2. *Irrational Exponents*

Synopsis: Math teachers confidently describe the steps that lead from positive integer powers to rational powers. (We fail to mention that each step requires some effort; nevertheless, the path we suggest should work—and does—as shown in Section C.1.)

When we start talking about taking limits of rational powers to get irrational powers, however, our voices trail off with disclaimers about how the details are "beyond the scope of the course." Indeed, this transition requires a surprising amount of mathematical machinery. The necessary ideas are discussed in Chapters 4 and 5.

Problems 5.3

1. Let X be a metric space. Let $f : X \to \mathbb{R}$ be a real-valued function. Prove that the following statements are equivalent:

    i. The function f is continuous.

    ii. For all $t, s \in \mathbb{R}$ with $t < s$, $f^{-1}((t,s))$ is open in X.

    iii. For all $t \in \mathbb{R}$, $f^{-1}((-\infty, t])$ and $f^{-1}([t, \infty))$ are both closed in X.

2. Algebraic combinations of real-valued functions.

    (a) Prove Theorem 5.3.1. (*Hint:* You can compute the necessary estimates directly, or you can be clever and get this theorem almost for free by exploiting the fact that you already have a similar theorem for real sequences.)

    (b) Use Theorem 5.3.1 to deduce Theorem 5.3.3.

3. Consider a sort of converse to Theorem 5.3.1: "If $\lim_{x \to a} (f(x) + g(x))$ exists, then $\lim_{x \to a} f(x)$ and $\lim_{x \to a} g(x)$ exist as well" (and so forth for the other cases). Are any of these statements true? Give a proof or a counterexample in each of the four cases.

4. Polynomials and rational functions.

    (a) Let $P : \mathbb{R} \to \mathbb{R}$ be a polynomial.
    $$P(x) = a_0 + a_1 x + a_2 x^2 + a_3 x^3 + \cdots + a_n x^n$$
    Prove that P is continuous on $\mathbb{R}$.

    (b) Let f be a rational function. That is, suppose there exist polynomials $P : \mathbb{R} \to \mathbb{R}$ and $Q : \mathbb{R} \to \mathbb{R}$ such that
    $$f(x) = \frac{P(x)}{Q(x)}$$
    Prove that f is continuous on its domain.

5. Continuity of the arithmetic operations.

    (a) Prove that the arithmetic operations $+$, $-$, and $\cdot$ are continuous functions from $\mathbb{R}^2$ to $\mathbb{R}$. Prove that $\div$ is a continuous function from $\mathbb{R}^2 \setminus \{(t, 0) : t \in \mathbb{R}\}$ to $\mathbb{R}$.

    (b) Which of the arithmetic operations are uniformly continuous?

    (c) Informally say what the results from parts (a) and (b) tell you about the arithmetic operations. (In other words, make statements such as "if *such and such* are sufficiently close together, then *so and so* are also close together.")

6. Let X be a metric space. Let $f : X \to \mathbb{R}$ and $g : X \to \mathbb{R}$ be real-valued Lipschitz functions with Lipschitz constants K and M, respectively.

    (a) Prove that $f + g$ is a Lipschitz function.

    (b) Prove that if f and g are bounded, then $f \cdot g$ is a Lipschitz function.

    (c) Show that the boundedness condition in part (b) is necessary. Do so by giving an example of two Lipschitz functions whose product is not a Lipschitz function.

7. Let X be a metric space. Let $f : X \to \mathbb{R}$ and $g : X \to \mathbb{R}$ be uniformly continuous functions.

    (a) Determine whether the sum of f and g must be uniformly continuous. Prove or give a counterexample.

    (b) Because Lipschitz functions are uniformly continuous (see Theorem 4.4.4), Problem 6 proves that the product of f and g need not be uniformly continuous. If you add the requirement that the two functions be bounded, must it be the case that $f \cdot g$ is uniformly continuous?

8. Suppose that a and $y \in \mathbb{R}$. Suppose also that $f : \mathbb{R}^2 \to \mathbb{R}$ is continuous at the point (a, y).

   (a) Show that $g : \mathbb{R} \to \mathbb{R}$ given by
   $$g(x) = f(a, x)$$
   is continuous at y and that $h : \mathbb{R} \to \mathbb{R}$ given by
   $$h(x) = f(x, y)$$
   is continuous at a.

   (b) Show by giving a counterexample that it is possible for g to be continuous at y and h to be continuous at a, and for f to be discontinuous at (a, y).

Chapter 6

Completeness

In the discussion of the axioms for the real number system, we used the least upper bound axiom to mathematically describe the difference between the rational numbers and the real numbers. Basically, we were able to observe that the "holes" left by the rationals could not be satisfactorily dealt with by either the field axioms or the order axioms. The least upper bound axiom is sometimes called the "completeness axiom" because it guarantees that $\mathbb{R}$ satisfies a condition that in general metric spaces is called **completeness**.

The idea here is that we want to talk about the existence (or lack thereof) of "pinprick" holes in a space. In the absence of an ordering, nothing parallel to the least upper bound axiom is possible. Instead, we need a completely new idea.

6.1 Cauchy Sequences

Intuitively, we say that a sequence (a_n) converges to a if the terms of the sequence (a_n) get closer and closer to a as they get farther out in the sequence. Of course, if the terms beyond some point are all close to a, they must also be close to one another. Sequences in which the terms get closer and closer together the farther out in the sequence they are called *Cauchy sequences*.

Definition 6.1.1 Let (a_n) be a sequence in a metric space X. Then (a_n) is said to be a **Cauchy sequence** provided that for every $\epsilon > 0$, there exists $N \in \mathbb{N}$ such that if $n, m > N$, then $d(a_n, a_m) < \epsilon$.

We can easily establish some simple facts.

Theorem 6.1.2 Let (a_n) be a sequence in a metric space X.

1. If (a_n) converges, then it is a Cauchy sequence.

2. If (a_n) is a Cauchy sequence, then every subsequence of (a_n) is a Cauchy sequence.

3. If (a_n) is a Cauchy sequence, then it must be bounded.

4. If (a_n) is a Cauchy sequence that has a *convergent* subsequence, then (a_n) itself must converge.

Proof: The proof is Problem 1 at the end of this section. $\square$

Problems 6.1

1. Prove Theorem 6.1.2, which lays out the basic characteristics of Cauchy sequences. (*Hint*: In the proof of the last part, you may find a use for Theorem 0.4.6.)

2. Let X be a metric space and (a_n) be a sequence in X. For each $n \in \mathbb{N}$, let $T_n = \{a_n, a_{n+1}, a_{n+2}, \ldots\}$. Prove that (a_n) is a Cauchy sequence if and only if $\lim_{n \to \infty} \text{diameter}(T_n) = 0$.

3. Let X and Y be metric spaces. Let $f : X \to Y$ be a function. Let (a_i) be a Cauchy sequence in X.

    (a) Prove that if f is uniformly continuous, then $(f(a_i))$ is a Cauchy sequence in Y.

    (b) Give an example of a continuous function f and a Cauchy sequence (a_i) that shows that if f is *not* uniformly continuous, then $(f(a_i))$ need not be a Cauchy sequence. [*Hint*: You need not seek anything very "pathological." Try working with functions $f : (0, 1) \to \mathbb{R}$.]

6.2 Complete Metric Spaces

One might think that if the terms of a sequence are getting closer and closer together, then the sequence "should" converge to something. However, it is easy to see that this is not so. For instance, the sequence

$$3, 3.1, 3.14, 3.141, 3.1415, 3.14159, \ldots$$

is getting closer and closer to the number that we know as π. In $\mathbb{R}$, the sequence converges to π; in the rational numbers, however, it doesn't converge at all. Why? Because there is a little "pinprick" hole in the rational numbers.

Also, if our space is $X = \mathbb{R}^2 \setminus \{(0,0)\}$ together with the metric "inherited" from the Euclidean metric in $\mathbb{R}^2$, then the sequence $(0,1), (0, 1/2), (0, 1/3), (0, 1/4) \ldots$ is a Cauchy sequence but doesn't converge. Once again, we see that a "pinprick" is the culprit.

Definition 6.2.1 A metric space X is said to be **complete** if every Cauchy sequence in X converges to an element of X.[1]

Some spaces are not complete—the rationals, for example. However, many of our favorite mathematical friends *are* complete. As mentioned earlier, completeness plays a role in general metric spaces that is similar to the role played by the least upper bound property in $\mathbb{R}$. Thus our first task is to show that $\mathbb{R}$ is complete.

Theorem 6.2.2 $\mathbb{R}$ is complete.

Proof: The proof is Problem 1 at the end of this section. □

We can generalize this result to n-dimensional space.

Corollary 6.2.3 $\mathbb{R}^n$ is complete.

Proof: The proof is Problem 2 at the end of this section. □

Like the least upper bound property, completeness guarantees the existence of certain kinds of points, something that is tremendously important in practice. Suppose that we need to find a solution to a particular equation.[2] We may

---

[1]. The name comes from the fact that our mathematical forebearers thought Cauchy sequences "ought" to converge. They felt that a space was "missing something" if it contained a Cauchy sequence that did not converge. Thus they used the word "incomplete" to describe it.

[2]. This is a very general sort of problem. We could be talking about an algebraic equation, but it might also be a differential equation, for example. In Excursion N you will study a complete metric space in which the points are themselves functions. But you aren't quite ready for that topic yet.

attack the problem numerically by successive approximation. In effect, we are constructing a Cauchy sequence that we hope will converge to the solution we want. If the space is complete, we are guaranteed that the limit will be there.

Completeness is also an important theoretical tool. You will soon see this method of "successive approximation" used in the proof of the Heine–Borel theorem in Section 7.3.

Problems 6.2

1. Prove that $\mathbb{R}$ is complete (Theorem 6.2.2). (*Hint*: Recall that bounded sequences of real numbers have some very nice properties.)

2. Prove that $\mathbb{R}^n$ is complete (Corollary 6.2.3). (*Hint*: Use the fact that $\mathbb{R}$ is complete and mess around with the appropriate estimates.)

3. Prove that every closed subspace of a complete space is complete.

4. Recall Problem 8 on page 70, which shows two ways to combine two metric spaces (X, d_X) and (Y, d_Y) into a new metric space $(X \times Y, d)$. Pick one of these strategies and prove that if X and Y are both complete metric spaces, then $X \times Y$ is also complete. How would your proof need to be modified to prove the same result for the other case?

5. Prove that the metric space ℓ_∞ (defined in Problem 9 on page 70–71) is complete.

6. Let X be a metric space, and let D be a dense subset of X. Suppose that Y is a complete metric space and that f is a uniformly continuous function from D to Y. Prove that there exists a uniformly continuous function $F : X \to Y$ such that $F|_D = f$. (*Hint*: If you are stuck, take a peek at Problem 7 at the end of Section 4.4. The problems are attacked in a similar way. Consider also Problem 3a at the end of Section 6.1.)

Chapter 7

Compactness

In this chapter, we will explore a special class of sets called compact sets. Our mathematics often becomes much simpler when we deal with finite sets. Here are some examples:

- Every real-valued function with a finite domain K has a maximum value and a minimum value.

- If K is a finite set, there exist points x and y such that $d(x,y) =$ diameter(K).

- Given any finite subset K of a metric space X and any point $x \in X \setminus K$, there is a point $y \in K$ such that $d(x,y) \leq d(x,k)$ for all $k \in K$. In other words, there is an element $y \in K$ that is "closest" to x.

- Every sequence with a finite range has a convergent subsequence.

Exercise 7.0.1 Show that none of these conditions holds for all sets. That is, for each of the preceding statements, come up with an example for which the statement minus the word "finite" fails. ∎

Compact sets share some of the nice mathematical properties of finite sets (including, under reasonable circumstances, those listed previously). For this and other reasons, compactness is a *very* useful condition. Unfortunately, it can seem fairly abstract at first, so be patient.

7.1 Compact Sets

Definition 7.1.1 Let X be a metric space, and let $S \subseteq X$. If $\{U_\alpha\}_{\alpha \in \Lambda}$ is a collection of open subsets of X and

$$S \subseteq \bigcup_{\alpha \in \Lambda} U_\alpha.$$

then $\{U_\alpha\}_{\alpha \in \Lambda}$ is an **open cover** for S.

Any subcollection of $\{U_\alpha\}_{\alpha \in \Lambda}$ whose union still contains S is called a **subcover** for S.

This idea of open covers is almost certainly new to you, so examine the following examples very carefully. Make sure you come away with a picture in your mind for each of the various covers.

Example 7.1.2 [Open Covers]

1. Let X be a metric space. Let $S \subseteq X$, and let r be a fixed positive real number. Then the collection $\{B_r(x)\}_{x \in S}$ is an open cover of S.

2. Let X be a metric space, let $S \subseteq X$ and let $a \in X$. Then the collection $\{B_n(a)\}_{n=1}^{\infty}$ is an open cover of S. (In fact, it covers all of X.)

3. Suppose that $S = (0, 1)$. Let $r_1, r_2, r_3 \ldots$ be an enumeration of the rational numbers in S. For each $i \in \mathbb{N}$, let

$$U_i = \left(r_i - \frac{1}{4}, r_i + \frac{1}{4}\right).$$

The collection $\{U_i\}_{i=1}^{\infty}$ is an open cover of S, and the subcollection

$$\{\left(\frac{1}{4} - \frac{1}{4}, \frac{1}{4} + \frac{1}{4}\right), \left(\frac{1}{2} - \frac{1}{4}, \frac{1}{2} + \frac{1}{4}\right), \left(\frac{3}{4} - \frac{1}{4}, \frac{3}{4} + \frac{1}{4}\right)\}$$
$$= \{\left(0, \frac{1}{2}\right), \left(\frac{1}{4}, \frac{3}{4}\right), \left(\frac{1}{2}, 1\right)\}$$

still covers all of the interval $(0, 1)$. The latter is a *subcover* of the former.

Definition 7.1.3 Let X be a metric space. A subset S of X is said to be **compact** provided that every open cover for S has a *finite* subcover. In other words, whenever S is a subset of the union of a collection of open subsets of X, then it is a subset of the union of *finitely many* of those sets.

A metric space X is said to be a **compact metric space** if it is compact as a subset of itself.

Exercise 7.1.4 Does Example 7.1.2, part 3, tell us that $(0,1)$ is a compact set? Why or why not? ∎

Exercise 7.1.5 Here are some examples that will help you work through the definition of compactness and gain a little insight.

- *There are compact sets.*
  All finite sets are compact. Explain why.

- *Not all sets are compact.*
  The open interval $I = (0, 1)$ in $\mathbb{R}$ is not compact. To see this, consider the following sequence of open sets: For $n = 1, 2, 3, \ldots$, $U_n = (\frac{1}{n}, 1)$. Show that this sequence covers all of I. Now show that it has no finite subcover.

- *There are compact sets that are not finite.*
  Let X be a metric space. Let (a_n) be a sequence of distinct terms in X converging to a. Show that the set $S = \{a, a_1, a_2, a_3, \ldots\}$ is compact. (*Hint*: Cover S with open sets. What can you say about the open set that contains a?)

∎

Exercise 7.1.5 may lead you to believe that all compact sets are sort of "skimpy"; not all are finite, but those that are infinite seem not to be very "thick," either. (You may be wondering whether they have to be countable sets.) Theorem 7.1.6 shows that this is not so.

Theorem 7.1.6 Let a and b be real numbers with $a < b$. Then the closed interval $[a, b]$ is compact.

Proof: Suppose we have an open covering $\mathcal{C} = \{U_\alpha\}_{\alpha \in \Lambda}$ of $[a, b]$. We begin by defining

$$\mathcal{S} = \{x \in [a, b] : [a, x] \text{ can be covered by finitely many of the sets in } \mathcal{C}\}.$$

It is our goal to show that $b \in \mathcal{S}$. Note two things:

- At least one open set in the covering contains a, so $\mathcal{S}$ is non-empty.

- $\mathcal{S}$, being a subset of $[a, b]$, is a bounded subset of $\mathbb{R}$.

Therefore, the least upper bound property of $\mathbb{R}$ tells us that $\mathcal{S}$ has a least upper bound y. Note that $a < y \le b$.

Next, we will show that $y \in \mathcal{S}$. To this end, suppose that U_α in $\mathcal{C}$ contains y. Note that U_α must also contain the interval $(x, y]$ for some $x \in \mathcal{S}$. (*Do you see why?*) By virtue of x's membership in $\mathcal{S}$, we know that there are sets

$U_1, U_2, \ldots, U_n$ in $\mathcal{C}$ such that $[a, x] \subseteq U_1 \cup U_2 \cup \ldots \cup U_n$. Then $[a, y] \subseteq U_1 \cup U_2 \cup \ldots \cup U_n \cup U_\alpha$, still a finite union. Thus y meets the criterion necessary for being an element of $\mathcal{S}$.

Moreover, the set U_α, being open, must contain some interval $(y - r, y + r)$ centered at y. That is, $[a, y + r/2] \subseteq U_1 \cup U_2 \cup \ldots \cup U_n \cup U_\alpha$. Because all elements in $\mathcal{S}$ must be smaller than or equal to y, the only way this can happen is for $(y, y + r/2]$ to lie *outside* the interval $[a, b]$. We conclude that $y = b$. □

The Idea Behind the Proof of Theorem 7.1.6

You have an open cover of the real interval $[a, b]$. (To make it easier to visualize our heuristic argument, suppose that all of the open sets in the cover are open intervals.) You want to know whether you can throw out all but finitely many of the open sets and still cover everything in the set. The basic idea, elaborated in the actual proof, is to "work from left to right." You want to see how far to the right you can go, covering with finitely many intervals. You know that you can cover the left endpoint a with some set in the cover. Then you choose another interval that covers "the next piece" of the interval, and you keep adding intervals, moving from left to right. (See Figure 7.1.) The problem with this strategy, is that if you do not choose your intervals—wisely—it is entirely possible to pick increasingly shorter intervals so that you can keep adding intervals forever and never reach b. Thus the proof needs to look at *all possible ways* to cover from the left endpoint a to some point x partway from a to b. In the end, the least upper bound property comes to the rescue.

Figure 7.1 You try to cover the interval going from left to right, adding one interval at a time.

Besides providing a proof that a large class of sets satisfies our definition of compactness, Theorem 7.1.6 lets us know that proving a set is compact (or even determining whether or not it is!) is likely to be a nontrivial exercise. My next goal is to establish some properties of compact sets that will help you to better understand what compact sets are like.

Theorem 7.1.7 Every compact subset of a metric space is bounded. In particular, compact metric spaces are bounded.

Proof: The proof is Problem 2 at the end of this section. □

Theorem 7.1.8 Every compact subset of a metric space is closed.

Proof: Let S be a compact subset of a metric space X. We will show that the complement of S is open. Let $x \in S^C$. Then by Problem 4 on page 78 we know that for each $y \in S$, there exist disjoint open sets U_y and V_y containing x and y, respectively. The set $\{V_y\}_{y \in S}$ clearly covers all of S and thus has a finite subcover $\{V_{y_1}, V_{y_2}, \ldots, V_{y_n}\}$. The set $U = U_{y_1} \cap U_{y_2} \cap \ldots \cap U_{y_n}$ is open, contains x, and does not intersect $V_{y_1} \cup V_{y_2} \cup \ldots \cup V_{y_n}$. Therefore, it does not intersect S either. We conclude that S is closed. □

Not every closed subset of a metric space is compact. The interval $[a, \infty)$ in $\mathbb{R}$, for instance, is closed but not bounded and so cannot be compact. Nevertheless, there is a useful theorem that is *sort of* a converse to Theorem 7.1.8.

Theorem 7.1.9 Every closed subset of a compact set is compact.

Proof: The proof is Problem 4 at the end of this section. □

Corollary 7.1.10 A subset C of $\mathbb{R}$ is compact if and only if it is closed and bounded.

Proof: The proof is Problem 5 at the end of this section. □

Remark: Corollary 7.1.10 might prompt us to ask a question: Are the compact subsets of a metric space identical with those that are closed and bounded? (This would certainly make it easier to identify them!) In general, the answer is no; Problem 14 at the end of this section shows that there are closed, bounded sets that are not compact. However, it is true that all closed, bounded subsets of $\mathbb{R}^n$ are compact, a fact that we will prove in Section 7.3.

And now, here is perhaps the most useful characterization of compact sets.

Theorem 7.1.11 Let X be a metric space, and let S be a subset of X. Then the following statements about S are equivalent:

1. S is compact.

2. Every sequence in S has a subsequence that converges to a point of S.

3. Every infinite subset of S has a limit point in S.

Proof

1 $\implies$ 2 Let (a_i) be a sequence in S. If (a_i) has a convergent subsequence, then that subsequence will converge to a point of S by Theorem 7.1.8. Therefore, we will assume that (a_i) has no convergent subsequence. In particular, no point of X can appear infinitely many times as a term of (a_i), so the range of (a_i) must infinite.

Let x be any point in S. Then there is an open subset U_x containing x that contains only finitely many of the a_i's. (See Problem 8 on page 86.) The collection $\{U_x\}_{x \in S}$ clearly covers S, but it has no finite subcover because any finite collection of these open sets can contain at most finitely many of the terms in the range of (a_i). This contradiction tells us that the assumption that (a_i) had no convergent subsequence was an incorrect assumption.

2 $\implies$ 3 This is Problem 7 at the end of this section.

3 $\implies$ 1 This is Problem 10 at the end of this section. It makes use of the results in Problems 8 and 9.

$\square$

Theorem 7.1.12 Every compact metric space is complete.

Proof: The proof is Problem 12 at the end of this section. $\square$

Problems 7.1

1. Let X be a metric space. Suppose that F is a closed subset of X and that K is a compact subset of X. Prove that $K \cap F$ is compact.

2. Prove that compact sets are bounded (Theorem 7.1.7). (*Hint*: There are several ways to prove this theorem, but one way is to think about Example 7.1.2, part 2.)

3. The proofs given in the text for Theorems 7.1.6 and 7.1.8 used Definition 7.1.3 to prove that a certain set is compact. Given all of our previous work with sequences and subsequences, it is actually easier to prove these things by using the fact that a set C is compact if and only if every sequence in C has a subsequence that converges to a point of C (Theorem 7.1.11). Find these alternative proofs.

4. In this problem, you are asked to prove Theorem 7.1.9 in two different ways.

    (a) Prove Theorem 7.1.9 using only the open cover definition for compactness.

    (b) Prove Theorem 7.1.9 using the fact that a set $\mathcal{S}$ is compact if and only if every sequence in $\mathcal{S}$ has a subsequence that converges to an element of $\mathcal{S}$.

5. Prove that closed and boundedness is equivalent to compactness in $\mathbb{R}$ (Corollary 7.1.10).

6. Let K be a compact metric space. Let B be a dense subset of K. Fix an arbitrary positive real number δ. Prove that there exists a finite subset $\{b_1, b_2, \ldots, b_n\}$ of B such that for every $x \in K$, there exists $i \leq n$ such that $d(x, b_i) < \delta$. (See Problem 4 at the end of Section 3.5 for the definition of dense.)

7. Prove that the second condition implies the third condition in Theorem 7.1.11.

8. Let X be a metric space in which every infinite subset has a limit point. Without appealing to Theorem 7.1.11, prove the following statements.

    (a) Let $\delta > 0$. Show that X cannot contain an infinite, δ-separated subset. (See Problem 3 at the end of Section 3.5 for the definition of δ-separated.)

    (b) Use part (a) to prove that X has a countable, dense subset.

9. **Countable Compactness:** A metric space in which every open cover has a countable subcover is sometimes called a **countably compact** space. Countable compactness is not as strong a condition as compactness, but countably compact spaces have some interesting properties. Clearly, any compact space is also countably compact and so shares these properties.

    Let X be a metric space. Prove that the following statements about X are equivalent.

    i. X has a countable base.

    > **Definition:** A collection $\mathcal{V}$ of open subsets of X is called a **base** for the topology on X if given any open set U and any element $x \in U$, there exists an element $V \in \mathcal{V}$ such that $x \in V \subseteq U$.

    ii. X is countably compact.

    iii. X has a countable dense subset.

10. Prove that the third condition implies the first condition in Theorem 7.1.11. (*Hint*: Proceed by contradiction. Take an open cover with no finite subcover. Appealing to Problems 8 and 9, begin by extracting a *countable* subcover. Use it to get a contradiction.)

11. Let K be a compact subset of a metric space X. Let (x_n) and (y_n) be sequences in K. Prove that there exists an increasing sequence (n_i) in $\mathbb{N}$ such that susequences (x_{n_i}) and (y_{n_i}) both converge.

12. Prove that compact metric spaces are complete (Theorem 7.1.12).

13. Use Theorem 7.1.12 to give yet another proof that $\mathbb{R}$ is complete.

14. This problem refers to the metric space ℓ_∞ defined in Problem 9 on pages 70-71. Prove that the closed ball of radius 1 about $\vec{0} = (0,0,0,0,\dots)$ is not compact. To do so, consider the sequence (of sequences) $\{\vec{e_1}, \vec{e_2}, \vec{e_3}, \dots\}$ in ℓ_∞ where $\vec{e_i} = (e_{i1}, e_{i2}, e_{i3}\dots)$ is given by

$$e_{ij} = \begin{cases} 0 & \text{if } i \neq j \\ 1 & \text{if } i = j \end{cases}.$$

Show that $(\vec{e_i})$ is, indeed, in $C_1(\vec{0})$ and that it is 1-separated. Conclude that the closed ball is not compact. (*Hint*: See Problem 8a.)

15. Let X be a compact metric space. Prove that every nested sequence of non-empty closed subsets of X has a non-empty intersection. That is, if $C_1 \supseteq C_2 \supseteq C_3 \supseteq \dots$ are non-empty closed sets, then

$$\bigcap_{i=1}^{\infty} C_i \neq \emptyset.$$

> **Remark:** Note that the nested interval theorem (Theorem 1.4.10) is an important special case of this more general principle.

16. Recall Problem 8 at the end of Chapter 2, which shows two ways to combine two metric spaces (X, d_X) and (Y, d_Y) into a new metric space $(X \times Y, d)$. For both these situations, prove that if X and Y are compact metric spaces, then $X \times Y$ is also compact.

17. On the notion of a compact metric space.

    (a) Prove the following theorem:

    Let Y be a metric space. Suppose that $S \subseteq X \subseteq Y$. Then S is compact as a subset of X if and only if S is compact as a subset of Y.

> This theorem implies something very important. The answer to the question "Is S compact?" does not need to be answered with the question "In what space?" The compactness (or lack of compactness) of S as a subset of X and Y is independent of which superspace is being considered.

(b) The point made in the box accompanying part (a) about compact sets is definitely *not* true of other set properties such as openness and closedness. Take $Y = \mathbb{R}^2$ and $X = \{(x, y) \in \mathbb{R}^2 : x^2 + y^2 \leq 1, x \geq 0, \text{ and } y \geq 0\}$. Give examples (drawing pictures will do) of the following:

- A set that is open in X but not in Y
- A set that is open in both X and Y
- A set that is closed in X but not in Y
- A set that is closed in both X and Y

(c) Why is it meaningless to define a metric space to be open if it is open as a subset of itself and closed if it is closed as a subset of itself?

> The moral here is that it makes sense to talk about compact metric spaces, whereas it does not make sense to talk about open or closed metric spaces.

7.2 Continuity and Compactness

Theorem 7.2.1 The continuous image of a compact set is compact. In other words, suppose that X and Y are non-empty metric spaces and that S is a compact subset of X. Then if $f : X \to Y$ is a continuous function, $f(S)$ is a compact subset of Y.

Proof: The proof is Problem 2 at the end of this section. □

Theorem 7.2.2 [Max-Min Theorem] Let X be a non-empty compact metric space, and let $f : X \to \mathbb{R}$ be a continuous function. Then f achieves both a maximum value and a minimum value. That is, there exist x and $y \in X$ such that for all $z \in X$,

$$f(x) \leq f(z) \leq f(y)$$

In this case we say that f *achieves* its minimum value at x and its maximum value at y. $f(x)$ and $f(y)$ are, respectively, the minimum and the maximum values of f.

Proof: By Theorem 7.2.1, the range of f is compact and therefore both closed and bounded. Because $X \neq \emptyset$, the range of f is non-empty. Thus, by the least upper bound axiom and by Theorem 1.4.5, the range of f has a least upper bound M and a greatest lower bound m.

By Theorem 3.4.10, there are sequences (a_i) and (b_i) in $\mathcal{R}an(f)$ converging to m and M, respectively. Because $\mathcal{R}an(f)$ is closed, this implies that m and M are in $\mathcal{R}an(f)$. That is, there exist points x and y in X such that $f(x) = m$ and $f(y) = M$. Because m is a lower bound for $\mathcal{R}an(f)$ and M is an upper bound for $\mathcal{R}an(f)$,

$$f(x) = m \leq f(z) \leq M = f(y) \text{ for all } z \in X.$$

$\square$

Theorem 7.2.3 Let X be a compact metric space, and let Y be any metric space. If $f : X \to Y$ is a continuous function, then f is uniformly continuous.

Proof: The proof is Problem 5 at the end of this section. $\square$

Problems 7.2

1. Let X be a compact metric space and Y be any metric space. Let $f : X \to Y$ be a continuous function. Without using Theorem 7.2.1, prove that f is bounded. (For the definition of bounded function, see Problem 4 at the end of Section 4.4.)

2. Prove that the continuous image of a compact set is compact (Theorem 7.2.1).

3. Let X be a metric space, let K be any non-empty compact subset of X, and let $x \in X$. Prove that there is a point $y \in K$ such that $d(x, y) \leq d(x, k)$ for all $k \in K$. That is, every non-empty compact set K has a point that lies "closest" to x. (*Hint*: See Problem 2 on page 112.)

> The "closest" point need not be unique. If K is the boundary of a circle in $\mathbb{R}^2$ and x is the center of the circle, every point in K satisfies the given condition.

4. Prove that if S is a compact subset of a metric space X, then there exist points x and y in S that are "farthest apart." That is, show that there exist points x and y in S such that $d(x,y) = \text{diam}(S)$.

5. Prove that a continuous function on a compact domain must be uniformly continuous (Theorem 7.2.3). There are at least two very different ways to do this problem:

    - You can prove the theorem directly by using an ϵ-δ argument. This is fairly tricky, but what you think should work does work; just keep at it.
    - It is a bit easier to prove the theorem by contrapositive. Look at Problem 3b at the end of Section 4.4.

6. Open and closed maps: This next problem puts Problem 5 on page 113 into a more general context.

    Definition: Let X and Y be metric spaces. A function $f : X \to Y$ is said to be a **closed map** if the image under f of every closed subset of X is a closed subset of Y. The function f is said to be an **open map** provided that the image under f of every open subset of X is an open subset of Y.

    (a) Let X be a compact metric space and Y be any metric space. Let $f : X \to Y$ be a continuous function. Prove that f is a closed map.

    (b) Let X be a compact metric space and Y be any metric space. Let $f : X \to Y$ be a bijective, continuous function. Prove that f is an open map.

    (c) The hypothesis that f be one-to-one in part (b) is necessary. To see why, show that $f : [-1, 1] \to [0, 1]$ given by $f(x) = x^2$ is not an open map.

    (d) Let X be a compact metric space and Y be any metric space. Let $f : X \to Y$ be a bijective, continuous function. Prove that f^{-1} is continuous. (A bijective continuous function whose inverse is also continuous is called a **homeomorphism**.[1])

7.3 Compactness in $\mathbb{R}^n$

In Problem 14 of Section 7.1, you showed that not all closed and bounded subsets of a metric space are compact. The counterexamples need not be bizarre

---

1. Homeomorphic spaces are "mathematically identical" in a way that is made precise in topology, a branch of mathematics that is closely related to both analysis and geometry.

sets; in the metric space ℓ_∞, for example, the closed ball of radius 1 centered at $\vec{0}$ is not compact.[2] However, the only subsets of $\mathbb{R}^n$ that are closed and bounded are the compact sets; a fact known as the Heine–Borel theorem. The proof is not easy, but the resulting ease with which we will be able to identify compact sets in $\mathbb{R}^n$ makes the effort well worthwhile.

The basic difficulty comes from the fact that there is no way to prove this theorem without digging down into the **geometric** structure of $\mathbb{R}^n$. The equivalence of closed and boundedness with compactness is not simply a metric property, or they would be equivalent in all metric spaces, including infinite-dimensional spaces such as ℓ_∞.

Exercise 7.3.1 Let $(x_1, x_2, \ldots, x_n) \in \mathbb{R}^n$, and let $\epsilon > 0$. Fix $m \in \mathbb{N}$ such that $m > \dfrac{\sqrt{n}}{\epsilon}$. Problem 8c in Section 1.4 tells us that we can find integers $a_1, a_2, \ldots, a_n$ such that

$$\frac{a_i}{m} < x_i \leq \frac{a_i + 1}{m} \text{ for each } i \leq n.$$

Show that

$$d\left((x_1, x_2, \ldots, x_n), \left(\frac{a_1}{m}, \frac{a_2}{m}, \ldots, \frac{a_n}{m}\right)\right) < \epsilon.$$

■

Exercise 7.3.1 demonstrates an important geometric fact about $\mathbb{R}^n$—namely, that we can cover it by a countably infinite set of evenly spaced open balls, all of the same diameter. To help you picture what is going on, think about the plane $\mathbb{R}^2$ (see Figure 7.2). Take each integer and divide it by the same natural number m. Then take all points in $\mathbb{R}^2$ whose coordinates are numbers of this form. The left drawing in Figure 7.2 shows that we get an evenly spaced "grid." (In the diagram, every denominator is 3.) Now we center a circle on each of the grid points. If the diameter of each of these circles is sufficiently large, the circles will cover the entire plane. Exercise 7.3.1 shows that if we make the grid sufficiently "fine," we can make the diameter of the circles as small as we like. Lemma 7.3.2 tells us that covering a *bounded* set requires only finitely many of these balls (see Figure 7.3).

---

2. This is because ℓ_∞ is infinite-dimensional. Even though a closed ball is "small" in one sense (its diameter is finite), in an infinite-dimensional space it is nevertheless "huge" because there are so many independent directions from each point.

7.3 Compactness in $\mathbb{R}^n$ 147

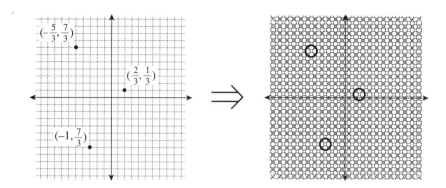

Figure 7.2 The diagram on the left shows a grid with grid-points $\frac{1}{3}$ unit apart. Three illustrative points are highlighted. The diagram on the right shows the plane covered by circles centered at the grid points. The three highlighted circles are centered at the points shown on the left.

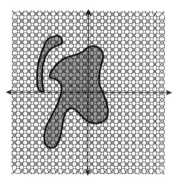

Figure 7.3 Any bounded subset of $\mathbb{R}^n$ can be covered by finitely many of the evenly spaced balls.

Lemma 7.3.2 Let S be a bounded subset of $\mathbb{R}^n$. Then for any $\epsilon > 0$, S is contained in the union of a finite number of balls of radius ϵ.

Proof: Let S be any bounded subset of $\mathbb{R}^n$, and let $\epsilon > 0$. Fix $m \in \mathbb{N}$ such that $m > \dfrac{n}{\epsilon}$.

Because S is bounded, there exists $R \in \mathbb{R}$ such that $S \subseteq \mathrm{B}_R(\vec{0})$. Thus, for an arbitrary element $(x_1, x_2, \ldots, x_n)$ of S, we have

$$|x_i| \leq \sqrt{x_1^2 + x_2^2 + \ldots + x_n^2} < R \text{ for each } i \leq n.$$

Referring to this same arbitrary element of S, we appeal to Problem 8c of Section 1.4 and choose integers $a_1, a_2, \ldots, a_n$ so that

$$\frac{a_i}{m} \leq x_i \leq \frac{a_{i+1}}{m}.$$

Thus, for each $i \leq n$,

$$\left|\frac{a_i}{m}\right| = \left|x_i + \left(\frac{a_i}{m} - x_i\right)\right|$$
$$\leq |x_i| + \left|\left(\frac{a_i}{m} - x_i\right)\right|$$
$$\leq |x_i| + \left|\frac{1}{m}\right| < R + \frac{1}{m}.$$

This, in turn, allows us to deduce that $|a_i| < Rm + 1$, an upper estimate that is independent of the choice of $(x_1, x_2, \ldots, x_n)$! Finally, referring to the calculation you did in Exercise 7.3.1, we observe that every element of S lies within ϵ of a point

$$\left(\frac{a_1}{m}, \frac{a_2}{m}, \ldots, \frac{a_n}{m}\right)$$

in which $a_1, a_2, \ldots, a_n$ are integers whose absolute values are less than $Rm+1$. Clearly, there are only finitely many such points! $\square$

Theorem 7.3.3 [Heine–Borel Theorem] Any closed, bounded subset of $\mathbb{R}^n$ is compact.

Proof: The proof is by contradiction. Suppose that $K \subseteq \mathbb{R}^n$ is closed and bounded but not compact. Then there exist open sets $\{U_\alpha\}_{\alpha \in \Lambda}$ such that

$$K \subset \bigcup_{\alpha \in \Lambda} U_\alpha$$

but K cannot be covered by any finite subcollection of the U_α's.

However, because K is bounded, Lemma 7.3.2 tells us that there are finitely many closed balls $B_1, B_2, \ldots, B_t$ of radius $\frac{1}{2}$ such that $K \subset \bigcup_{i=1}^{t} B_i$. Then consider the sets $K \cap B_1, K \cap B_2, \ldots, K \cap B_t$. These are closed, bounded sets whose union is K. Furthermore, at least one of these sets is not contained in the union of finitely many of the U_α's—call it K_1. Note that $\operatorname{diam}(K_1) \leq 1$. Now apply Lemma 7.3.2 to K_1 with $\epsilon = \frac{1}{4}$ to get K_1 contained in a union of finitely many closed balls of radius $\frac{1}{4}$. Repeat the preceding argument to get a closed subset K_2 of K_1 with diameter less than $\frac{1}{2}$ that is not covered by a finite subcollection of the U_α's.

Repeating the argument with $\epsilon = \frac{1}{6}, \frac{1}{8}, \frac{1}{10}, \ldots$, we obtain a sequence of closed sets $K_1 \supset K_2 \supset K_3 \ldots$, none of which is covered by a finite subcollection of the collection $\{U_\alpha\}_{\alpha \in \Lambda}$ and so that $\operatorname{diam}(K_i) \leq \frac{1}{i}$ for $i \in \mathbb{N}$.

Because all the K_i's are non-empty, we can choose a sequence $(\mathbf{p}_k)$ such that for all i, $\mathbf{p}_i \in K_i$. Claim: $(\mathbf{p}_i)$ is a Cauchy sequence. To see this, we let $\epsilon > 0$. Choose $N \in \mathbb{N}$ such that $\frac{1}{N} < \epsilon$. Suppose m and $n \in \mathbb{N}$ with $m \geq n > N$. Then because $K_m \subset K_n$, both $\mathbf{p}_n$ and $\mathbf{p}_m$ are in K_n, and

$$d(\mathbf{p}_n, \mathbf{p}_m) \leq \mathrm{diam}(K_n) \leq \frac{1}{n} < \frac{1}{N} < \epsilon.$$

Because $\mathbb{R}^n$ is complete, we know there exists $\mathbf{p}_0 \in \mathbb{R}^n$ such that $\mathbf{p}_n \to \mathbf{p}_0$. Because $(\mathbf{p}_i)$ is a sequence in K and K is closed, $\mathbf{p}_0 \in K$. Thus $\mathbf{p}_0 \in U_{\alpha^*}$ for some $\alpha^* \in \Lambda$. Given that U_{α^*} is open, there exists $r > 0$ such that $B_r(\mathbf{p}_0) \subseteq U_{\alpha^*}$. Choose $M \in \mathbb{N}$ large enough so that $\frac{1}{M} < \frac{r}{2}$ and $d(\mathbf{p}_0, \mathbf{p}_M) < \frac{r}{2}$. Then, for any point $\mathbf{p} \in K_M$,

$$d(\mathbf{p}_0, \mathbf{p}) \leq d(\mathbf{p}_0, \mathbf{p}_M) + d(\mathbf{p}_M, \mathbf{p}) < \frac{r}{2} + \frac{1}{M} < 2\left(\frac{r}{2}\right) = r.$$

Thus $\mathbf{p} \in B_r(\mathbf{p}_0)$, which proves that $K_M \subset B_r(\mathbf{p}_0) \subset U_{\alpha^*}$. That is, K_M is covered by a *single* one of the U_α's. This contradicts the fact that K_M could not be covered by any finite number of the U_α's! □

Remark: The argument just given used only two facts about $\mathbb{R}^n$: $\mathbb{R}^n$ is complete, and every bounded subset of $\mathbb{R}^n$ can be covered by finitely many closed balls of radius ϵ (ϵ arbitrary!). Thus the equivalence of the conditions "closed and bounded" and "compact" holds in any metric space that satisfies these two conditions.

Corollary 7.3.4 Every bounded sequence in $\mathbb{R}^n$ has a convergent subsequence.

Proof: The proof is in Problem 4 at the end of this section. □

Problems 7.3

1. Let X be a metric space. Let $f : \mathbb{R}^n \to X$ be a continuous function. Let K be a closed, bounded subset of $\mathbb{R}^n$. Prove that $f(K)$ is bounded in X. Show, by giving an example, that the continuous image of a bounded set need not (in general) be a bounded set.

2. Show that the closed ball of radius 1 centered at $\vec{0}$ in ℓ_∞ cannot be covered by finitely many closed balls of radius $\frac{1}{4}$. (*Hint:* Think about the result of Problem 14 in Section 7.1.)

3. Let K be a bounded, non-empty subset of $\mathbb{R}^n$. Let $f : K \to \mathbb{R}$ be a uniformly continuous function. Prove that f is bounded.

   (*Note that this problem generalizes Problem 5 at the end of Section 4.4. What was the essential property, shared by $\mathbb{R}$ and $\mathbb{R}^n$, that made the proofs work? Can you generalize the theorem further?*)

4. Prove that every bounded sequence in $\mathbb{R}^n$ has a convergent subsequence (Corollary 7.3.4).

5. Give an alternate proof of the Heine–Borel theorem by proving that every sequence in a closed, bounded subset K of $\mathbb{R}^n$ has a subsequence that converges to an element of K.

Chapter 8

Connectedness

8.1 The Intermediate Value Theorem

Definition 8.1.1 A real number z is **between** the real numbers x and y if either $x \leq z \leq y$ or $y \leq z \leq x$.

One of the more useful theorems in the theory of real functions is the intermediate value theorem (IVT). In its simplest and most familiar form this theorem says the following:

Theorem 8.1.2 [Intermediate Value Theorem] Let a and b be real numbers, and let $f : [a, b] \to \mathbb{R}$ be a continuous function. Then if γ is between $f(a)$ and $f(b)$, there exists $c \in [a, b]$ such that $f(c) = \gamma$.

Proof: The proof is Problem 1 at the end of this section. $\square$

Example 8.1.3 The IVT is used all of the time. Here are two examples.

- The IVT is the principle behind the bisection method for finding roots. If $f : \mathbb{R} \to \mathbb{R}$ is a continuous function that is positive at $x = -2$ and negative at $x = 0$, then f must have a root somewhere between -2 and 0. Bisect the segment. The function f has a root in either $[-2, -1]$ or $[-1, 0]$; the IVT tells you which one. Repeat the procedure as often as necessary to approximate the root to any desired degree of accuracy.

- Every algebra student learns to solve real inequalities of the form $f(x) \leq 0$ by first solving the equation $f(x) = 0$ and then "testing" the sign of the function at a single point in between each pair of zeros. The sign of the function at that point gives the sign of the function on that entire subinterval. This works because a continuous function has to pass through the x axis to change signs—the IVT!

Although the IVT on the real line can be proved directly by appealing to properties of the real numbers (see Problem 1 on page 153), we wish to explore the fundamental mathematical principles that underly this important theorem. These are:

- $[a, b]$ is a connected set.

- The continuous image of a connected set is connected. Thus $f([a, b])$ is connected.

- You will show in Problem 4a at the end of Section 8.2 that a subset U of $\mathbb{R}$ is connected if and only if the following holds: If x and y are in U, then every point that lies between x and y is also in U. From this, it follows that if γ is between $f(a)$ and $f(b)$, then it must be an element of the connected set $f([a, b])$. That is, there is some real number c between a and b such that $f(c) = \gamma$.

Notice the frequency with which the word "connected" appears in the preceding list. We may have a good intuitive understanding of this word. (See Figure 8.1.) Section 8.2, however, describes the idea of connectedness in a mathematically rigorous way.

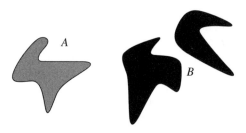

Figure 8.1 A (shown in gray) is connected, while B (shown in black) is disconnected.

Problems 8.1

1. Prove the IVT for a continuous function $f : [a, b] \to \mathbb{R}$ as follows. Suppose that γ is between $f(a)$ and $f(b)$. Let $c = \sup\{x \in [a, b] : f(x) \leq \gamma\}$. Show that $f(c) = \gamma$. (*Don't forget to show that c actually exists and that it is in $[a, b]$.*)

2. Let $K \subseteq \mathbb{R}$. A function $f : K \to \mathbb{R}$ is said to be **strictly monotonic** if it is either strictly increasing or strictly decreasing. That is, one of the following two conditions holds:

$$x, y \in K \text{ with } x < y \text{ implies that } f(x) < f(y)$$

or

$$x, y \in K \text{ with } x < y \text{ implies that } f(x) > f(y)$$

   (a) Let I be an interval[1] in $\mathbb{R}$, and let $f : I \to \mathbb{R}$ be a continuous function. Prove that f is one-to-one if and only if f is strictly monotonic.

   (b) Show that the result from part (a) is false if we remove the result that f be continuous.

   (c) Show that the result from part (a) is false if we remove the requirement that I be an interval. (In other words, the *connectedness* of the domain is a crucial element here.)

3. Let I be an interval in $\mathbb{R}$, and let $f : I \to \mathbb{R}$ be a one-to-one, continuous function. Then prove that $f^{-1} : f(I) \to \mathbb{R}$ is also continuous.

8.2 Connected Sets

Intuitively, a set is connected if it is composed of a single piece. It is disconnected if it is broken up into two or more disjoint pieces. Unfortunately, this intuitive description is woefully inadequate from a mathematical standpoint. To see why, consider the real numbers, which are clearly connected; thus any definition that we give for connectedness should be satisfied by $\mathbb{R}$. Write $\mathbb{R} = (-\infty, 0) \cup [0, \infty)$ ... There! $\mathbb{R}$ is made up of two disjoint pieces. For that matter, we can take any set with at least two elements and "break" it into two or more disjoint pieces. This is not, of course, what we meant to say, but it shows that our "definition" needs some work!

---

1. *I* may be open, closed, half-open, bounded, unbounded—it doesn't matter.

Let's try again: A set is disconnected if it is broken into several pieces that are somehow "separated" from one another—they cannot be "close together" like $(-\infty, 0)$ and $[0, \infty)$. What might we mean by this? Let C be a disconnected set made up of the "separated pieces" A and B. No point of A can be a limit point of B, and no point of B can be a limit point of A.

Exercise 8.2.1 Explain why it is necessary to specify *both* that B contain no limit point of A *and* that A contain no limit point of B. One or the other is insufficient to define connectedness. ∎

Recall: If X is a metric space and $S \subseteq X$, then S inherits a metric from X and becomes a metric space in its own right. Problem 7 in Section 3.1 demonstrates that $U \subseteq S$ is open in the metric space S if and only if there exists $W \subseteq X$ such that W is open in X and $U = W \cap S$. We emphasize the fact that U may not be open in X by saying that U is *relatively* open in S.

It is easy to see, using complementation, that $C \subseteq S$ is relatively closed in S if and only if there exists $K \subseteq X$ such that K is closed in X and $C = K \cap S$.

Theorem 8.2.2 Let X be a metric space and $S \subseteq X$. Suppose that A and B are non-empty, disjoint subsets of S such that $S = A \cup B$. The following conditions are then equivalent:

1. No point of A is a limit point of B, and no point of B is a limit point of A.

2. A is both a closed and an open subset of the subspace S.

3. B is both a closed and an open subset of the subspace S.

Proof: It is clearly only necessary to show that 1 is equivalent to 2. The equivalence of 1 and 3 then follows from the symmetry between the sets A and B.

$1 \implies 2$ Because no element of A is a limit point of B, for each element $a \in A$, there is an open ball about a that does not intersect B. The union U of all such open balls is open in X, contains A, and does not intersect B. Thus $A = S \cap U$ and A is relatively open in S.

This same argument establishes that B is relatively open in S, and (because A is B's complement in S) we can conclude that A is relatively closed in S.

$2 \implies 1$ Suppose that A is both open and closed as a subset of the metric space S. Then there exists an open subset U of X such that $A = S \cap U$. In particular, $U \cap B = \emptyset$. We conclude that no element of A is a limit point of B. Next we note that B is open in S because its relative complement A is closed in S. Therefore, the preceding argument tells us that no point of B can be a limit point of A.

$\square$

Definition 8.2.3 Sets that are both open and closed are called **clopen** sets.

Definition 8.2.4 A metric space X is said to be **disconnected** if it can be written as the disjoint union of two non-empty clopen sets. X is **connected** if it is not disconnected.

A subset S of a metric space X is said to be connected if the subspace S is connected, S is a disconnected subset of X if the subspace S is disconnected.

Corollary 8.2.5 Let X be a metric space. The following statements are then equivalent:

1. X is connected.

2. The only subsets of X that are both open and closed are $\emptyset$ and all of X.

3. There do not exist non-empty, disjoint, open sets A and B such that $X = A \cup B$.

4. There do not exist non-empty, disjoint, closed sets A and B such that $X = A \cup B$.

Proof: The proof is Problem 1 at the end of this section. $\square$

Theorem 8.2.6 Let X and Y be metric spaces, and let $f : X \to Y$ be a continuous function. If X is connected, so is $f(X)$.

Proof: The proof is by contraposition. Suppose that $f(X)$ is the disjoint union of two (relatively) open sets A and B. Then there exist open sets U and V of Y such that $A = U \cap f(X)$ and $B = V \cap f(X)$. Then, $f^{-1}(A) = f^{-1}(U)$ and $f^{-1}(B) = f^{-1}(V)$ (this is Problem 12 on page 26). Because f is continuous, $f^{-1}(U)$ and $f^{-1}(V)$ are both open in X. Furthermore, $f(X) = A \cup B$, so

$$X = f^{-1}(f(X)) = f^{-1}(A \cup B) = f^{-1}(A) \cup f^{-1}(B) = f^{-1}(U) \cup f^{-1}(V)$$

Now A and B are non-empty and disjoint; thus $f^{-1}(U)$ and $f^{-1}(V)$ are non-empty and disjoint as well. Therefore X is the union of disjoint, non-empty open sets. It is not connected.

□

Problems 8.2

1. Prove Corollary 8.2.5, which sets out several equivalent conditions for describing a connected metric space.

2. Unions and intersections of connected sets.

    (a) Let A and B be connected subsets of a metric space X. Show that if $A \cap B \neq \emptyset$, then $A \cup B$ is connected.

    (b) Show that it is not necessarily true that if two connected sets have a non-empty intersection, then their intersection will be connected. (A picture will do.)

3. Problem 4 at the end of Section 5.2 defines the intermediate value property for a function on a real interval. We might generalize this property as follows: Let X be a metric space and let $f : X \to \mathbb{R}$ be a real-valued function. We say that f has the *generalized intermediate* value property provided that for all $a, b \in X$ and $\alpha \in \mathbb{R}$ with α between $f(a)$ and $f(b)$, there exists $c \in X$ such that $f(c) = \alpha$.

    (a) Generalized intermediate value theorem. Suppose that X is a connected metric space and $f : X \to \mathbb{R}$ is a continuous function. Prove that f has the generalized intermediate value property.

    As it happens, the converse is also true. This gives us an alternative way of defining connectedness.

    (b) Prove that a set S in a metric space X is connected if and only if every continuous function $f : S \to \mathbb{R}$ has the generalized intermediate value property.

4. Connectedness in $\mathbb{R}$:

    (a) Let S be a subset of $\mathbb{R}$. Prove that the following statements are equivalent:

        i. S is connected.

ii. If x and y are in S, then every point that lies between x and y is also in S.
iii. S is an interval.

(b) Prove that $\mathbb{R}$ is a connected set.

(c) Deduce that if X is a connected metric space and $f : X \to \mathbb{R}$ is a continuous function, then $f(X)$ is an interval.

(d) Show by giving examples that it is possible to have a subset Y of $\mathbb{R}$ and function $f : Y \to \mathbb{R}$ for which $f(Y)$ is an interval and yet satisfies any one of the following conditions:

- f is continuous and Y is disconnected.
- f is discontinuous and Y is connected.
- f is discontinuous and Y is disconnected.

5. Let A be a connected subset of a metric space X.

(a) Show that $\overline{A}$ is connected.

(b) Show that if D is a subset of X such that $A \subseteq D \subseteq \overline{A}$, then D is also connected.

> **Remark:** The fact that the closure of a connected set is connected shows that some things are connected that we might think wouldn't be. Let A be the graph of the function $f(x) = \sin(\frac{1}{x})$ for $x > 0$ (shown in Figure 8.2). It might take you a bit of work to prove it, but I presume you believe that this is a connected subset of $\mathbb{R}^2$. Then its closure is $\overline{A} = A \cup \{(0, x) \in \mathbb{R}^2 : -1 \leq x \leq 1\}$. You have just proved that this is a connected set!
>
> If this fact strikes you as just *wrong*, what you are really saying is that you think the definition of connectedness is inadequate. What makes you nervous about this example? Can you think of a way to strengthen the definition of connectedness so that this disturbing example is excluded? If the result doesn't trouble you, how would you convince the doubters that the definition *is* adequate? What are the forceful arguments on both sides?

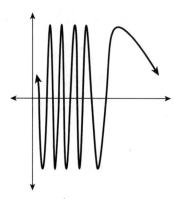

Figure 8.2 The graph of $f(x) = \sin\left(\dfrac{1}{x}\right)$ for $x > 0$.

Chapter 9
Differentiation of Functions of One Real Variable

In the Preliminary Remarks section *"What Is Analysis?"* I said that analysis had its origins in an attempt to give a careful mathematical account of the theory underlying the calculus. Although it has outgrown these early trappings, this is still one of its jobs.

9.1 Regarding Domains

Most introductory calculus courses discuss the basic theory of the derivative fairly well, if informally. Although few calculus students can define a limit precisely, most can tell you that the derivative f' of a function f at the point x can be defined by

$$f'(x) = \lim_{y \to x} \frac{f(y) - f(x)}{y - x}.$$

They can also draw diagrams that illustrate what is going on in this limiting process. Our work to this point allows us to fill in many of the mathematical gaps immediately, which we will do in Section 9.2. But it is useful to discuss some nuances first.

The focal point in the study of differentiation is, of course, the limit of a difference quotient. Peripheral choices, however, also affect the way that the theory "plays itself out." Interestingly, the choice of domain is quite significant. Suppose we consider a class of functions $f : K \to \mathbb{R}$, where $K \subseteq \mathbb{R}$. The assumptions we make about the domain K have consequences for the generality of the results we ultimately obtain and, in particular, for the class of functions to which those results will eventually speak. The assumptions we make about K are, therefore, crucial.

Some considerations are easy. For example, it is impossible to talk about differentiating a function at a point of its domain that is not also a limit point of the domain. (We cannot define the appropriate limit, otherwise.) So K should certainly *have* limit points—probably lots of them. Elaborating on this trivial observation becomes strangely complicated, and there doesn't seem to be a lot of agreement about how to proceed. After thinking about it for some time, I decided that the best way to proceed was to ask myself what sorts of functions I want to be able to differentiate.

At the very least, I want to be able to differentiate the standard elementary functions: polynomials, rational functions, sines and cosines, exponentials, and logarithms, and their respective inverse functions. I also want to consider sums, products, quotients, and compositions of those functions. What sorts of domains do I get? Here are some examples:

- The set $\mathbb{R}$ is the domain for polynomials, sines and cosines, and exponentials.

- Rational functions have domains of the form $\mathbb{R} \setminus S$, where S is a finite set of real numbers.

- The domains of the logarithm and some close relatives require that we consider open intervals of the form (a, ∞) and $(-\infty, a)$.

- If we compose the square root function with a polynomial, we get a function whose domain is a finite union of closed intervals (they may be bounded, unbounded, or some combination of the two). The "standard" inverses of the sine and cosine are also closed, bounded intervals.

- Quotients of sines and cosines, from which we get functions such as the tangent and the secant, have domains of the form $\mathbb{R} \setminus S$, where S is an infinite discrete subset of $\mathbb{R}$.

These domains share an important characteristic. Each is a union of countably, many nondegenerate,[1] disjoint intervals. Let K be such a set. Notice that every point in K is a limit point of K. Therefore, if K is the domain of a real-valued function f, we can sensibly ask whether f is differentiable at each point of K. So this seems like a pretty good place to start! Some theorems will require that we restrict our domain further (e.g., to a connected domain or a compact domain), in which case we will add hypotheses that make this restriction clear.

---

1. An interval is said to be *degenerate* if it is empty or contains only one point. For instance, the intervals $[a, b]$ and (a, b) are degenerate when $b \leq a$. We can sensibly define these sets, but they are not really intervals in the sense that we usually intend the word. From now on, we will deal with only nondegenerate intervals.

To simplify the language later on, we make the following definition:

Definition 9.1.1 A subset K of $\mathbb{R}$ is called a **D-domain** (short for "differentiation domain") if K either is all of $\mathbb{R}$ or is a union of countably many non-degenerate disjoint intervals. (These intervals may be open or closed, bounded or unbounded, or some combination thereof.)

Be aware that this term is nonstandard and is made up for our purposes here.

9.2 The Derivative

Let K be a D-domain. Let $f : K \to \mathbb{R}$ be a function. Let $x \in K$. We want to know whether f is differentiable at x. Although we do not usually name the function explicitly, the limit in question is the limit of the function

$$DQ_x : K \setminus \{x\} \longrightarrow \mathbb{R}$$

given by

$$DQ_x(y) = \frac{f(y) - f(x)}{y - x}.$$

The function DQ_x is called the **difference quotient of f at the point x**. To get the derivative of f at a limit point x of K, we take the limit of the difference quotient as y approaches x.

Definition 9.2.1 Let K be a D-domain, and let $f : K \to \mathbb{R}$ be a function. If $x \in K$, we say that **f is differentiable at the point x** provided that

$$\lim_{y \to x} \frac{f(y) - f(x)}{y - x}$$

exists.

If f is differentiable at every point in a set S, we say that **f is differentiable on S**. In this case, we can define the function $f' : S \to \mathbb{R}$ by

$$f'(x) = \lim_{y \to x} \frac{f(y) - f(x)}{y - x}.$$

If S is the set of all points at which f is differentiable, this function is called the **derivative of f**.

Finally, we say that **f is differentiable** provided that f is differentiable at every point of its domain.

> Let K be a D-domain. Let $f : K \to \mathbb{R}$ be a function, and let $x \in K$. Our previous knowledge of limits immediately gives us three different ways of saying what it means for f to be differentiable at x. The following statements are equivalent:
>
> - The function f is differentiable at x with derivative $f'(x)$.
>
> - For every $\epsilon > 0$, there exists $\delta > 0$ such that if $y \in K$ and $0 < |y - x| < \delta$, then
> $$\left| f'(x) - \frac{f(y) - f(x)}{y - x} \right| < \epsilon.$$
>
> - For every sequence (y_n) in $K \setminus \{x\}$ that converges to x, the sequence
> $$\left(\frac{f(y_n) - f(x)}{y_n - x} \right) \text{ converges to } f'(x).$$
>
> Given that these statements are equivalent, we will (as always) use them interchangeably without comment.

Our task in this chapter is not to retread all of differential calculus, going through crossing the t's and dotting the i's. Acknowledging that the basic principles of differential calculus are already understood, this chapter discusses the theoretical threads that tie those ideas together. In that vein, it is now time to widen our theoretical perspective. You have probably heard that one way to think about differentiability for a function f at a point a is to imagine "zooming in" on the graph, always keeping the point $(a, f(a))$ at the center of our view. We say that the function is differentiable if and only if the graph looks more and more like a straight (but not vertical) line as we zoom in. The line that we see in this circumstance is called the *tangent line* to the curve at the point.

Calculus teachers make much of this graphical interpretation of differentiability, but rarely give any mathematical explanation for it. This perspective is completely appropriate for a first-semester calculus course, because the theoretical formulation is not very enlightening on first glance. Nevertheless, the idea can be made quite rigorous and simplifies some pieces of the theory. More importantly, this alternative view allows us to generalize the notion of differentiation to functions of two or more variables. In this vein, we might think of it as a more "mature" look at the idea of differentiability.

Here, finally, is a mathematically rigorous way of saying that if we look through a microscope at a very tiny piece of the graph of a differentiable function, we will see a straight line.

9.2 The Derivative

Theorem 9.2.2 Let K be a D-domain, let $f : K \to \mathbb{R}$ be a function, and let $x \in K$. The function f is differentiable at x if and only if there exist a real number $f'(x)$ and a function $r : K \to \mathbb{R}$ satisfying

$$\lim_{y \to x} \frac{r(y)}{y - x} = 0$$

such that for all $y \in K$

$$f(y) = f'(x)(y - x) + f(x) + r(y).$$

Proof: The proof is Problem 3 at the end of this section. □

Exercise 9.2.3 Explain the connection between the intuitive notion of "zooming in" on the function and the theoretical formulation given in Theorem 9.2.2. (*Hint*: It might be useful to think of the letter r as standing for "remainder." The letter e, for "error," is also frequently used to denote this function.) ∎

Lemma 9.2.4 Let K be a D-domain. Suppose that $r : K \to \mathbb{R}$ satisfies

$$\frac{r(y)}{y - x} \to 0 \text{ as } y \to x.$$

Then $r(y) \to 0$ as $y \to x$.

Proof: The proof is Problem 4 at the end of this section. □

When you were considering the relationship between "zooming in" and the characterization given in Theorem 9.2.2, you may have wondered why the theorem requires that $\frac{r(y)}{y-x} \to 0$ rather than just $r(y) \to 0$ as $y \to x$. Lemma 9.2.4 tells us that the requirement in the theorem is the stronger of these two conditions.

If we just required the weaker $r(y) \to 0$ as $y \to 0$, what would this imply about the function f at the point x? (*Hint*: take the limit of both sides of the expression as $y \to x$. What happens?)

Corollary 9.2.5 Let K be a D-domain, and let $f : K \to \mathbb{R}$ be differentiable at $x \in K$. Then f is continuous at x.

Proof: The proof is Problem 5 at the end of this section. □

We can also use the characterization given in Theorem 9.2.2 to prove the standard algebraic rules for differentiation.

Theorem 9.2.6 Let K be a D-domain, and let $f : K \to \mathbb{R}$ and $g : K \to \mathbb{R}$ be functions. Suppose that f and g are differentiable at $x \in K$. Then the following statements hold:

- **The Constant Multiple Rule:** If $k \in \mathbb{R}$, the function kf is differentiable at x and $(kf)'(x) = kf'(x)$.

- **The Sum and Difference Rules:** The functions $f + g$ and $f - g$ are differentiable at x. Furthermore,
$$(f+g)'(x) = f'(x) + g'(x), \text{ and } (f-g)'(x) = f'(x) - g'(x).$$

- **The Product Rule:** The function fg is differentiable at x and
$$(fg)'(x) = f'(x)g(x) + f(x)g'(x).$$

Proof: Here we will prove only the sum rule. The rest of the cases are left to you in Problem 8 at the end of this section.

Before we begin, we appeal to Theorem 9.2.2 to obtain expressions
$$f(y) = f'(x)(y-x) + f(x) + r(y), \text{ and}$$
$$g(y) = g'(x)(y-x) + g(x) + e(y),$$
such that $\dfrac{r(y)}{y-x}$ and $\dfrac{e(y)}{y-x}$ both approach 0 as y approaches x.

Now we consider the value of $f + g$ at an arbitrary $y \in K$.
$$(f+g)(y) = f(y) + g(y)$$
$$= [f'(x)(y-x) + f(x) + r(y)] + [g'(x)(y-x) + g(x) + e(y)]$$
$$= [f'(x) + g'(x)](y-x) + f(x) + g(x) + r(y) + e(y)$$
$$= [f'(x) + g'(x)](y-x) + (f+g)(x) + r(y) + e(y)$$

This gives us an expression of the form required by Theorem 9.2.2. The remainder in the expression is $R(y) = r(y) + e(y)$. We want to consider the limit as y approaches x of
$$\frac{R(y)}{y-x} = \frac{r(y) + e(y)}{y-x} = \frac{r(y)}{y-x} + \frac{e(y)}{y-x}$$

Because each of the two terms goes to zero as $y \to x$, so does their sum. $\square$

Theorem 9.2.7 [Chain Rule] Let K and K' be D-domains. Suppose that $f : K \to \mathbb{R}$ and $g : K' \to \mathbb{R}$ are functions with $f(K) \subseteq K'$. Suppose further that f is differentiable at $x \in K$ and g is differentiable at $f(x)$. Prove that $g \circ f$ is differentiable at x and that
$$(g \circ f)'(x) = g'(f(x))f'(x).$$

Proof Outline:

A thorough proof outline is given here. Problem 11 at the end of this section asks you to fill in the details and write a complete, polished proof.

i. First we use Theorem 9.2.2 to get expressions
$$f(y) = f'(x)(y-x) + f(x) + r(y), \text{ and}$$
$$g(y) = g'(f(x))(y - f(x)) + g(f(x)) + e(y),$$
such that $\dfrac{r(y)}{y-x} \to 0$ as $y \to x$, and $\dfrac{e(y)}{y-f(x)} \to 0$ as $y \to f(x)$.

ii. Use these expressions to show that
$$g \circ f(y) = g'(f(x))f'(x)(y-x) + g \circ f(x) + g'(f(x))r(y) + e(f(y)).$$
Note that the remainder required by Theorem 9.2.2 is
$$R(y) = g'(f(x))r(y) + e(f(y)).$$

iii. The first term of
$$\frac{R(y)}{y-x} = g'(f(x))\frac{r(y)}{y-x} + \frac{e(f(y))}{y-x}$$
goes to zero, so this leaves us with the second term.

iv. Fix a sequence (y_n) in K converging to x. We need to show that the sequence
$$\left(\frac{e(f(y_n))}{y_n - x}\right) \to 0 \text{ as } n \to \infty.$$

v. Observe that $e(f(x)) = 0$. Thus we need only worry about the terms of the sequence for which $f(y_n) \neq f(x)$. We simplify our notation by extracting these terms and renumbering, so that for all $n \in \mathbb{N}$, $f(y_n) \neq f(x)$. (*What if all but finitely many terms satisfy $f(y_n) = f(x)$?*)

vi. Multiply and divide by $f(y_n) - f(x)$ to get
$$\frac{e(f(y_n))}{y_n - x} = \frac{e(f(y_n))}{f(y_n) - f(x)} \frac{f(y_n) - f(x)}{y_n - x}.$$

vii. Because the second factor is a sequence of difference quotients for f at x, it approaches $f'(x)$ as $n \to \infty$.

viii. For the other factor, set $z_n = f(y_n)$; then
$$\frac{e(f(y_n))}{f(y_n) - f(x)} = \frac{e(z_n)}{z_n - f(x)}.$$

Given that f is continuous at x, $z_n = f(y_n) \to f(x)$ as $n \to \infty$. It follows that
$$\lim_{n \to \infty} \frac{e(f(y_n))}{y_n - f(x)} = 0, \text{ by definition of } e.$$

This finishes the proof. $\square$

Problems 9.2

1. In calculus courses, one frequently sees the difference quotient written in a slightly different form. This can sometimes be convenient. Let K be a D-domain, let $f : K \to \mathbb{R}$ be a function, and let $x \in K$.

    (a) Show that f is differentiable at x with derivative $f'(x)$ if and only if $\lim_{h \to 0} \dfrac{f(x+h) - f(x)}{h}$ exists and is equal to $f'(x)$.

    (b) Find characterizations for this difference quotient parallel to the second and third characterizations given in the box on page 162.

2. Differentiation at an endpoint of the domain.

    (a) Let $K \subseteq \mathbb{R}$ be a D-domain, and let $a \in K$. Suppose that there exists $\delta > 0$ such that $(a, a+\delta) \cap K = \emptyset$. (That is, a is the right endpoint of one of the intervals that make up K.) Using our definition, it is perfectly sensible to say that a function $f : K \to \mathbb{R}$ is differentiable at a. Using the language of the definition and a picture to help you, give a graphical interpretation of this statement.

    (b) Use the insight you gained from part (a) to define the (sometimes handy) terms **right-handed derivative** and **left-handed derivative**.

    (c) If a is an interior point of a D-domain K, and $f : K \to \mathbb{R}$ is a function, what is the relationship between the left-handed derivative, the right-handed derivative, and the derivative at the point a?

> **Remark:** Some authors treat differentiability at an endpoint of the domain as a separate concept, but our definition allows us to treat differentiation at an endpoint exactly the same way we treat the derivative at an interior point. There are some theorems in which we will need to specify that the point of interest is an interior point of the domain. In those cases, we will simply add this specification as a hypothesis. In the absence of such a hypothesis, you know that the theorem applies equally well to endpoints of the domain with the appropriate one-sided derivative.

3. Prove that the "zoom in and find a straight line" formulation for differentiation is equivalent to the usual difference quotient definition (Theorem 9.2.2). (*Hint*: Don't work too hard!)

4. Referring to the remainder in the characterization of differentiation, show that
$$\frac{r(y)}{y-x} \to 0 \text{ as } y \to x \text{ implies that } r(y) \to 0 \text{ as } y \to x.$$
(Lemma 9.2.4.)

5. Prove that differentiable functions are continuous (Corollary 9.2.5).

6. Consider the characterization of the derivative given in Theorem 9.2.2. Given this formulation, it may not be obvious that a function $f : K \to \mathbb{R}$ can have only one derivative at the point x. Assume that there exist real numbers s and t and functions $r : K \to \mathbb{R}$ and $e : K \to \mathbb{R}$ such that
$$\frac{r(y)}{y-x} \to 0 \text{ and } \frac{e(y)}{y-x} \to 0 \text{ as } y \to x$$
and such that for all $y \in K$,
$$f(y) = s(y-x) + f(x) + r(y) \text{ and } f(y) = t(y-x) + f(x) + e(y).$$
Prove that $s = t$.

7. As you know, some continuous functions are not differentiable.

   (a) Show that $f(x) = |x|$ is continuous everywhere and differentiable everywhere except at $x = 0$.

   (b) The function $f(x) = |x|$ fails to be differentiable because its graph has a "corner" at $x = 0$. Show that $f(x) = x^{\frac{1}{3}}$ also fails to be differentiable at the origin. (*Hint*: What feature of the graph causes this difficulty? Use this knowledge to construct your proof.)

   (c) Construct a function that is continuous everywhere and fails to be differentiable at every integer.

   > How much farther can we carry this idea? In Excursion K, "Everywhere Continuous, Nowhere Differentiable," you will see that some continuous functions fail to be differentiable at *every* real number.

8. Complete the proof of Theorem 9.2.6. Feel free to appeal to the notation given in the proof of the sum rule, if you wish.

   (a) Use Theorem 9.2.2 to prove that the constant multiple rule holds.

(b) Use part (a) and the sum rule to prove that the difference rule holds.

(c) Use Theorem 9.2.2 to prove that the product rule holds.

9. Let $f : (-\infty, 0) \cup (0, \infty) \to \mathbb{R}$ be the function $f(x) = \frac{1}{x}$. Use Theorem 9.2.2 to prove that f is differentiable.

10. The quotient rule: Let f and g be as in Theorem 9.2.6. Use the result given in Problem 9, the product rule, and the chain rule to prove that if $g(x)$ is never zero, the function $\frac{f}{g}$ is differentiable at x and that

$$\left(\frac{f}{g}\right)'(x) = \frac{f'(x)g(x) - f(x)g'(x)}{(g(x))^2}$$

11. Fill in the details of the proof outline given for the chain rule (Theorem 9.2.7) and write a complete, polished proof of the result.

12. Every differentiable function is continuous, but the derivative need not be continuous. It is hard to picture such a function, but there are examples. Here's one:

$$f(x) = \begin{cases} x^2 \sin\left(\frac{1}{x}\right) & \text{if } x \neq 0 \\ 0 & \text{if } x = 0 \end{cases}$$

(a) Show that f is continuous at $x = 0$.

(b) Show that f is differentiable at $x = 0$.

(c) Show that f' is not continuous at $x = 0$.

In your proof, feel free to assume standard algebraic and calculus facts, including facts about the sine function. You may find Theorem 5.1.5 (the sandwich theorem) useful.

13. Here's a typical calculus problem: Define a function f by

$$f(x) = \begin{cases} 3kx^2 - 5kx + 2k & \text{if } x < 1 \\ 2x - 2 & \text{if } x \geq 1 \end{cases}$$

Notice that f is continuous at $x = 1$ for all choices of k. Find the value of k that makes f differentiable at $x = 1$. (See Figure 9.1.)

Many calculus students will try the following argument: Clearly, f is differentiable at all values of x not equal to 1, and

$$f'(x) = \begin{cases} 6kx - 5k & \text{if } x < 1 \\ 2 & \text{if } x > 1 \end{cases}$$

Thus we need to find k so that $6kx - 5k = 2$ when $x = 1$. This gives us a value of $k = 2$.

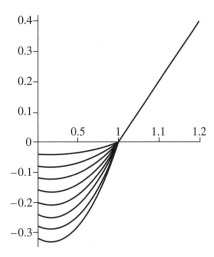

Figure 9.1 Graphs of the function f for various values of k.

(a) The correct answer *is* $k = 2$, but what do you think of the argument? What assumptions are being made? Is an unstated general principle at work here?

> This question does not have a "correct" answer; it certainly doesn't have an *easy* correct answer! It is meant to be food for thought. Write a paragraph or two that talks about the assumptions being made by the student and indicates where his or her reasoning is suspect or breaks down altogether. Part (b) of the problem gives you a hint about one possible way to formulate the problem more precisely, but even *it* is phrased sloppily and asks you to be careful about stating hypotheses that make the general principle hold.

(b) Consider a more general situation in which f is a piecewise-defined function based on two differentiable functions f_1 and f_2:

$$f(x) = \begin{cases} f_1(x) & \text{if } x < a \\ f_2(x) & \text{if } x \geq a \end{cases}$$

Suppose that f is continuous at a. Is it, in general, true that f is differentiable at $x = a$ if $f_1'(a) = f_2'(a)$? State a theorem precisely, and prove it.

9.3 What Does the Derivative Tell Us about the Function?

Let a and b be real numbers. Suppose that $f : (a, b) \to \mathbb{R}$ is a differentiable function. It is easy to see that if f is constant on (a, b), then the derivative of f is zero at every point of (a, b). (Set up the difference quotients; they are all zero. Presto!)

Now consider the converse:

If for all $c \in (a, b)$, $f'(c) = 0$, then f is constant on (a, b).

In other words, if the slope of the graph of f is zero everywhere, then the graph of f is flat. This is certainly a plausible, if not downright obvious, statement. But how would we prove it?

The Naive (but Doomed) Approach: Suppose that $x \in (a, b)$ is fixed but arbitrary and that the limit of the difference quotient for f at x approaches 0. Informally, for y "close" to x,

$$\frac{f(y) - f(x)}{y - x} \approx 0$$

Given that the denominator is close to zero, the fraction can be close to zero only if the numerator is as well. That is, $f(y) \approx f(x)$. Of course, approximate equality is not equality. Nevertheless, this chain of reasoning raises the hope that we might be able to obtain something like

$$\text{for all } \epsilon > 0, \; |f(y) - f(x)| < \epsilon$$

from which we could conclude that $f(y) = f(x)$. Unfortunately, this idea doesn't work, for two important reasons:

- Our choice of ϵ governs how close y must be to x. We can't fix y ahead of time and then let ϵ vary. Shrinking ϵ will confine us to values of y that come ever closer to x.

- Even if we could somehow overcome this hurdle, our most optimistic scenario would still get us $f(y) = f(x)$ only for y "close" to x. This is a local, not a global statement about the function values of f.

A statement about the limit of difference quotients for f gives us, at best, approximate numerical information about the values of f near x. But the theorem requires exact numerical information that holds over the whole domain.

When we have information about f, setting up the difference quotient and then taking the limit gives us information about f'. The difficulty arises when we have information about f' and want information about f. We can't "untake" the limit and easily obtain a relation between values of f at widely separated points. We need a new theoretical tool to do this. This tool, called the mean value theorem, is arguably the second most important theorem in the theory of the calculus, after the fundamental theorem of calculus (Theorem 11.6.2).

Theorem [Mean Value Theorem] Let $K \subseteq \mathbb{R}$ be a D-domain, and let $f : K \to \mathbb{R}$ a function. Assume that $a, b \in \mathbb{R}$ with $[a, b] \subseteq K$. Suppose that f is continuous on the closed interval $[a, b]$ and differentiable on the open interval (a, b). Then there exists $c \in (a, b)$ such that

$$f'(c) = \frac{f(b) - f(a)}{b - a}.$$

Remark: The word "mean" in mean value theorem is used in the sense of "average."

$$\frac{f(b) - f(a)}{b - a}$$

is the average (or mean) rate of change of the function f on the interval $[a, b]$. The mean value theorem states that there must be at least one point in the interior of the interval whose instantaneous rate of change equals the average rate of change of f over the entire interval. The theorem is illustrated in Figure 9.2.

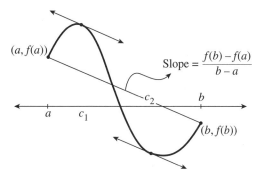

Figure 9.2 The mean value theorem.

The proof of the mean value theorem is the subject of Section 9.4.

9.4 Proving the Mean Value Theorem

The proof of the mean value theorem (Theorem 9.4.2) uses the local extreme theorem. This familiar theorem is closely associated with the theory of optimization, but we prove it here so that we have it available for proving the mean value theorem. First, we need some definitions.

Definition 9.4.1 Let $f : X \to \mathbb{R}$ be a function. Let $z \in X$. Then $f(z)$ is said to be

- a **global maximum** (or simply a **maximum**) for f on X if $f(z) \geq f(x)$ for all $x \in X$.

- a **local maximum** for f on X if there exists $r > 0$ such that, for all $x \in B_r(z)$, $f(z) \geq f(x)$.

Parallel definitions hold for **global minimum** (or **minimum**), and **local minimum**.

If f attains its maximum or minimum value at z, then $f(z)$ is a **global extreme value** (or **extremum**) for f. The real number z is called a **global extreme point** for f.

If $f(z)$ is either a local maximum or a local minimum value for f, then $f(z)$ is a **local extreme value** (or **local extremum**) for f. The real number z is called a **local extreme point** for f.

The plurals of maximum, minimum, and extremum are maxima, minima, and extrema, respectively. We use the adjective *global* for emphasis when discussing both global and local extrema.

Theorem 9.4.2 [Local Extreme Theorem] Let $K \subseteq \mathbb{R}$ be a D-domain, and let $f : K \to \mathbb{R}$ be a function. Suppose that f has a local extreme value at c. If f is differentiable at c and c is in the interior of K, then $f'(c) = 0$.

Proof: The proof is Problem 1 at the end of this section. □

Although we could now prove the mean value theorem directly, it is traditional first to state and prove a special case, from which the general case follows. The special case, called Rolle's theorem, is important and useful enough to be called a theorem in its own right.

Theorem 9.4.3 [Rolle's Theorem] Let $K \subseteq \mathbb{R}$ be a D-domain and $f : K \to \mathbb{R}$ be a function. Assume that $a, b \in \mathbb{R}$ with $[a, b] \subseteq K$. Suppose that the following conditions hold:

- f is continuous on the closed interval $[a, b]$.

- f is differentiable on the open interval (a, b).

- $f(a) = f(b)$.

Then there exists $c \in (a, b)$ such that $f'(c) = 0$.

Proof: The proof is Problem 2 at the end of this section. □

9.4 Proving the Mean Value Theorem

Exercise 9.4.4 Draw a picture that illustrates Rolle's theorem. Use the diagram to explain why Rolle's theorem is a special case of the mean value theorem. ∎

Finally, with the insight from Exercise 9.4.5, we will be ready to prove the mean value theorem.

Exercise 9.4.5 Let $f : \mathbb{R} \to \mathbb{R}$ be a differentiable function. Fix any two points a and b in $\mathbb{R}$. Let g be the straight line joining $(a, f(a))$ and $(b, f(b))$. Now define a new function $h : \mathbb{R} \to \mathbb{R}$ by

$$h(x) = f(x) - g(x).$$

What is the relationship between the graphs of f and h? (Try experimenting with a graphing program.) ∎

Theorem 9.4.6 [Mean Value Theorem] Let $K \subseteq \mathbb{R}$ be a D-domain, and let $f : K \to \mathbb{R}$ be a function. Assume that $a, b \in \mathbb{R}$ with $[a, b] \subseteq K$. Suppose that f is continuous on the closed interval $[a, b]$ and differentiable on the open interval (a, b). Then there exists $c \in (a, b)$ such that

$$f'(c) = \frac{f(b) - f(a)}{b - a}.$$

Proof: The proof is Problem 3 at the end of this section. □

Corollary 9.4.7 Let K be a D-domain, and let $f : K \to \mathbb{R}$ be a function. Let $I \subseteq K$ be an interval. Then f is constant on I if and only if $f'(x) = 0$ for all $x \in I$.

Proof: The proof is Problem 4 at the end of this section. □

Corollary 9.4.7 has its own very important corollary.

Corollary 9.4.8 Let K be a D-domain and let $f : K \to \mathbb{R}$ and $g : K \to \mathbb{R}$ be functions. Let $I \subseteq K$ be an interval. Then there exists $C \in \mathbb{R}$ such that $f(x) = g(x) + C$ for all $x \in I$ if and only if $f'(x) = g'(x)$ for all $x \in I$.

Proof: The proof is Problem 5 at the end of this section. □

Recall the concept of an antiderivative.

Definition 9.4.9 Let K be a D-domain. The function $F : K \to \mathbb{R}$ is said to be an **antiderivative** for $f : K \to \mathbb{R}$ provided that $F'(x) = f(x)$ for all $x \in K$. (If it is convenient, we can also restrict the domains of F and f to some subset I of K and say that **F is an antiderivative of f on I**.)

It is easy to see that any function f with an antiderivative must have infinitely many antiderivatives. (If F is an antiderivative for f, then so is $G(x) = F(x) + C$ for any real number C.) Corollary 9.4.8 says that this family of functions exhausts the set of antiderivatives for f.

> **Why Is Corollary 9.4.8 Interesting?**
>
> We know that $F(x) = x^2 + 7$ and $G(x) = x^2 - \pi$ are antiderivatives for $f(x) = 2x$, but how we know that some complicated monster like
>
> $$H(x) = \frac{\arctan(x^2+4)\ln\left(\frac{x}{x^2+4}\right)}{x^4+4}$$
>
> *is not*? It is not a forgone conclusion that, when we take the derivative of H, a monumental amount of algebra will not render plain old $2x$. Of course, in the case of this *particular* complicated monster, we can carry out the experiment and see that, indeed, no such tricky simplification occurs. But what about all the other complicated monsters that lurk out there? We certainly cannot rule them out one by one—we need Corollary 9.4.8. This fact has important consequences for integration and for solving differential equations.

Problems 9.4

1. Prove the local extreme theorem (Theorem 9.4.2).

   Hint: (For the case where c is a local minimum.) Since c is in the interior of K, we know there exists $\delta > 0$ such that $(c-\delta, c+\delta)$ is entirely contained in K and for each x in this interval $f(x) \geq f(c)$. Consider separately

   $$\lim_{x \to c^+} \frac{f(x) - f(c)}{x - c} \quad \text{and} \quad \lim_{x \to c^-} \frac{f(x) - f(c)}{x - c}.$$

2. Prove Rolle's theorem (Theorem 9.4.3). (*Hint*: This proof follows fairly readily from the local extreme theorem.)

3. Prove the mean value theorem (Theorem 9.4.6). (*Hint*: Look at Exercise 9.4.5. Show that the function h satisfies the conditions of Rolle's theorem.)

4. Prove that a differentiable function is constant if and only if its derivative is zero everywhere (Corollary 9.4.7). (*Hint*: Apply the mean value theorem.)

5. Prove that any two antiderivatives for the same function must differ by a constant (Corollary 9.4.8).

6. Let $f : \mathbb{R} \to \mathbb{R}$ be a differentiable function with a constant derivative. Prove that f is a straight line function (that is, there exist real numbers a and b such that $f(x) = ax + b$ for all $x \in \mathbb{R}$).

7. Consider the following (approximate) converse to the mean value theorem:

   Let $f : [a, b] \to \mathbb{R}$ be a continuous function that is differentiable on (a, b). Let $x \in (a, b)$. Then there exist t_1 and t_2 in (a, b) such that
   $$f'(x) = \frac{f(t_1) - f(t_2)}{t_1 - t_2}.$$

   Show by giving a counterexample that this statement is false in general.

8. The following is a generalization of the mean value theorem that is useful in certain contexts:

   Cauchy Mean Value Theorem. Let a and b be real numbers with $a < b$. Let K be a D-domain with $[a, b] \subseteq K$. Let $f : K \to \mathbb{R}$ and $g : K \to \mathbb{R}$ be continuous on $[a, b]$ and differentiable on (a, b). Assume that if $x \in (a, b)$, $g'(x) \neq 0$. Then there exists $c \in (a, b)$ such that
   $$\frac{f(b) - f(a)}{g(b) - g(a)} = \frac{f'(c)}{g'(c)}.$$

   (a) Prove the Cauchy mean value theorem. (*Hint*: Use a trick similar to the one you used for the proof of the mean value theorem. Be sure to show that $g(b) \neq g(a)$!)

   (b) Show that the Cauchy mean value theorem is a generalization of the mean value theorem by showing that if $g(x) = x$, then the Cauchy mean value theorem reduces to the mean value theorem.

9. **Lipschitz functions and the derivative.** Let $a, b \in \mathbb{R}$, and let $f : [a, b] \to \mathbb{R}$ be a function. One way to think about Lipschitz conditions (see page 115) is to observe that saying f is Lipschitz is equivalent to saying that there is a positive real number M such for all y, x in the domain of f,
$$\left| \frac{f(y) - f(x)}{y - x} \right| \leq M.$$

Because the quantities on the left are the absolute values of difference quotients for f, it seems likely that the Lipschitz condition is somehow related to the differentiability of f. The purpose of this problem is to further explore this observation.

(a) Let M be a positive real number and suppose that f is differentiable. Prove that f is Lipschitz with constant M if and only if for all $x \in K$, $|f'(x)| \leq M$.

This result, of course, allows us to deduce that functions with bounded derivatives are always uniformly continuous (see Theorem 4.4.4). But the converse is not true.

(b) Give an example that shows that there are uniformly continuous functions with unbounded derivatives. (*Hint*: Recall Theorem 7.2.3 and think about functions you know that have unbounded derivatives.)

It is tempting to believe that all Lipschitz functions have to be differentiable. But this is not true, either.

(c) Give an example of a Lipschitz function that is not differentiable.

10. Suppose that K is a D-domain, and that $t \in \mathbb{R}$ such that $t > 1$. Let $f : K \to \mathbb{R}$ be a function that satisfies the condition $|f(x) - f(y)| \leq M|x - y|^t$, for all $x, y \in K$.

(a) Prove that f is differentiable everywhere on K. (*Hint*: You should be able to calculate the derivative at an arbitrary point.)

(b) Using the result you found in part (a), completely describe the function f.

9.5 Monotonicity and the Mean Value Theorem

One of the first things that calculus students internalize is that a differentiable function f is increasing on an interval (a, b) if and only if f' is non-negative on (a, b), and similarly for a decreasing function and a non-positive derivative. To be a bit careful about all of this, let's state some definitions.

Definition 9.5.1 Let $A \subseteq K \subseteq \mathbb{R}$, and let $f : K \to \mathbb{R}$ be a function. We say that f is **increasing** on A if $x_1 < x_2$ implies that $f(x_1) \leq f(x_2)$ for every x_1 and x_2 in A. f is **strictly increasing** on A if $x_1 < x_2$ implies that $f(x_1) < f(x_2)$ for every x_1 and x_2 in A.

Similar definitions apply for f **decreasing** and f **strictly decreasing**.

If f is either increasing on A or decreasing on A, we say that f is **monotonic** on A. The function f is **strictly monotonic** on A if it is either strictly increasing on A or strictly decreasing on A.

(If we use the terms increasing, decreasing, or monotonic, respectively, without qualification, we mean that the function is increasing, decreasing, or monotonic on its entire domain.)

Theorem 9.5.2 Let $K \subseteq \mathbb{R}$ be a D-domain, and let $f : K \to \mathbb{R}$ be a function. Suppose that I is an interval and that $I \subseteq K$. Suppose that f is differentiable on I. Then

- f is increasing on I if and only if $f' \geq 0$ on I.
- f is decreasing on I if and only if $f' \leq 0$ on I.

Proof: The proof is Problem 1 at the end of this section. □

> Many of the theorems in this chapter, including Theorem 9.5.2, confine our attention to a subinterval of the domain. Consider Theorem 9.5.2. Is such a restriction necessary? Why or why not? You should consider this same issue for other theorems as well.

Figure 9.3 gives the idea of the following theorem:

Theorem 9.5.3 Let K be a D-domain, and let $f : K \to \mathbb{R}$ be a function. Suppose that f is differentiable on $[a, b]$ where $a < b$ and $[a, b] \subseteq K$. If for all $x \in (a, b)$, $f'(x) \neq 0$, then f is one-to-one on $[a, b]$.

Proof: The proof is Problem 3 at the end of this section. □

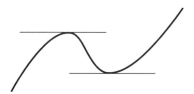

Figure 9.3 Intuitively, it seems that nonconstant, differentiable functions that are not one-to-one have to rise and fall and, therefore, must have places where their derivative is zero.

Theorem 9.5.3 now allows us to prove an amazing theorem of which the following lemma is a crucial special case:

Lemma 9.5.4 Let K be a D-domain, and let $f : K \to \mathbb{R}$ be a function. Suppose that f is differentiable on $[a, b]$ where $a < b$ and $[a, b] \subseteq K$. If $f'(a)$ and $f'(b)$ are opposite in sign, then there exists $c \in (a, b)$ such that $f'(c) = 0$.

Proof: The proof is Problem 4 at the end of this section. □

Lemma 9.5.4 tells us something fundamental about the nature of derivative functions: If a function is the derivative of another function on an interval $[a, b]$, it cannot go from negative to positive, or vice versa, without crossing through zero. Darboux's theorem generalizes this idea.

Let I be an interval in $\mathbb{R}$. Recall that a function $f : I \to \mathbb{R}$ is said to have the *intermediate value property*, provided that if a and b are in I and α is any real number between $f(a)$ and $f(b)$, then there exists $c \in I$ such that $f(c) = \alpha$. (See Problem 4 at the end of Section 5.2.) The intermediate value *theorem* tells us that every continuous function f defined on an interval in $\mathbb{R}$ has the intermediate value property. Darboux's theorem (Theorem 9.5.5) tells us that derivatives do as well.

Theorem 9.5.5 [Darboux's Theorem] Let K be a D-domain, and let a and b be real numbers with $a < b$ and $[a, b] \subseteq K$. Suppose that $f : K \to \mathbb{R}$ is differentiable on $[a, b]$. Then f' satisfies the intermediate value property on $[a, b]$.

Proof: The proof is Problem 5 at the end of this section. □

Exercise 9.5.6 Consider the graphs shown here

1. Explain why the function in the top graph cannot be the derivative of any function.

2. What about the function in the bottom graph? Could it be the derivative of some function? Why or why not?

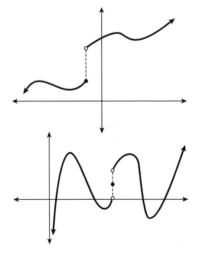

■

9.5 Monotonicity and the Mean Value Theorem

Why Is Darboux's Theorem Amazing?

Notice that if f' is continuous on $[a, b]$, then Darboux's theorem is a corollary to the intermediate value theorem. The remarkable thing is that Darboux's theorem does *not* assume the continuity of f'. Under the circumstances, it may be tempting to think that Darboux's theorem implies that derivative functions are automatically continuous, but this isn't the case. A derivative function may exist everywhere in an interval and yet be discontinuous. (Remember Problem 12 at the end of Section 9.2?) Darboux's theorem tells us that, if this is the case, then the derivative function must behave fairly wildly in the neighborhood of the discontinuity. (See Problem 6a at the end of this section.)

Problems 9.5

1. Prove Theorem 9.5.2, which states the connection between the sign of the derivative and the monotonicity of the function.

2. What about *strict* monotonicity?

   (a) Show by giving an example that f can be strictly decreasing on an interval and yet have a derivative of zero at one or more places.

   (b) What about the converse? If $f' < 0$ on (a, b), must f be strictly decreasing on (a, b)?

3. Prove Theorem 9.5.3, which says, roughly, that if the derivative of a function is never zero on an interval, then the function is one-to-one on the interval.

4. Prove Lemma 9.5.4, the special case of Darboux's theorem that says that if the derivative of a differentiable function has opposite signs at the two ends of an interval, then there must be some place in the interior of the interval where the derivative is zero. (*Hint*: Prove the contrapositive. You may want to consider Problem 2 at the end of Section 8.1.)

5. Prove Darboux's theorem (Theorem 9.5.5). [*Hint*: Suppose that s and t are distinct points in $[a, b]$ and that r is between $f'(s)$ and $f'(t)$. Consider the function $h(x) = f(x) - rx$.]

6. Let K be a D-domain, and let $f : K \to \mathbb{R}$ be differentiable on the interval $[a, b] \subseteq K$ ($a < b$).

   (a) Let $x \in (a, b)$. Show that if $\lim_{y \to x^+} f'(y)$, and $\lim_{y \to x^-} f'(y)$ both exist, then f' must be continuous at x.

   > Recall that a function g is said to have a **removable discontinuity** at $x \in (a, b)$, if $\lim_{y \to x} g(y)$ exists but is not equal to $g(x)$. It has a **jump discontinuity** at $x \in (a, b)$ if $\lim_{y \to x^+} g(y)$, and $\lim_{y \to x^-} g(x)$ both exist but are not equal. All other discontinuities are called **essential** discontinuities.
   >   Darboux's theorem implies that derivative functions are either continuous or "badly" discontinuous. Any discontinuity in a derivative function must be an essential discontinuity. That is, the function given in Problem 12 at the end of Section 9.2 is typical of discontinuous derivative functions.

   (b) We all know that the function
   $$f(x) = \begin{cases} 1 & \text{if } x > 0 \\ -1 & \text{if } x < 0 \end{cases}$$
   is the derivative of $f(x) = |x|$. Why doesn't this contradict Darboux's theorem?

9.6 Inverse Functions

Let $K \subset \mathbb{R}$, and let $f : K \to \mathbb{R}$ be a function. Strictly speaking, f is invertible if and only if f is both one-to-one and onto, but inverses are so useful that we talk about them in other contexts as well. If f is one-to-one but not onto, we can still define the inverse of f on $f(K)$ (e.g., the natural logarithm as the inverse of the exponential function). Even if f is neither one-to-one nor onto, we can restrict our attention to a subset S of K where f *is* one-to-one and talk about its inverse on $f(S)$. [For example, $f(x) = x^2$ and $g(x) = \sqrt{x}$ are inverses only if we restrict the domain of f to the positive reals. Consider also the sine and the arcsine.]

When we invert a function, we hope that some of the nice properties of the function transfer to the inverse. And, indeed, our intuition tells us that this is so. Suppose that $K \subseteq \mathbb{R}$ and $f : K \to \mathbb{R}$ is one-to-one. We know from previous experience that the graph of f and the graph of its inverse are mirror images across the line $y = x$. If we think intuitively of continuous functions as functions without any breaks in them, it seems believable that the inverse of a continuous function should be continuous (reflecting a graph across a line should not "break it"). Example 9.6.1 shows that this is, in fact, not true without qualification.

Example 9.6.1 Let $K = [0,1] \cup (2,3]$. Define $f : K \to \mathbb{R}$ by

$$f(x) = \begin{cases} x & \text{if } 0 \leq x \leq 1 \\ x - 1 & \text{if } 2 < x \leq 3 \end{cases}$$

The function f is one-to-one and continuous; its inverse has a jump discontinuity. To see this, draw graphs of f and f^{-1}.

The difficulty in Example 9.6.1 was caused by the fact that the domain was not connected. As it turns out, a connected domain does give us the result we need. This was laid out in Problem 3 on page 153. That result is restated here as a theorem for your reference.

Theorem 9.6.2 Let I be an interval in $\mathbb{R}$, and let $f : I \to \mathbb{R}$ be a one-to-one continuous function. Then $f^{-1} : f(I) \to \mathbb{R}$ is also continuous.

The domain need not be connected, however—a closed, bounded domain also gives us what we want. This was Problem 5 on page 113 and is restated here.

Theorem 9.6.3 Let K be a closed, bounded subset of $\mathbb{R}$, and let $f : K \to \mathbb{R}$ be a one-to-one, continuous function. Then $f^{-1} : f(K) \to \mathbb{R}$ is also continuous.

What about differentiability? Suppose we have a differentiable function f and we wish to invert it. What can we say about f^{-1}? First note that if K is a D-domain and f is continuous on K, then $f(K)$ is a D-domain. (*Why?*) Thus we can, without difficulty, talk about differentiating the inverse of a one-to-one function.

Let us continue with our intuitive musings about reflecting a graph across a line. We have said that a function f is differentiable if and only if, as we "zoom in" on the graph of f, we see something that looks more and more like a straight (but not vertical) line. Thus, if the graph of f has a "corner" at $(x, f(x))$, then $f'(x)$ will not exist. It will also fail to be differentiable at x if its tangent line at $(x, f(x))$ is vertical. We can readily believe if the graph of f doesn't have a sharp corner, the graph of f^{-1} won't either. The same reasoning doesn't hold for vertical tangents, however. If f has a horizontal tangent line, the mirror image of its graph across $y = x$ *will* have a vertical tangent line. Our intuition tells us that if f is differentiable *and its derivative is never zero*, then f^{-1} should be differentiable as well.

Theorem 9.6.4 [Inverse Function Theorem] Let K be a closed, bounded subset of $\mathbb{R}$, and let $f : K \to \mathbb{R}$ be a one-to-one continuous function that is differentiable at $x \in K$. If $f'(x) \neq 0$, then the inverse function $f^{-1} : f(K) \to \mathbb{R}$ is differentiable at $f(x)$ and

$$\left(f^{-1}\right)'(f(x)) = \frac{1}{f'(x)}.$$

Proof: The proof is the problem at the end of this section. $\square$

Problems 9.6

1. Prove the inverse function theorem (Theorem 9.6.4). (*Hint*: Things will go much more smoothly if you use a sequential approach to proving the limit exists.)

9.7 Polynomial Approximation and Taylor's Theorem

We frequently want to approximate a complicated function with a simpler one. Polynomial approximations are some of the most important. Among those, the Taylor polynomial approximations are the easiest to find and use.

Suppose we have a real-valued function f defined on the closed interval $[a, b]$. For the sake of argument, assume this function is "nicely behaved," in the sense that it has derivatives to any order that we need. We would like to find a polynomial function that will approximate f near some point $s \in [a, b]$, and we want to know how accurate the approximation is.

Taylor's theorem is a generalization of the mean value theorem that speaks to this very question. To understand Taylor's theorem and its connection to the mean value theorem, we want to think about the mean value theorem in slightly different terms than we have before.

Consider the function $f : [a, b] \to \mathbb{R}$ and $s \in [a, b]$, as mentioned earlier. The crudest possible polynomial approximation for f near s is to assume that

$$f(t) \approx f(s) \text{ for all } t.$$

In other words, we approximate f by a horizontal line, as shown in Figure 9.4. In this case, the error made by the approximation depends on two factors: how

9.7 Polynomial Approximation and Taylor's Theorem

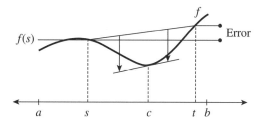

Figure 9.4 Degree 0: Error $= f(t) - f(s) = f'(c)(t-s)$.

far t is from s and how far f is from being horizontal. If the derivative of f were zero everywhere, the approximation would be exact. The larger $|f'|$ is, the larger the error. The mean value theorem tells us that, for a particular value t, there is a c between s and t such that the error made when we approximate $f(t)$ by $f(s)$ is the product of $f'(c)$ and $t - s$:

$$f'(c) = \frac{f(t) - f(s)}{t - s} \implies f(t) = f(s) + f'(c)(t-s)$$

Of course, we all know that a horizontal line approximation is not generally very satisfactory, so we elaborate. The main idea of differential calculus is that a function that is differentiable at s can be approximated reasonably well by a straight line in the region around s. (See Figure 9.5.) Again, the error will depend on how far t is from s, but this time our approximation takes into consideration the fact that f is not a horizontal line. Instead, the error will depend on how far f is from being a straight line. In other words, it will depend on how much f curves. As you probably know, the second derivative is a measure of curvature. The simplest form of Taylor's theorem tells us that there is a point c between s and t such that the error made by the best local linear approximation is precisely the product of $\frac{(t-s)^2}{2}$ and $f''(c)$. To see how this works, we will prove Taylor's remainder theorem in the special case of linear approximations. The nth-degree generalization is left to you.

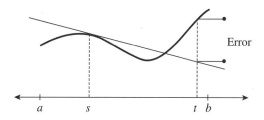

Figure 9.5 Degree 1: Error $= f(t) - f(s) - f'(s)(t-s) = \frac{f''(c)}{2}(t-s)^2$.

Theorem 9.7.1 [Taylor's Remainder Theorem for Linear Approximations] Let a and b be real numbers with $a < b$, and let $f : [a, b] \to \mathbb{R}$ be differentiable on $[a, b]$ and twice differentiable on (a, b). Assume that t and s are any two points in $[a, b]$. Then there exists a real number c lying between t and s such that

$$f(t) = f(s) + f'(s)(t-s) + \frac{f''(c)}{2!}(t-s)^2.$$

Proof: Let M be the real number that satisfies

$$f(t) = f(s) + f'(s)(t-s) + \frac{M}{2!}(t-s)^2.$$

For each $x \in [a, b]$, define

$$A(x) = f(x) - \left(f(s) + f'(s)(x-s) + \frac{M}{2!}(x-s)^2 \right).$$

Note that $A(t) = 0$ by our choice of M; $A(s)$ is also 0. Thus Rolle's theorem tells us that there exists x_1 between s and t such that $A'(x_1) = 0$. Because

$$A'(x) = f'(x) - f'(s) - M(x-s)$$

we see that $A'(s) = 0$, too.

We appeal to Rolle's theorem once again to get c between x_1 and s such that
$$A''(c) = 0.$$
But $A''(x) = f''(x) - M$, so $A''(c) = 0$ which implies that $M = f''(c)$. Because any number that lies between x_1 and s also lies between t and s, we have the desired result. □

The standard tangent line approximation matches the value of f at s and the derivative of f at s. We can do better with a quadratic equation by further matching the second derivative of f at s, thus taking into consideration the curvature of f at s. The error, then, will depend on the rate of change of the curvature, or the third derivative, of f. And so it goes.

Definition 9.7.2 The polynomial

$$P_n(x) = f(s) + f'(s)(x-s) + \frac{f''(s)}{2!}(x-s)^2 + \cdots + \frac{f^{(n)}(s)}{n!}(x-s)^n$$

is called the nth-degree Taylor polynomial approximation of f based at s or simply the nth Taylor polynomial of f at s.

9.7 Polynomial Approximation and Taylor's Theorem

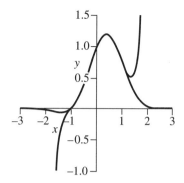

Figure 9.6 $f(x) = (x+1)e^{-x^2}$ and its tenth-degree Maclaurin polynomial.

The case when $s = 0$ is sufficiently important that it has its own name. The nth Taylor polynomial of f at $s = 0$ is called **the nth Maclaurin polynomial of f**:

$$M_n(x) = f(s) + f'(s)x + \frac{f''(s)}{2!}x^2 + \ldots + \frac{f^{(n)}(s)}{n!}x^n$$

Note that P_n matches the value of f at s, and also all the values of its derivatives up to the nth derivative at s.

Taylor polynomials do a remarkably good job of approximating most familiar transcendental functions—not only on a *microscopic* scale but also on a larger scale. The higher the degree of the polynomial, the larger the region over which the approximation is good. (See Figure 9.6.)

Let P_n be the nth Taylor polynomial of f at s. As you would expect, the error made by $P_n(t)$ in approximating $f(t)$ depends on the distance between g and t and also on the size of the $(n+1)^{\text{st}}$ derivative of f on the interval between s and t. Taylor's remainder theorem tells us that there is a point c between s and t such that the error made when $f(t)$ is approximated by $P_n(t)$ is precisely

$$f(t) - P_n(t) = R_n(t) = \frac{f^{(n+1)}(c)}{(n+1)!}(t-s)^{n+1}$$

where R stands for "remainder."

Theorem 9.7.3 [Taylor's Theorem] Let a and b be real numbers with $a < b$, and let $f : [a, b] \to \mathbb{R}$ be a function with derivatives $f', f'', f''', \ldots, f^{(n)}$ all existing and continuous on $[a, b]$. Suppose that $f^{(n+1)}$ exists everywhere on (a, b). Assume further that t and s are any two points in $[a, b]$. Then there exists a real number c lying between t and s such that

$$f(t) = f(s) + f'(s)(t-s) + \frac{f''(s)}{2!}(t-s)^2 + \cdots + \frac{f^{(n)}(s)}{n!}(t-s)^n + \frac{f^{(n+1)}(c)}{(n+1)!}(t-s)^{n+1}$$

Proof: The proof is Problem 1 at the end of this section. □

Notice that the choice of c depends on *both* s and t. So Taylor's theorem may seem to be of limited usefulness, given that we have to know $f(t)$ to find c. If we can bound the size of the various derivatives in some global way, however, then we can get an upper bound for the size of the error that doesn't require prior knowledge of $f(t)$. In general this solution is fine, because an upper bound for the error is all that is ordinarily needed (or possible), anyhow. This is demonstrated by the following, extremely useful, corollary.

Corollary 9.7.4 Let a and b be real numbers with $a < b$, and let K be a positive real number. Let $f : [a,b] \to \mathbb{R}$ be a function with derivatives f', f'', f''', $\ldots$, f^n all existing and continuous on $[a,b]$. Suppose that $f^{(n+1)}$ exists everywhere on (a,b) and that

$$|f^{(n+1)}(x)| \leq K \text{ for all } x \in (a,b).$$

Assume further that t and s are any two points in $[a,b]$. Then

$$|R_n(t)| \leq \frac{K}{(n+1)!}|t-s|^{n+1}.$$

Proof: The proof is Problem 2 at the end of this section. □

Problems 9.7

1. Prove Taylor's theorem (Theorem 9.7.3).

2. Prove Corollary 9.7.4.

3. Consider the functions $f(x) = e^x$ and $f(x) = \sin(x)$. (For this problem, you may assume the familiar facts about exponentials, sines and cosines, and their derivatives.) Let P_n be the nth Taylor polynomial for f based at $s = 0$. As before, the error is given by $R_n = f - P_n$.

   For each of the functions, use Corollary 9.7.4 to do the following things:

   (a) Determine a value of n such that $|R_n(3)| \leq 0.001$. (Crude upper estimates are fine. For instance, feel free to note that $e \leq 3$.)

(b) Fix x. Assuming the "well-known" fact[2] that
$$\lim_{n \to \infty} \frac{|x|^{n+1}}{(n+1)!} = 0,$$
show that $\lim_{n \to \infty} |R_n(x)| = 0$. What does this allow you to conclude?

---

2. The proof of this fact is sketched in Section D.4 in Excursion D (Theorem D.4.3).

Chapter 10

Iteration and the Contraction Mapping Theorem

One of the most important tools in modern analysis is the iteration of functions. In this chapter, you will learn some elementary facts about it. Our main goal is to explore the contraction mapping theorem, which is an important theoretical tool in analysis. But iteration is also amazingly useful in applied mathematics, where it often plays a central role. Furthermore, iteration underlies some of our most beautiful and fascinating "pure" mathematics—fractals and the theory of chaos. Once you have studied it here, you will almost certainly find it cropping up in other places.

10.1 Iteration and Fixed Points

Definition 10.1.1 Let X be a set, and let $f : X \to X$ be a function. If x_0 is any element of X, we can define a sequence (x_n) in the set X by repeatedly applying the function f beginning at the point x_0. That is,

$$x_1 = f(x_0), \ x_2 = f(x_1), \ x_3 ,= f(x_2) \ \ldots$$

This process is called **iteration**. The resulting sequence is called the **iterated map on f based at x_0** or the **orbit of x_0 under iteration by f**. (See Figure 10.1.) If the function f is understood, we may just refer to the iterated map at x_0 or the orbit of x_0.

Remark: The iterated map on f based at x_0 can also be written as

$$x, f(x), f(f(x)), f(f(f(x))), f(f(f(f(x)))), \ldots$$

Thus we adopt a natural shorthand:

$$x, f(x), f^2(x), f^3(x), f^4(x), \ldots$$

190 Chapter 10 ■ Iteration and the Contraction Mapping Theorem

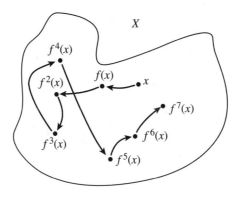

Figure 10.1 The first few terms of the orbit of x under iteration by f.

Exercise 10.1.2 [A Simple Example] Let $A = \{a, b, c, d\}$. Consider the function $f : A \to A$, which is depicted (in two different ways!) in Figure 10.2.

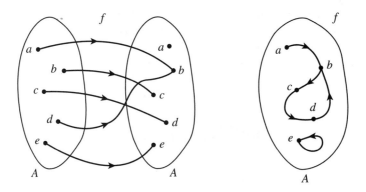

Figure 10.2 $f : A \to A$: two views.

1. Find the iterated maps on f based at the points a and c.

2. We call the point c a *periodic point* and the point a an *eventually periodic point*. Why do these names make sense?

■

Although iteration is a very simple procedure, it can lead to very complicated results. Exercise 10.1.3 should begin to suggest some of the surprising behavior of iterated functions.

Exercise 10.1.3 [An Experiment] Consider the family of quadratic functions of the form

$$f_k(x) = kx(1-x) \text{ where } k > 1 \text{ is a constant}$$

(See Figure 10.3.) These functions are called the *logistic maps*. Iteration on these maps is a fascinating enterprise.

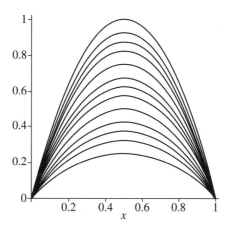

Figure 10.3 The logistic maps on $[0,1]$ for $1 < k \leq 4$.

1. Begin by verifying that for $1 \leq k \leq 4$, the range of the logistic map is contained in $[0,1]$. (Think about the values at the endpoints and the position of the vertex.) This allows us to think about iterating f starting with a point in $[0,1]$.

2. For the values $k = 2$, 2.5, 3.2, 3.5, and 4, try iterating the logistic maps beginning at several points in $[0,1]$. What do you observe about the "long-term" behavior of the resulting sequences? In other words, can you make any conjectures about the limiting behavior of the iterated maps?

You will want to generate a fair number of iterates, so you will need to use a computer to help. Visualizing the orbits using "cobweb diagrams" can also be useful. This process is discussed in the box on page 192. ■

Graphical Analysis: Cobweb Diagrams

Looking at long lists of terms to guess the long-term behavior of a sequence is cumbersome and often misleading. For real functions, a graph can help us to visualize the behavior of an iterated map. Figures 10.4 and 10.5 illustrate the following procedure:

Step 1.: Choose a point x_0 at which to begin the iteration. On a graph of f and the diagonal line $y = x$, draw a vertical line from the point x_0 on the x-axis to the graph of the function f. The y-coordinate of this point is the first term in the iterated map on f based at x_0.

Step 2.: To obtain the second term in the iterated map, we need to evaluate f at $f(x_0)$. To accomplish this graphically, we draw a horizontal line from the point $(x_0, f(x_0))$ to the point $(f(x_0), f(x_0))$ on the diagonal line.

Step 3.: Draw a vertical line from the diagonal to the graph of f. The point where this vertical line meets the graph is $(f(x_0), f^2(x_0))$, whose y-coordinate is the second term of the iterated map.

Step 4.: Repeat Steps 2 and 3 as many times as necessary to visualize the long-term behavior of the iterated map.

Cobweb diagrams are difficult to draw by hand, because they are extremely sensitive to any variation from perfectly horizontal and vertical lines. Computers, of course, are tailor-made for producing them.

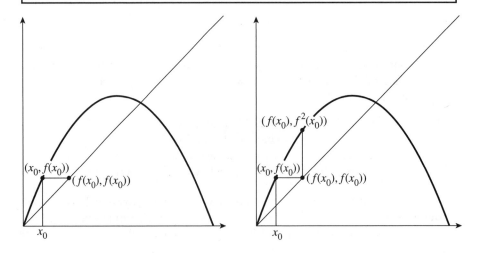

Figure 10.4 Graphical analysis of iteration.

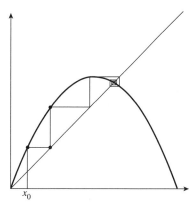

Figure 10.5 Cobweb diagram.

Definition 10.1.4 Let X be a set, and let $A \subseteq X$. Suppose that $f : A \to X$ is a function, and a is an element of A. If $f(a) = a$, we say that a is a **fixed point** of f.

Exercise 10.1.5 [Finding Fixed Points]

1. Find the fixed point of the function given in Exercise 10.1.2.

2. Find the fixed points of the function $f : \mathbb{R} \to \mathbb{R}$ given by $f(x) = x^3 - x$.

3. Find the fixed points of the logistic maps $f_k(x) = kx(1 - x)$.

■

Example 10.1.6 For functions $f : \mathbb{R} \to \mathbb{R}$, it is easy to identify fixed points using a graph. Suppose a is a fixed point of f. Then $f(a) = a$, so the point $(a, f(a))$ is a place where the graph of f intersects the line $y = x$. Plotting f and the diagonal line together make it easy to see whether (and where) f has fixed points. Figure 10.6 shows the fixed points of $f(x) = \frac{1}{10}(x^5 - 5x^3 + x^2 + 5)$. What are they, approximately?

You probably noticed when you iterated $f(x) = 2x(1-x)$ in Exercise 10.1.3 that the iterated maps on f based at points in $[0, 1]$ all converge to $\frac{1}{2}$, which is also the fixed point of the map. As it turns out, this behavior is quite common.

Theorem 10.1.7 Let X be a metric space, and let $x_0 \in X$. Suppose that f is a continuous function from X to itself and that the iterated map on f based at x_0 converges to some point $y \in X$. Then y is a fixed point of f.

Proof: The proof is Problem 6 at the end of this section. □

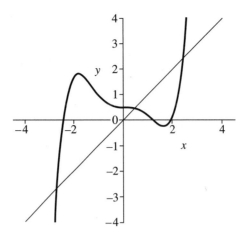

Figure 10.6 The function $f(x) = \frac{1}{10}(x^5 - 5x^3 + x^2 + 5)$ has three fixed points.

We can immediately deduce the following corollary.

Corollary 10.1.8 Let X be a metric space. If $f : X \to X$ is a continuous function with no fixed point, then none of the iterated maps on f converge.

It is important to understand what Theorem 10.1.7 does *not* say:

- The mere existence of a fixed point does not guarantee convergence of the iterated maps, even for a very well-behaved function. For example, the logistic map $f_{3.2}(x) = 3.2x(1-x)$ has fixed points at $\frac{11}{16}$ and at 0, yet none of the nonconstant iterated maps of $f_{3.2}$ converge.[1]

- The lack of a fixed point guarantees the nonconvergence of iterated maps only for functions that are continuous. Problem 7 at the end of this section gives a function with no fixed point whose iterated maps all converge.

---

1. As you saw in Exercise 10.1.3, iterated maps based at points of $(0, 1)$ seem eventually to oscillate between two values. The exact values are
$$a_1 = \frac{21 + \sqrt{21}}{32} \text{ and } a_2 = \frac{21 - \sqrt{21}}{32}.$$
It is easy to verify that $f_{3.2}(a_1) = a_2$ and $f_{3.2}(a_2) = a_1$. These points are said to have "period two." Any book on discrete dynamical systems will tell you more about this interesting phenomenon. See, for instance, [HOL].

Attractors and Repellors

Exercise 10.1.9 [An Experiment] Consider the function
$$f(x) = \frac{1}{10}(x^5 - 5x^3 + x^2 + 5)$$
on $\mathbb{R}$. Figure 10.6 shows that this function has fixed points at approximately -2.6, 0.5, and 2.4.

1. Try iterating the function starting with values in the interval $(-2.6, 2.4)$. (Be sure to include values spread out around the interval.) What happens?

2. Try iterating f with values around $x = -2.7$ (to the left of the leftmost fixed point) and around $x = 2.5$ (to the right of the rightmost fixed point). What happens?

■

Exercise 10.1.9 suggests that the iterated maps of f, when they converge, always converge to 0.5. Even when the base point is quite close to -2.6 or 2.4, the iterated map moves quickly away. This motivates the following terminology.

Definition 10.1.10 Let X be a metric space, and let U be an open subset of X. Let $f : U \to X$ be a function with a fixed point at $a \in X$. We say that a is an **attracting fixed point of f** (or just an **attractor**) if there exists an open ball B centered at a such that all iterated maps on f that are based at points of B converge to a.

The fixed point a is called a **repelling fixed point of f** (or **repellor**) if there is an open ball B centered at a such that all iterated maps on f based at points of B must eventually leave B. That is, for all $b \in B$, there exists $n \in \mathbb{N}$ such that $f^n(b) \notin B$.

There are also fixed points that are neither attractors nor repellors. (See Problem 9 at the end of this section.)

Theorem 10.1.11 [Attractors and Repellors in $\mathbb{R}$] Let U be an open subset of $\mathbb{R}$, and let $f : U \to \mathbb{R}$ be a differentiable function. Suppose that $p \in \mathbb{R}$ is a fixed point of f and that f' is continuous at p.

1. If $|f'(p)| < 1$, then there exists an open interval $I \subseteq U$ containing p such that whenever $x \in I$, $f^n(x) \to p$.

2. If $|f'(p)| > 1$, then there exists an open interval $I \subseteq U$ containing p such that for all $x \in I \setminus \{p\}$, $f^n(x) \notin I$ for some $n \geq 1$.

Proof: The proof is Problem 10 at the end of this section. □

Problems 10.1

1. Use a computer to investigate the iterated maps of the following real functions. Briefly report your procedure and your results. Describe the behavior of the iterated maps as completely as you can. (Your descriptions will, of course, be educated guesses based on the results of the experiment. No proofs are expected.)

    (a) $f(x) = \dfrac{x-4}{3}$

    (b) $f(x) = 2x + 1$

    (c) $f(x) = x^3$

    (d) $f(x) = x^{\frac{1}{3}}$

    (e) $f(x) = x^2 + 2x$ [*Hint*: Consider separately base points in the intervals $(-\infty, -2)$, $(-2, 0)$, $(0, \infty)$. What happens if $x_0 = -2$ or $x_0 = 0$?]

    (f) $f(x) = x^2 - 1$ [*Hint*: Consider separately base points in the intervals $(-\infty, -t)$, $(-t, t)$, (t, ∞), where t is approximately 1.62.]

2. Let C be a proper subset of A. Suppose that $f : A \to C$ is a one-to-one function. Prove that A contains a sequence of distinct terms.[2]

3. Consider the real-valued functions $f(x) = x^2 + 1$ and $g(x) = \dfrac{x^2-1}{2x}$. Neither of these functions has a fixed point. What happens when you iterate them at various values? (Use a computer to experiment.)

4. Let a and b be real numbers with $a < b$. Suppose that $f : [a, b] \to [a, b]$ is a continuous function. Prove that f has a fixed point. (*Hint*: Consider the function $g = f(x) - x$ and use the intermediate value theorem.)

5. Let a and b be real numbers with $a < b$. Suppose that $f : [a, b] \to \mathbb{R}$ is a continuous function that satisfies $f([a, b]) \supset [a, b]$. Prove that f has a fixed point. (*Hint*: Same hint as Problem 4.)

---

2. *Iteration and infinite sets:* Let A be a set. A is infinite if it contains a sequence of distinct terms. It is easy to see that an infinite set A can be put into one-to-one correspondence with a *proper* subset of itself. If $A = \{a_1, a_2, \ldots\} \cup B$ (where no a_i is in B), then we can map A onto $C = \{a_2, a_3, \ldots\} \cup B$ in a one-to-one fashion as follows:

$$f(x) = \begin{cases} a_{i+1} & \text{if } x = a_i \text{ for some } i \in \mathbb{N} \\ b & \text{if } x \in B \end{cases}$$

This is a one-to-one correspondence because all of the a_i's are different and none of them are in B.

Problem 2 asks you to prove the converse of this fact. An iterated map is central to the proof.

6. Prove that if an iterated map on a continuous function converges, it must converge to a fixed point (Theorem 10.1.7).

7. Consider the function
$$f(x) = \begin{cases} \frac{1}{2}x + 1 & \text{if } x < 2 \\ 3 - x & \text{if } x \geq 2 \end{cases}$$

   (a) Show that f has no fixed point.
   (b) Show that every iterated map on f converges to 2.
   (c) Look at a graph. Can you explain what is happening?

8. In Exercise 10.1.5, you showed that $f_k(x) = kx(1-x)$ has fixed points at 0 and at $x_k = \frac{k-1}{k}$. Prove that x_k is an attractor when $1 \leq k \leq 3$ and that it is a repellor when $k > 3$. The fixed point at 0 is always a repellor, except when $k = 1$. (In this case, $\frac{k-1}{k} = 0$; f_0 has only one fixed point.)

9. **Neither attractors nor repellors.** In this problem you will consider the real functions $f(x) = -x$ and $g(x) = e^x - 1$.

   (a) Find the fixed points of f and g, and verify that they are neither attractors nor repellors.
   (b) What do you notice about the derivatives of the functions at the fixed points given in the examples? (Compare this to the conditions given in Theorem 10.1.11.)
   (c) The fixed points of these functions fail to be attractors or repellors for different reasons. Give a geometric explanation of what is happening in each case. (A clear, well-labeled diagram with a sentence or two of explanation will be fine for each function.)

10. Prove Theorem 10.1.11, which explores the connection between the derivative of a function at a fixed point and whether that fixed point is attracting or repelling.

11. **Super-attracting fixed points.** Let U be an open subset of $\mathbb{R}$, and let $f : U \to \mathbb{R}$ be a differentiable function. Suppose that p is a fixed point of f and that $f'(p) = 0$. Then p is called a super-attracting fixed point for f.

   (a) Suppose that p is a super-attracting fixed point for the differentiable function f. Prove that for all $\epsilon > 0$, there exists $\delta > 0$ such that
   $$\text{if } 0 < |p - x| < \delta, \text{ then } |p - f(x)| < \epsilon |p - x|$$

(b) Show that a super-attracting fixed point is an attracting fixed point. (*Careful*! Theorem 10.1.11 does not apply.)

(c) In view of the fact proved in parts (a) and (b), explain why it makes sense to call p "super"-attracting.

12. Period Doubling. We start with a definition:

    Definition: Let X be a metric space, let $f : X \to X$ be a function, and let $n \in \mathbb{N}$. Suppose that $a \in X$ satisfies $f^n(a) = a$. Then we say that a is a **period-n point**. If a is a period-n point but $f(a) \neq a$, $f^2(a) \neq a$, ..., $f^{n-1}(a) \neq a$, then we say that a has **prime period-n**.

    Note that if k divides n, any period-k point is also a period-n point but not a *prime* period-n point. Note also that if a has period n, then it has prime period-n if and only if $a, f(a), f^2(a), \ldots, f^{n-1}(a)$ are distinct.

    Let I be a closed interval in $\mathbb{R}$, and let $f : I \to I$ be a continuous function.

    (a) Suppose that $f(a) = b$ and $f(b) = a$ with $a \neq b$. Prove that f has a fixed point in (a, b). In other words, every continuous, real-valued function that has a prime period-2 point has a fixed point as well. (*Hint*: Problem 5 may prove useful.)

    (b) Let $n \in \mathbb{N}$. Prove that if f has a prime period-2^n point, then it must have prime periodic points of periods $\{1, 2, 4, 8, \ldots, 2^n\}$. (Hint: If x is a prime period-2^{n+1} point for f, then it is a prime period 2^n point for f^2.

10.2 The Contraction Mapping Theorem

Definition 10.2.1 Let X and Y be metric spaces. Let $f : X \to Y$ be a function. We say that f **contracts distances** if there exists a positive number $k < 1$ such that

$$d(f(x), f(y)) \leq k\, d(x, y) \quad \text{for all} \quad x, y \in X.$$

The number k is called the **contraction constant of f**. We may say that f k-contracts distances or that it contracts distances with constant k.

Exercise 10.2.2 Let X be a metric space, and let S be any subset of X. Suppose that $f : S \to X$ contracts distances. Prove that f can have at most one fixed point in S. ∎

Remark: Notice that a function that contracts distances is a Lipschitz function (see page 115); thus, it's, in particular, uniformly continuous (Theorem 4.4.4).

10.2 The Contraction Mapping Theorem

Definition 10.2.3 Let X be a metric space. A function $f : X \to X$ that contracts distances is called a **contraction of X**. If the contraction constant of f is k, then f is called a k-*contraction*.

> **Why the Term *Contraction*?**
>
> If we picture a function from a set to itself as "shifting" the points of the set about within the set, we can envision a contraction as a function that brings everything closer together by at least some fixed factor. (Any two image points in the set are closer together than were their pre-images.) In effect, you can picture f as *shrinking or contracting* the metric space X. See Figure 10.7.

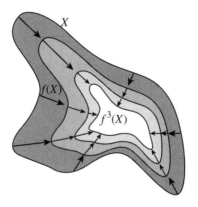

Figure 10.7 *A useful picture*: A contraction "shrinks" the domain. Iterating the function continues the "shrinking." *Picture's limitation*: the images may not be "the same shape" as the original.

Exercise 10.2.4 Let X be a metric space, and let $k \in (0,1)$. Suppose that $f : X \to X$ is a k-contraction of X with fixed point p.

1. Prove that f moves every point in X closer to p. That is, show that for all $x \in X$ with $x \neq p$, $d(f(x),p) < d(x,p)$.

2. Extending the reasoning you used in part (1), show that if $x \in X$, then the orbit of x under iteration by f converges to p.

■

Figure 10.7 suggests that iterating a contraction on a metric space X will shrink it over and over again. The diagram makes it appear as though all the orbits will be converging toward a single (fixed) point—and so they do, if there is a fixed point for them to move toward. There will be such a fixed point if the metric space is complete. This is the content of the contraction mapping theorem.

Theorem 10.2.5 [Contraction Mapping Theorem] Let X be a complete metric space. Let $f : X \to X$ be a k-contraction of X. Then f has a unique fixed point x_0, and for all $x \in X$, the iterated map on f based at x converges to x_0.

Proof: You will be asked to prove this theorem (with some direction) in Problem 5 at the end of this section. □

> The contraction mapping theorem tells us not only that a contraction on complete metric space has a unique fixed point, but that that fixed point is (in a spectacular way!) an attracting fixed point.

The following generalization of the contraction mapping theorem says that $f : X \to X$ need not itself contract distances. If X is complete and some *iterate* of f is a contraction on X, then f has a unique fixed point in X and all iterated maps of f converge to it.

Theorem 10.2.6 Let X be a complete metric space. Let $f : X \to X$ be a function. Suppose that there exists $N \in \mathbb{N}$ such that the Nth iterate f^N of f is a k-contraction of X. Then f has a unique fixed point x_0, and, for all $x \in X$, the iterated map on f based at x converges to x_0.

Proof: You will be asked to prove this theorem (with some direction) in Problem 7 at the end of this section. □

Suppose we relax the condition that our map contract distances by at least a fixed factor and assume only that it decreases distances. Can we then deduce the existence of a fixed point? In general, no—not unless we also assume that X is compact instead of only complete.

Theorem 10.2.7 Suppose X is a compact metric space and $f : X \to X$ is a function that satisfies

$$d(f(x), f(y)) < d(x, y) \quad \text{for all} \quad x, y \in X \text{ with } x \neq y.$$

Then f has a unique fixed point x_0 and every iterated map on f converges to x_0.

Proof: The proof of this theorem is outlined for you in Problem 9 at the end of this section. □

Why You Should Care about Fixed Points

This chapter began by saying that the iteration of functions is "one of the most important tools in modern analysis." By now you are beginning to see that iteration helps us detect the presence of fixed points. It can also help us find or approximate them. What you may not yet be able to fathom is why we would want to do so. Here's a simple example.

Example 10.2.8 Suppose that we want to solve the equation

$$0.005x^4 + 0.0281x^3 - 0.051x^2 - 0.3x + 0.3 = 0.$$

We can recast this as a fixed point problem by adding x to both sides of the equation:

$$(0.005x^4 + 0.0281x^3 - 0.051x^2 - 0.3x + 0.3) + x = x.$$

Note that x is a root of the original equation if and only if it is a fixed point of the function given by the left side of the modified equation:

$$g(x) = (0.005x^4 + 0.0281x^3 - 0.051x^2 - 0.3x + 0.3) + x.$$

If we iterate g starting at $x = 0$, we get a sequence that converges to $r \approx 0.94059$. Because g is continuous, we know r must be a fixed point for g and, therefore, a root of our original equation.

You might ask yourself whether the scheme in Example 10.2.8 always works. In fact, it does, provided that the function is continuous and that the orbit converges. (It may not.) So we have to ask, when does this iterative procedure converge? Clearly, the function being iterated must have a fixed point (the original equation must have a root), but the fixed point must also be an *attracting* fixed point, ... and so on. We immediately run up against the issues discussed earlier in this chapter.

Many numerical approximation methods involve iteration. We use analysis to study the efficacy of such methods. *When* do they converge? How *fast* do they converge? This requires an understanding of iteration and fixed point theory. See, for instance, Excursion L on Newton's Method.

You will see even more of the power of fixed point theory if you delve into the excursions on the implicit function theorem (Excursion M) and solutions to differential equations (Excursion O).

You are now ready for Excursion L:

Excursion L. *Newton's Method*

Synopsis: This excursion discusses the dynamics of the well-known root-finding method. It requires a good working knowledge of both Chapter 9 and Chapter 10.

Problems 10.2

1. **Straight-line contractions.**

   (a) Let a and b be real numbers. Suppose that $f : \mathbb{R} \to \mathbb{R}$ is given by $f(x) = ax + b$ and that it k-contracts distances. What, if anything, can be said about a and b?

   (b) Give an example of a straight-line function $f : [-1, 1] \to \mathbb{R}$ that satisfies the following conditions:
   - f contracts distances with constant $k < 1$.
   - The domain and range of f have a non-empty intersection.
   - f has no fixed point.

2. Consider the function $f(x) = \frac{1}{2}(x - 3)$ on $[0, 1]$.

   (a) Prove that f for $x, y \in [0, 1]$,
   $$|f(x) - f(y)| \leq \frac{1}{2}|x - y|.$$

   (b) Prove that f has no fixed point.

   (c) Why don't parts (a) and (b) of this problem contradict the contraction mapping theorem?

3. Some functions are not contractions, yet still have fixed points.

   (a) Any function $f : \mathbb{R} \to \mathbb{R}$ that has more than one fixed point cannot be a contraction. Give an example of such a function and justify the statement that it cannot be a contraction.

   (b) Give an example of a function $f : \mathbb{R} \to \mathbb{R}$ that has a unique fixed point but is not a contraction. Justify your answer.

4. **Contractions and differentiability.** Let $I \subseteq \mathbb{R}$ be an interval.

   (a) Let $f : I \to \mathbb{R}$ be a differentiable function. Suppose that
   $$|f'(x)| \leq k < 1 \quad \text{for all} \quad x \in I.$$
   Prove that f k-contracts distances. If, in addition, $f(I) \subseteq I$, we can conclude that f is a k-contraction.

   (b) Let $f : I \to \mathbb{R}$ be a differentiable function, and let p be a fixed point of f that lies in the interior of the interval I. Suppose that f' is continuous at p with $|f'(p)| < 1$. Show that there exists a closed interval J centered at p such that f is a contraction on J.

10.2 The Contraction Mapping Theorem

(c) Show by giving an example that a contraction $f : I \to I$ need not be differentiable.

(d) Suppose that $f : I \to \mathbb{R}$ is continuously differentiable and k-contracts distances. Show that $|f'(x)| \leq k$ for all $x \in I$.

5. In this problem you will complete the arguments needed to prove the contraction mapping theorem (Theorem 10.2.5). Let X be a metric space, and let $f : X \to X$ be a k-contraction.

   (a) Let $x \in X$. Prove that, for all $n \in \mathbb{N}$,
   $$d(f^n(x), f^{n+1}(x)) \leq k^n \, d(x, f(x))$$

   (b) Prove that if X is complete, then every iterated map on f converges. (*Hint*: You will need to use standard facts about geometric series to do this problem. You may use these facts without proof in this exercise, but—if you wish for further information—all necessary ideas are laid out in Excursion H.)

   (c) Using parts (a) and (b) as lemmas, write a proof of the contraction mapping theorem.

6. An alternative proof of the contraction mapping theorem in a metric space of finite diameter. Let X be a metric space with diameter $D < \infty$. Let $f : X \to X$ be a k-contraction of X.

   (a) Prove that the diameter of $f(X)$ is no more than kD.

   (b) Show that $X \supseteq f(X) \supseteq f^2(X) \supseteq f^3(X) \supseteq f^4(X) \ldots$.

   (c) What can you say about the diameters of X, $f(X)$, $f^2(X)$, $f^3(X)$, $f^4(X)$, ... ?

   (d) Use the facts proved in parts (a)–(c) to give an alternate proof of the contraction mapping theorem for a complete metric space of finite diameter.

7. In this problem you will prove Theorem 10.2.6. Along the way, you will obtain some independent, partial results.

   Let X be a metric space, and let $f : X \to X$ be a function. Let N be a fixed natural number.

   (a) Suppose that f^N has a unique fixed point $x^* \in X$. Prove that x^* is also a fixed point of f.

   (b) Assume that X is a complete metric space. Use the result from part (a) to prove that if f^N is a k-contraction of X, then f has a unique fixed point in X.

(c) Assume that x^* is a fixed point of f^N and that every iterated map on f^N converges to x^*. Prove that every iterated map on f converges to x^*.

(d) Use the results from parts (a)–(c) to help you prove Theorem 10.2.6.

8. Give an example of a function $f : \mathbb{R} \to \mathbb{R}$ that satisfies

$$d(f(x), f(y)) < d(x, y) \quad \text{for all} \quad x, y \in \mathbb{R}.$$

but has no fixed point.

> Problem 8 shows that the function given in the contraction mapping theorem must really contract distances by some uniform factor and not just shrink them. Alternatively, it demonstrates that the hypothesis of compactness is truly necessary in Theorem 10.2.7.

9. The proof of Theorem 10.2.7 is outlined below. Your job is to fill in the details and write a complete proof.

   Part 1. Showing that f has a unique fixed point.

   Step 1. Consider the function $F : X \to \mathbb{R}$ given by $F(x) = d(x, f(x))$. Prove that F is continuous. Conclude that F achieves a minimum value at some point x_0 in X.

   Step 2. Prove that x_0 is a fixed point of X.

   Step 3. Observe that the uniqueness of the fixed point follows by the same argument you used in Exercise 10.2.2.

   Part 2. Showing that every iterated map of f converges to x_0. Fix $y \in X$ and consider the iterated map $(f^n(y))$ based at y.

   Step 1. Show that the sequence

   $$d(y, x_0), d(f(y), x_0), d(f^2(y), x_0), d(f^3(y), x_0), \ldots$$

   is a decreasing sequence of real numbers that is bounded below. Deduce that this sequence must converge to some non-negative real number δ. We hope to show that $\delta = 0$.

   Step 2. Show that any convergent subsequence of $(f^n(y))$ converges to a point whose distance from x_0 is precisely δ.

   Step 3. Show that $(f^n(y))$ has at least one convergent subsequence, $(f^{n_i}(y))$.

Step 4. Prove that $(f^{n_i+1}(y))$ also converges. (What does it converge to?)

Step 5. Use the results from the previous steps to prove that $\delta = 0$.

$\square$

10.3 More on Finding Attracting Fixed Points

The contraction mapping theorem is a wonderful tool, but it often happens that not quite all of the hypotheses hold. For example, suppose the function contracts distances only on some subset of its domain. Can we restrict the domain of our function and obtain a contraction? Not necessarily.

Let S be a subset of a metric space X, and let $f : S \to X$ be a function that contracts distances. The function f need not have a fixed point, even if X is complete. After all, S and $f(S)$ might be completely disjoint! But even if S and $f(S)$ do overlap, there might not be a fixed point. (See Problem 1 at the end of Section 10.2.)

However, if $f : S \to X$ does have a fixed point that lies in the interior of its domain, the domain can be restricted so that we get a contraction. This is easy to see.

Exercise 10.3.1 Let X be a metric space and $S \subseteq X$. Let $f : S \to X$ be a function that contracts distances. Suppose p is a fixed point for f lying in the interior of S. Prove that there exists $r > 0$ such that f restricted to $\mathrm{C}_r(p)$ is a contraction. ∎

Exercise 10.3.1 doesn't help us much unless we know that f has a fixed point. Nevertheless, it does tell us that if there *is* a fixed point in the interior of S, then we should be able to restrict the domain of f to get a contraction.

If we have reason to suspect that f has a fixed point in the interior of S, how can we make use of this idea without knowing what the fixed point is? We will prove two theorems that will help us. But first we want to consider the intuition that underlies them.

Consider a contraction on a complete space. The contraction moves every point in the space toward the fixed point so that the distance is decreased by at least a factor of k. Thus points near the fixed point are moved very little, while points much farther away are moved more. If we can find a point that is "nearly fixed" by the contraction, we know it must be close to the fixed point.

When we say that a function f contracts distances, we place a strong restriction on the action of f. Let us make an elementary observation.

Exercise 10.3.2 Let X be a metric space. Suppose that $f : X \to X$ is a k-contraction. Let $b \in X$ and $r > 0$. Prove that $f(B_r(b)) \subseteq B_{kr}(f(b))$. (See Figure 10.8.) ∎

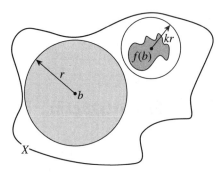

Figure 10.8 The dark gray region on the right of the diagram represents $f(B_r(b))$. If f is a k-contraction, then no point in this image can be farther from $f(b)$ than kr.

Continue to suppose that $f : X \to X$ is a k-contraction. As Figure 10.8 shows, if $f(b)$ is relatively far away from b, then $B_r(b)$ may be entirely disjoint from its image, precluding the existence of a fixed point in $B_r(b)$. However, the fact that $k < 1$ implies that $kr < r$. Thus, if $f(b)$ is fairly close to b, then the entire image of $B_r(b)$ will lie inside $B_r(b)$. (See Figure 10.9.)

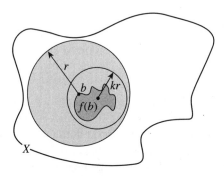

Figure 10.9 If a k-contraction f doesn't move b too much, then $f(B_r(b)) \subseteq B_r(b)$.

Theorem 10.3.3 Let X be a metric space, b be an element of X, and $k < 1$ be a real number. Suppose that we have a function $f : C_r(b) \to X$ that k-contracts distances, and assume that

$$d(b, f(b)) \leq r - kr.$$

Then $f(C_r(b)) \subseteq C_r(b)$.

If X is complete, it follows from the contraction mapping theorem that f has a unique fixed point in $C_r(b)$.

Proof: The proof is Problem 1 at the end of this section. □

A bit of fudging with strict inequalities allows us to show that a similar theorem holds if our domain is an open ball.

Corollary 10.3.4 Let X be a complete metric space and $b \in X$. Suppose that $f : B_r(b) \to X$ k-contracts distances for some real number $k < 1$. If $d(b, f(b)) < r(1 - k)$, then f has a unique fixed point in $B_r(b)$.

Proof: The proof is Problem 2 at the end of this section. □

Problems 10.3

1. Prove Theorem 10.3.3.

2. Use Theorem 10.3.3 to prove Corollary 10.3.4.

Chapter 11
The Riemann Integral

11.1 What Is Area?

The ancient Greeks were interested in areas. They knew how to compute the area of figures with straight edges. They wondered how to calculate the areas of figures with curved edges. They took it for granted that figures such as those shown in Figure 11.1 *have* areas and, therefore, that it makes sense to try to compute them; we would tend to agree. Modern integration theories are the mathematical descendants of such musings. Our interest is deepened because we now know that the search for area has far-reaching consequences that go beyond geometry.

To address area questions with mathematical precision, we need a usable definition. Strangely, creating such a definition turns out to be harder than we might expect. Naively, we proceed as follows. What is area? Area is a number; this number tells us how much "two-dimensional stuff" it takes to "fill up" a figure. This is a pretty long way from a mathematical definition, but look at the ingredients. Given a figure (a subset of the plane), we need to assign to it a

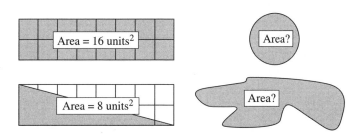

Figure 11.1 Finding Area.

non-negative real number—its area. This has the flavor of a function from the set of subsets of $\mathbb{R}^2$ to the real numbers. (This sort of function is sometimes called a **set function** because sets have numbers assigned to them.) To obtain a set function that gives area, we simply need to describe a *procedure* by which we can start with a figure and arrive at a real number that may reasonably be said to describe the figure's area. That is, we define area by computing it. We are right back to the questions posed by the Greeks!

Question to Ponder: The idea that we define area by computing it may seem circular at first glance. (That is, how can we compute what we can't define?) As you work through this chapter, you should consider whether you think there is a problem with circularity. Does the development seem correct to you from this standpoint?

11.2 The Riemann Integral

Modern integration theories have their origins in the attempt to make the concept of area mathematically precise. A number of different integration theories exist; we will study the Riemann integral.[1] Riemann's approach to the integral was modeled on the *method of exhaustion* used by the ancient Greeks.

From introductory calculus, we are already familiar with the following scheme for approximating the area under the graph of a positive function f on the interval $[a, b]$:

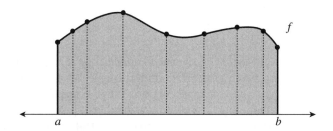

We divide $[a, b]$ into subintervals, which in turn divides the area under the graph into vertical strips. Under "suitable" conditions, each of these strips is approximately rectangular. Thus, we approximate the area of each strip by the area of an appropriate rectangle (which we *can* calculate). Finally, we approximate the area under the graph of f between a and b by adding up the areas of all the rectangles. We get an "exact value" for the area by means of an appropriate limiting process. This is all familiar territory, so we jump right in with the necessary definitions.

---

1. Other commonly used theories are the Riemann–Stieljes theory (see [AP3] or [RUD]), the Kurzweil–Henstock theory (also called the generalized Riemann integral; see [McL], and the Lebesgue theory (see [H&W] or [RUD]).

11.2 The Riemann Integral

Definition 11.2.1 Let a and b be real numbers with $a < b$. Then any set of the form
$$P = \{x_0, x_1, x_2, \ldots, x_{n-1}, x_n\}$$
that satisfies $a = x_0 < x_1 < x_2 < \cdots < x_{n-1} < x_n = b$ is called a **partition** of $[a, b]$. The intervals $[x_{i-1}, x_i]$ are called the **subintervals** of $[a, b]$ determined by P. The **mesh of the partition** is the length of its *longest* subinterval. It is denoted by $\|P\|$. That is:
$$\|P\| = \max_{1 \leq i \leq n} (x_i - x_{i-1}).$$

If f is a real-valued function whose domain contains the interval $[a, b]$, by a **Riemann sum for f corresponding to a given partition P**, we mean a sum of the form
$$\mathcal{R}(f, P) = \sum_{i=1}^{n} f(x_i^*)(x_i - x_{i-1})$$
where $x_{i-1} \leq x_i^* \leq x_i$ for each $i \leq n$.

The points x_i^* are sometimes called **sampling points** because they are the places at which the Riemann sum "samples" the value of the function for the subinterval (see Figure 11.2).

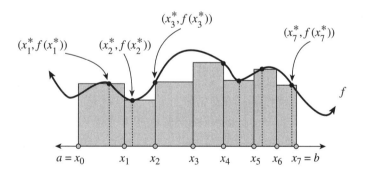

Figure 11.2 A Riemann sum with seven subintervals. In this example, $x_3^* = x_2$ and $x_4^* = x_4$. In all other cases, the sampling points are in the interiors of the subintervals.

Definition 11.2.2 Let $a, b \in \mathbb{R}$ with $a < b$. Let f be a real-valued function whose domain contains the interval $[a, b]$. We say that f is **Riemann integrable** on $[a, b]$ if there exists a real number I such that for all $\epsilon > 0$, there exists $\delta > 0$ such that $|\mathcal{R}(f, P) - I| < \epsilon$ whenever $\mathcal{R}(f, P)$ is a Riemann sum for f corresponding to a partition of $[a, b]$ of mesh less than δ. We denote I by $\int_a^b f$ and call it **the Riemann integral** of f over $[a, b]$.[2]

In keeping with the notation used for other limiting processes, we write

$$\lim_{\|P\| \to 0} \mathcal{R}(f, P) = \int_a^b f.$$

We read this expression as "the limit as the mesh of the partition goes to zero of the Riemann sums of f."

Terminology and Usage: If we say that f is a real-valued function that is Riemann integrable on the interval $[a, b]$, we are tacitly assuming that $[a, b]$ is a subset of the domain of f. It is not, in general, necessary to be more specific about the domain of f, and I won't be unless it is absolutely necessary. This will simplify the language and make our results applicable to as wide a class of functions as possible.

In Definition 11.2.2, we talk about *the* Riemann integral of f rather than *a* Riemann integral of f. This terminology is justified by Theorem 11.2.3.

Theorem 11.2.3 [Uniqueness of the Integral] Let a and b be real numbers with $a < b$. Suppose that f is a real-valued function that is a Riemann integrable on $[a, b]$. Then the Riemann integral of f is unique.

Proof: The proof is Problem 1 at the end of this section. □

Exercise 11.2.4 Let C be a real number. Suppose that f is the constant function $f(x) = C$. If our definition is doing its job, it certainly ought to be true that f is Riemann integrable and that $\int_a^b f = C(b-a)$. Show that this is, indeed, so. (*Hint*: Start by computing a Riemann sum. What happens?) ∎

Example 11.2.5 Let a and b be real numbers with $a < b$. Let $\xi \in [a, b]$. Let C be a non-zero real number. Set

$$f(x) = \begin{cases} 0 & \text{if } x \neq \xi \\ C & \text{if } x = \xi \end{cases}.$$

Then f is Riemann integrable and $\int_a^b f = 0$.

---

2. If the variable is unclear from the context, we may write out the full expression $\int_a^b f(x)\,dx$.

Proof: Let $\epsilon > 0$. Choose $\delta > 0$ such that
$$\delta < \frac{\epsilon}{2|C|}.$$

Let $P = \{x_0, x_1, x_2, \ldots, x_{n-1}, x_n\}$ be a partition of $[a,b]$ with $\|P\| < \delta$. Choose $x_1^*, x_2^*, \ldots, x_n^*$ arbitrarily such that $x_{i-1} \leq x_i^* \leq x_i$. Then
$$\mathcal{R}(f, P) - \sum_{i=1}^{n} f(x_i^*)(x_i - x_{i-1}).$$

All the terms in this summand are zero except those for which $x_i^* = \xi$. Thus $\mathcal{R}(f, P)$ can take on only three possible forms (see Figure 11.3.):

$$\mathcal{R}(f, P) = \begin{cases} 0 & \text{if } \xi \neq x_i^* \text{ for any } i \\ C(x_i - x_{i-1}) & \text{if } \xi = x_i^* \text{ for exactly one } i \\ C(x_{i+1} - x_i) + C(x_i - x_{i-1}) & \text{if } \xi = x_{i+1}^* = x_i^* \text{ for some } i \end{cases}.$$

In general, then,
$$|\mathcal{R}(f, P)| \leq 2|C|\,\|P\|.$$

Adding to this the fact that
$$\|P\| < \delta < \frac{\epsilon}{2|C|},$$

we conclude that
$$|\mathcal{R}(f, P)| < \epsilon.$$

□

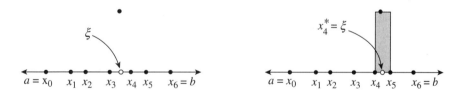

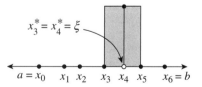

Figure 11.3 If f is non-zero only at one point, its Riemann sums can take on only one of three possible forms.

We have talked about integration theory as a means to define an "area function" from the subsets of $\mathbb{R}^2$ to $\mathbb{R}$. An interesting question arises in this context: Does every subset of the plane have an area, or are there subsets of $\mathbb{R}^2$ for which the concept of area is meaningless? Example 11.2.6 illustrates that, under Riemann's scheme, there are subsets of $\mathbb{R}^2$ to which we *cannot* assign an area.

Example 11.2.6 You will show in Problem 3 at the end of this section that the function f given by

$$f(x) = \begin{cases} 0 & \text{if } x \text{ is irrational} \\ 1 & \text{if } x \text{ is rational} \end{cases}$$

is *not* Riemann integrable on $[a, b]$.[3]

Remark: The function given in Example 11.2.6 is called the *characteristic function of the rational numbers*. More generally, if X is any set and $S \subseteq X$, then the characteristic function of S is the function $\chi_S : X \to X$ given by

$$\chi_S(x) = \begin{cases} 0 & \text{if } x \notin S \\ 1 & \text{if } x \in S \end{cases}.$$

There are still a couple of loose ends to tie up when it comes to defining the integral.

Definition 11.2.7 Let a and b be real numbers with $a < b$. Suppose f is a Riemann integrable function on $[a, b]$. Then for any $c \in [a, b]$, we define

$$\int_c^c f = 0.$$

Furthermore, we define

$$\int_b^a f = -\int_a^b f.$$

That is, the integral of f from b to a is defined if and only if the integral of f from a to b is defined and the integrals are related as shown above.

Finally, we give some preliminary information about the nature of Riemann integrable functions.

---

3. There are more general integration theories in which this function is integrable.

Theorem 11.2.8 Let a and b be real numbers with $a < b$. Suppose that f is a real-valued function that is Riemann integrable on $[a, b]$. Then f is bounded on $[a, b]$.

Proof: The proof is outlined in Problem 7 at the end of this section. $\square$

Theorem 11.2.9 [Translation Invariance of the Integral] Let a and b be real numbers with $a < b$. Let f be a real-valued function that is Riemann integrable on $[a, b]$. For a fixed $k \in \mathbb{R}$, let g_k be the function

$$g_k(x) = f(x + k),$$

where the domain of g is appropriately chosen so that x is in the domain of g_k if and only if $x + k$ is in the domain of f. Then g_k is Riemann integrable on $[a - k, b - k]$ and

$$\int_{a-k}^{b-k} g_k = \int_a^b f.$$

Proof: The proof is Problem 8 at the end of this section. $\square$

Exercise 11.2.10 Give a geometric interpretation of Theorem 11.2.9. ∎

You are now ready for Excursion I.

Excursion I. *Probing the Definition of the Riemann Integral*

Synopsis: On first glance, the definition of the Riemann integral may seem unnecessarily complicated. Why don't we just use regular partitions? Why do we allow for arbitrary sampling points? (For instance, why not just use the right endpoint or the left endpoint as our sampling point?) Excursion I examines this question and shows that our elementary notions of area really do require the more general definition. It is recommended as a supplement to this section.

Problems 11.2

1. Prove that the Riemann integral, when it exists, is unique (Theorem 11.2.3).

2. Suppose that c and d are distinct, non-zero real numbers. Fix $\xi \in [a,b]$. For $x \in [a,b]$, let
$$g(x) = \begin{cases} d & \text{if } x \neq \xi \\ c & \text{if } x = \xi \end{cases}.$$
Prove that g is Riemann integrable on $[a,b]$. What is $\int_a^b g$?

3. Let f be the characteristic function of the rational numbers. (See Example 11.2.6.)

    (a) Show that f is not Riemann integrable on $[0,1]$.

    (b) How would you need to modify your proof to show that the function is not Riemann integrable on an arbitrary interval $[a,b]$?

4. Let a, b, c, and d be real numbers with $a < c < d < b$. Let k be any non-zero real number. Define $f : \mathbb{R} \to \mathbb{R}$ as follows:
$$f(x) = \begin{cases} 0 & \text{if } x < c \\ k & \text{if } c \leq x \leq d \\ 0 & \text{if } d < x \end{cases}.$$

    (a) Prove that f is Riemann integrable on $[a,b]$. What is its integral?

    (b) How would your answer to part (a) change if every $<$ in the definition of f were changed to $\leq$, and vice versa?

5. Let a and b be real numbers with $a < b$. Suppose that f is Riemann integrable on $[a,b]$. Let ξ be any real number in the interval $[a,b]$, and let r be any real number different from $f(\xi)$. Define a function g as follows:
$$g(x) = \begin{cases} f(x) & \text{if } x \neq \xi \\ r & \text{if } x = \xi \end{cases}.$$
Prove that g is Riemann integrable on $[a,b]$ and that $\int_a^b g = \int_a^b f$.

6. Let D be a dense subset of $[a,b]$. Suppose that f is integrable on $[a,b]$ and $f(x) = 0$ for all $x \in D$. Prove that
$$\int_a^b f = 0.$$

7. This problem outlines a proof that Riemann integrable functions are bounded (Theorem 11.2.8). Your job is to fill in the details and write a complete, polished argument.

    Let a and b be real numbers with $a < b$. Suppose that f is a real-valued function whose domain contains the interval $[a, b]$. The proof is by contraposition, so assume throughout that f is unbounded on $[a, b]$.

    (a) Prove that there exists a sequence (y_n) in $[a, b]$ that converges to $y \in [a, b]$ such that for every $n \in \mathbb{N}$, $|f(y_n)| > n$.

    (b) Let P be any partition of $[a, b]$. Use the result from part (a) to prove that the set of Riemann sums of f corresponding to P is an unbounded set of real numbers.

    (c) Use the result from part (b) to show that f cannot be Riemann integrable on $[a, b]$.

8. Prove that the Riemann integral is translation invariant (Theorem 11.2.9).

9. Let a and b be real numbers with $a < b$. Let f be Riemann integrable on $[a, b]$. Let k be any real number, and let g be the function given by $g(x) = f(x) + k$.

    (a) Prove that g is Riemann integrable on $[a, b]$ and that
    $$\int_a^b g = k(b-a) + \int_a^b f.$$

    (b) Give a geometric interpretation of the result you proved in part (a).

11.3 Arithmetic, Order, and the Integral

Now that the definition is in place, we need to prove that the integral interacts in some useful ways with the arithmetic and order on the real numbers.

Theorem 11.3.1 [The Integral and Arithmetic] Let a, b, and k be real numbers. Suppose that $a < b$. Let f and g be real-valued functions that are Riemann integrable on the interval $[a, b]$. Then the following statements hold:

1. The function $f + g$ is Riemann integrable on the interval $[a, b]$, and
$$\int_a^b (f+g) = \int_a^b f + \int_a^b g.$$

2. The function kf is Riemann integrable on the interval $[a, b]$, and

$$\int_a^b kf = k \int_a^b f.$$

3. The function $f - g$ is Riemann integrable on the interval $[a, b]$, and

$$\int_a^b (f - g) = \int_a^b f - \int_a^b g.$$

Proof: The proof is Problem 1 at the end of this section. $\square$

Theorem 11.3.2 [The Integral and Order] Let a, b, m, and M be real numbers with $a < b$ and $m \leq M$. Let f and g be real-valued functions that are Riemann integrable on $[a, b]$. Then the following statements hold:

1. If $f(x) \geq 0$ for all $x \in [a, b]$, then $\int_a^b f \geq 0$.

2. If $f(x) \leq g(x)$ for all $x \in [a, b]$, then $\int_a^b f \leq \int_a^b g$.

3. If $m \leq f(x) \leq M$ for all $x \in [a, b]$, then

$$m(b - a) \leq \int_a^b f \leq M(b - a).$$

Proof: The proof is Problem 4 at the end of this section. $\square$

Problems 11.3

1. Prove Theorem 11.3.1, which relates the integral to arithmetic on $\mathbb{R}$.

2. Let g be the characteristic function of the irrational numbers.

$$g(x) = \begin{cases} 0 & \text{if } x \text{ is rational} \\ 1 & \text{if } x \text{ is irrational} \end{cases}$$

Use Theorem 11.3.1 to prove that g is not Riemann integrable.

3. Products of Riemann integrable functions:

    (a) Prove that there is no "product equivalent" for the first part of Theorem 11.3.1. Do this by giving two real-valued functions f and g that are Riemann integrable on a real interval $[a, b]$ for which it is not the case that
    $$\int_a^b fg = \int_a^b f \int_a^b g.$$

    (b) Find functions f and g, neither identically zero, and distinct real numbers a and b such that
    $$\int_a^b fg = \int_a^b f \int_a^b g.$$

4. Prove Theorem 11.3.2, which relates the integral to the order on $\mathbb{R}$.

5. Let a and b be real numbers with $a < b$. Let $f : [a, b] \to \mathbb{R}$ be Riemann integrable on $[a, b]$.

    (a) Suppose that $f(x) \geq 0$ for all $x \in [a, b]$ and that f is continuous on $[a, b]$. If $f > 0$ somewhere in $[a, b]$, prove that $\int_a^b f > 0$. (*Hint*: The second part of Theorem 11.3.2 and Problem 4 from Section 11.2 may be helpful to you.)

    (b) Is the statement in part (a) still true if f is not required to be continuous? Prove or give a counterexample.

6. Suppose that, as in Theorem 11.3.2, a and b are real numbers with $a < b$, and that f and g are real-valued functions that are Riemann integrable on $[a, b]$. Suppose further that $f(x) \leq g(x)$ for all $x \in [a, b]$ and that a strict inequality holds for at least one point in $[a, b]$.

    (a) Prove that $\int_a^b f < \int_a^b g$ need not hold.

    (b) Suppose that f and g are continuous functions. Is it, then, true that $\int_a^b f < \int_a^b g$? Prove or give a counterexample.

11.4 Families of Riemann Sums

Let $a, b \in \mathbb{R}$ with $a < b$; let $[a, b] \subseteq K \subseteq \mathbb{R}$; and let $f : K \to \mathbb{R}$ be a function. Given any partition $P = \{x_0, x_1, \ldots, x_n\}$ of $[a, b]$, we know that there are infinitely many different ways we can sample the subintervals to obtain a Riemann sum:
$$\mathcal{R}(f, P) = \sum_{i=1}^n f(x_i^*)(x_i - x_{i-1}).$$

Although we frequently use the notation $\mathcal{R}(f, P)$ to represent a single Riemann sum, $\mathcal{R}(f, P)$ is really an entire family of Riemann sums. This section explores these families of Riemann sums. What is the relationship between the various families? How do these families help us to determine whether or not the function f is integrable on $[a, b]$?

The Riemann "Envelope": Upper and Lower Sums

Suppose we consider two Riemann sums for f on the same partition P:

$$\sum_{i=1}^{n} f(x_i^*)(x_i - x_{i-1}) \quad \text{and} \quad \sum_{i=1}^{n} f(\tilde{x}_i)(x_i - x_{i-1}).$$

How far apart are they? If we assume that f is bounded (as are all Riemann integrable functions), we can give a partial answer to this question fairly easily.

Definition 11.4.1 Let $a, b \in \mathbb{R}$ with $a < b$; let $[a, b] \subseteq K \subseteq \mathbb{R}$; and let $f : K \to \mathbb{R}$ be bounded on $[a, b]$. Fix a partition $P = \{x_0, x_1, \ldots, x_n\}$ of $[a, b]$. Because f is bounded on $[a, b]$, it is bounded on each subinterval of $[a, b]$. Thus, for each $i \leq n$, we can define

$$M_i = \sup\{f(x) : x \in [x_{i-1}, x_i]\} \quad \text{and} \quad m_i = \inf\{f(x) : x \in [x_{i-1}, x_i]\}$$

Using these upper and lower bounds, we can define the **upper sum of f on P**

$$\mathcal{U}(f, P) = \sum_{i=1}^{n} M_i(x_i - x_{i-1}),$$

and the **lower sum of f on P**

$$\mathcal{L}(f, P) = \sum_{i=1}^{n} m_i(x_i - x_{i-1}).$$

$\mathcal{U}(f, P)$ and $\mathcal{L}(f, P)$ may or may not be Riemann sums.

Exercise 11.4.2

1. Give an example of a bounded function f on $[-1, 1]$ for which all upper and lower sums are Riemann sums.

2. Give an example of a bounded function f on $[-1, 1]$ and a partition P of $[-1, 1]$ for which neither the lower sum nor the upper sum is a Riemann sum.

Though they may not be Riemann sums, the upper and lower sums for f on P are upper and lower bounds for the family of Riemann sums of f on P.

Exercise 11.4.3 Let a and b be real numbers with $a < b$. Suppose that $[a,b] \subseteq K \subseteq \mathbb{R}$. Let $f : K \to \mathbb{R}$ be bounded on $[a,b]$, and let P be a partition of $[a,b]$. Let $\mathcal{R}(f, P)$ be any Riemann sum of f on P. Show that

$$\mathcal{L}(f,P) \leq \mathcal{R}(f,P) \leq \mathcal{U}(f,P).$$

∎

We can actually say more. If f is a function that is bounded on $[a,b]$, we can show that given any partition P of $[a,b]$, there are Riemann sums for f on P that are arbitrarily close to the upper sum for f on P and Riemann sums for f on P that are arbitrarily close to the lower sum for f on P.

Lemma 11.4.4 Let a and b be real numbers with $a < b$. Let $[a,b] \subseteq K \subseteq \mathbb{R}$, and suppose that $f : K \to \mathbb{R}$ is bounded on $[a,b]$. Let P be a partition of $[a,b]$, and let $\epsilon > 0$. Then there exist Riemann sums $\mathcal{R}_1(f,P)$ and $\mathcal{R}_2(f,P)$ such that

$$\mathcal{U}(f,P) \leq \mathcal{R}_1(f,P) + \epsilon \quad \text{and} \quad \mathcal{L}(f,P) \geq \mathcal{R}_2(f,P) - \epsilon.$$

Proof: The proof is Problem 1 at the end of this section. □

Thus we can think of the upper and lower sums as an upper and lower "envelope" for the family of Riemann sums on P. The interval between them gives us the range of values in the family $\mathcal{R}(f, P)$.

Intuition and Aims: If f is an integrable function and $\|P\|$ is very small, the family of Riemann sums of f on P will cluster about the number $\int_a^b f$. The family will, therefore, give a small range of values; in particular, the difference between the upper and the lower sums should be tiny. Our major goal in this section is to show that the converse of this statement is also true. If a small enough mesh for P implies an arbitrarily small range of values for the family of Riemann sums on P (or, equivalently, a very small difference between the upper and lower sums), then f is Riemann integrable.

Refinements

Before we proceed with this project, we need to consider some related issues. Suppose we start with a partition $P = \{x_0, x_1, x_2, \ldots, x_n\}$ of an interval $[a,b]$. Next, we throw in some extra points to get another "finer" partition Q of $[a,b]$:

$$Q = \{x_0, y_1^1, y_2^1, \ldots, y_{k_1}^1, x_1, y_1^2, y_2^2, \ldots, y_{k_2}^2, x_2, \ldots, x_{n-1}, y_1^n, y_2^n, \ldots, y_{k_n}^n, x_n\}.$$

Definition 11.4.5 Let $a, b \in \mathbb{R}$ with $a < b$. Let P and Q be partitions of $[a, b]$. If $P \subseteq Q$, then Q is said to be a **refinement** of P.

Let f be a real-valued function that is bounded on an interval $[a, b]$, and $P \subseteq Q$ partitions of $[a, b]$. What is the relationship between a Riemann sum of f on P and a Riemann sum of f on Q? For the sake of argument, suppose that we are dealing with a "well-behaved" function that is positive everywhere on $[a, b]$ so that the Riemann sums approximate the actual area between the graph of f and the x-axis. Now, Q gives us more and narrower subintervals than P. Thus our intuition tells us that a Riemann sum of f on Q should be a better approximation to the area under the graph of f on $[a, b]$ than a Riemann sum on P. Strictly speaking, this statement is false, as you will show in Problem 2 at the end of this section, but it does carry the germ of an important idea and to that idea we turn next.

Although an individual Riemann sum on a refinement Q of P does not necessarily improve on our estimate of the integral of f by a Riemann sum on P. Theorem 11.4.6 shows that the range of values of the family $\mathcal{R}(f, Q)$ is contained in the range of values of $\mathcal{R}(f, P)$. Thus, if these are approximating a single number (the integral of f), the *family* $\mathcal{R}(f, Q)$ will provide a "tighter" approximation than the *family* $\mathcal{R}(f, P)$.

Theorem 11.4.6 Let a and b be real numbers with $a < b$. Let $[a, b] \subseteq K \subseteq \mathbb{R}$, and let $f : K \to \mathbb{R}$ be bounded on $[a, b]$. If P is a partition of $[a, b]$ and Q is a refinement of P, then

$$\mathcal{L}(f, P) \leq \mathcal{L}(f, Q) \leq \mathcal{U}(f, Q) \leq \mathcal{U}(f, P).$$

Proof: The proof is Problem 3 at the end of this section. $\square$

Cauchy Criteria for the Existence of the Integral

Earlier, we said that we will be considering what happens to the range of values in $\mathcal{R}(f, P)$ as $\|P\|$ shrinks to zero. We hope to say that if a small mesh for P guarantees a small range of values for $\mathcal{R}(f, P)$, then f must be integrable. This is true, but we will have to work fairly hard to prove it. Nevertheless, the result will be a versatile and easy-to-use "Cauchy criterion" for integrability. The hard work here will pay off with nice theoretical results later. We begin with two technical lemmas. The proof of the first is fairly straightforward and is left to you.

Lemma 11.4.7 Let a and b be real numbers with $a < b$. Let $[a, b] \subseteq K \subseteq \mathbb{R}$, and let $f : K \to \mathbb{R}$ be a function. Let P be a partition of $[a, b]$, and let $\epsilon > 0$

be a real number. Suppose that whenever $\mathcal{R}_1(f,P)$ and $\mathcal{R}_2(f,P)$ are Riemann sums of f on P, then
$$|\mathcal{R}_1(f,P) - \mathcal{R}_2(f,P)| < \epsilon.$$
Several facts follow more or less immediately from this assumption:

1. f is bounded on $[a,b]$.

2. $\mathcal{U}(f,P) - \mathcal{L}(f,P) \leq \epsilon$.

3. If Q is any refinement of P, and $\mathcal{R}(f,P)$ and $\mathcal{R}(f,Q)$ are Riemann sums, then
$$|\mathcal{R}(f,P) - \mathcal{R}(f,Q)| \leq \epsilon.$$

Proof: The proof is Problem 5 at the end of this section. □

This next lemma may seem a bit strange at first, but it is the technical piece on which the proof of the Cauchy criterion hinges. Be sure you have a picture of the result in your mind before you proceed to the proof. Several readings of the statement may be necessary before you fully appreciate what it is saying.

Lemma 11.4.8 Suppose that $a < b$ with $[a,b] \subseteq K \subseteq \mathbb{R}$, and that the function $f : K \to \mathbb{R}$ is bounded on $[a,b]$. Let $\epsilon > 0$, and let $n \in \mathbb{N}$. Then there exists $\delta > 0$ such that if $P = \{x_0, x_1, x_2, \ldots, x_n\}$ is any partition of $[a,b]$ with $n+1$ elements (n subintervals), and Q is any partition of $[a,b]$ with $\|Q\| < \delta$, then
$$0 \leq \mathcal{U}(f,Q) - \mathcal{U}(f,Q^*) < \epsilon \text{ and } 0 \leq \mathcal{L}(f,Q^*) - \mathcal{L}(f,Q) < \epsilon$$
where $Q^* = Q \cup P$.

Proof: Suppose that $M > 0$ such that $|f(x)| < M$ for all $x \in [a,b]$. Fix $\epsilon > 0$ and $n \in \mathbb{N}$. Note that if $n = 1$, the theorem is trivial (*Why?*), so we assume this is not the case. Choose
$$\delta < \frac{\epsilon}{4(n-1)M}.$$

Fix P, Q, and Q^* as required by the theorem. Write $Q^* = \{z_0, z_1, z_2, \ldots, z_m\}$.

Given that $Q \subseteq Q^*$, Theorem 11.4.6 tells us that $0 \leq \mathcal{U}(f,Q) - \mathcal{U}(f,Q^*)$. We need to show that this difference is no larger than ϵ.

To this end, suppose that $i < j$ and that the partition points $z_i, z_j \in Q$, but the partition points $z_{i+1}, z_{i+2}, z_{i+3}, \ldots, z_{j-1}$ are in $P \setminus Q$. Then $[z_i, z_j]$ is a subinterval given by the partition Q, and the contribution made by this subinterval to $\mathcal{U}(f,Q)$ is
$$N^*(z_j - z_{i-1}) \text{ where } N^* = \sup\{f(x) : x \in [z_i, z_j]\}.$$

The intervals $[z_i, z_{i+1}], [z_{i+1}, z_{i+2}], [z_{i+2}, z_{i+3}], \ldots, [z_{j-1}, z_j]$ are subintervals given by Q^*. For each $k \in \mathbb{N}$ such that $i < k \leq j$, we set

$$M_k = \sup\{f(x) : x \in [z_{k-1}, z_k]\}.$$

Then the contribution made by these subintervals to $\mathcal{U}(f, Q^*)$ is

$$M_{i+1}(z_{i+1} - z_i) + M_{i+2}(z_{i+2} - z_{i+1}) + \ldots + M_j(z_j - z_{j-1}).$$

The difference between these two contributions will be zero if $i+1 = j$, because the two will be the same. If $i+1 < j$, it will be

$$(N^* - M_{i+1})(z_{i+1} - z_i) + (N^* - M_{i+2})(z_{i+2} - z_{i+1}) + \cdots + (N^* - M_j)(z_j - z_{j-1}).$$

This is the contribution made to $\mathcal{U}(f, Q) - \mathcal{U}(f, Q^*)$ by the interval $[z_i, z_j]$. For each of these terms, we have

$$|(N^* - M_k)(z_k - z_{k-1})| \leq (|N^*| + |M_k|)|z_k - z_{k-1}|$$
$$\leq 2M\delta$$

because $\delta > \|Q\| \geq \|Q^*\|$.

How many terms do we have in $\mathcal{U}(f, Q) - \mathcal{U}(f, Q^*)$? The "worst-case scenario" occurs when we have exactly $n - 1$ elements in $P \setminus Q$ and we never have more than one element of $P \setminus Q$ between two elements of Q. This gives no more than $2(n - 1)$ non-zero terms in $\mathcal{U}(f, Q) - \mathcal{U}(f, Q^*)$. (*Be sure you see why each of these things holds!*) Each of the terms in $\mathcal{U}(f, Q) - \mathcal{U}(f, Q^*)$ is smaller than $2M\delta$ in absolute value. Therefore,

$$0 \leq \mathcal{U}(f, Q) - \mathcal{U}(f, Q^*) \leq 2(n - 1)(2M\delta) = 4(n - 1)M\delta < \epsilon.$$

A parallel argument shows that (for this same choice of δ),

$$0 \leq \mathcal{L}(f, Q^*) - \mathcal{L}(f, Q) < \epsilon.$$

□

Theorem 11.4.9 Let $K \subseteq \mathbb{R}$. Suppose that $a < b$ and that $[a, b] \subseteq K$. Let $f : K \to \mathbb{R}$ be a function. Then the integrability of f on $[a, b]$ is equivalent to the following statement:

Cauchy Criterion for the Existence of the Integral: For every $\epsilon > 0$, there exists $\delta > 0$ such that if P is a partition of $[a, b]$ with $\|P\| < \delta$ and $\mathcal{R}_1(f, P)$ and $\mathcal{R}_2(f, P)$ are Riemann sums of f on P, then

$$|\mathcal{R}_1(f, P) - \mathcal{R}_2(f, P)| < \epsilon.$$

Proof

$\Longrightarrow$ The proof that if f is Riemann integrable, then its Riemann sums satisfy the Cauchy criterion, is straightforward and is left to you as Problem 6 at the end of this section.

$\Longleftarrow$ We assume that the Riemann sums for f satisfy the Cauchy criterion. Then f is bounded on $[a,b]$ by Lemma 11.4.7, part 1, so all results requiring the boundedness of f on $[a,b]$ apply. Choose a sequence (δ_n) of positive real numbers with δ_n corresponding to $\epsilon = \frac{1}{n}$ in the Cauchy criterion given on the preceding page. Next, find a sequence (P_n) of partitions of $[a,b]$ such that for all $n \in \mathbb{N}$, $\|P_n\| < \delta_n$ and P_{n+1} is a refinement of P_n. Finally, obtain a sequence $\mathcal{R}(f, P_n)$ of Riemann sums corresponding to the partitions (P_n). (The sampling points for these sums may be chosen arbitrarily.) Because $P_1 \subseteq P_2 \subseteq P_3 \subseteq P_4 \subseteq \ldots$, Lemma 11.4.7, part 3, tells us that the sequence of Riemann sums is a Cauchy sequence of real numbers and thus convergent. Suppose that it converges to x.

Let $\epsilon > 0$. Choose $N_1 \in \mathbb{N}$ such that if $n > N_1$, then $|\mathcal{R}(f, P_n) - x| < \frac{\epsilon}{4}$. Next, choose an integer $N > N_1$ such that $\frac{1}{N} < \frac{\epsilon}{4}$. Fix n to be the number of partition points in P_N. Finally, choose $\delta > 0$ that satisfies the conclusions of Lemma 11.4.8 for n and $\frac{\epsilon}{4}$. Let Q be any partition of mesh less than δ, and set $Q^* = Q \cup P_N$. If we can show that for any Riemann sum $\mathcal{R}(f,Q)$ of f on Q

$$|\mathcal{R}(f,Q) - x| < \epsilon,$$

then we will be able to conclude that $x = \int_a^b f$.

To this end, fix arbitrary Riemann sums $\mathcal{R}(f,Q), \mathcal{R}(f,Q^*)$. Then

$$|\mathcal{R}(f,Q) - x| \leq |\mathcal{R}(f,Q) - \mathcal{R}(f,Q^*)| + |\mathcal{R}(f,Q^*) - \mathcal{R}(f,P_N)| + |\mathcal{R}(f,P_N) - x|,$$

where $\mathcal{R}(f, P_N)$ is the Riemann sum chosen earlier in the proof.

Consider the last two terms of this sum:

$$|\mathcal{R}(f, P_N) - x| < \frac{\epsilon}{4} \text{ because } N > N_1.$$

Recall that any two Riemann sums of f on P_N are within $\frac{1}{N}$ of each other and that $\frac{1}{N} < \frac{\epsilon}{4}$. Because Q^* is a refinement of P_N, Lemma 11.4.7, part 3, tells us that

$$|\mathcal{R}(f, Q^*) - \mathcal{R}(f, P_N)| \leq \frac{\epsilon}{4}.$$

We will be finished if we can show that $|\mathcal{R}(f,Q) - \mathcal{R}(f,Q^*)| < \frac{\epsilon}{2}$.

Let us consider the cases $\mathcal{R}(f,Q) \geq \mathcal{R}(f,Q^*)$ and $\mathcal{R}(f,Q) < \mathcal{R}(f,Q^*)$ separately. For both cases, you should keep in mind three important facts:

- Q^* is a refinement of P_N, so

$$\mathcal{U}(f,Q^*) - \mathcal{L}(f,Q^*) \leq \mathcal{U}(f,P_N) - \mathcal{L}(f,P_N).$$

- Any two Riemann sums of f on P_N are within $\frac{\epsilon}{4}$ of one another, so Lemma 11.4.7, part 2, tells us that

$$\mathcal{U}(f, P_N) - \mathcal{L}(f, P_N) \leq \frac{\epsilon}{4}.$$

- Q and Q^* satisfy the conclusions of Lemma 11.4.8 for $\frac{\epsilon}{4}$, so

$$\mathcal{U}(f, Q) - \mathcal{U}(f, Q^*) < \frac{\epsilon}{4} \text{ and } \mathcal{L}(f, Q^*) - \mathcal{L}(f, Q) < \frac{\epsilon}{4}.$$

Suppose first that $\mathcal{R}(f, Q) \geq \mathcal{R}(f, Q^*)$. Then

$$\begin{aligned}
|\mathcal{R}(f, Q) - \mathcal{R}(f, Q^*)| &= \mathcal{R}(f, Q) - \mathcal{R}(f, Q^*) \\
&\leq \mathcal{U}(f, Q) - \mathcal{L}(f, Q^*) \\
&= [\mathcal{U}(f, Q) - \mathcal{U}(f, Q^*)] + [\mathcal{U}(f, Q^*) - \mathcal{L}(f, Q^*)] \\
&< \frac{\epsilon}{4} + [\mathcal{U}(f, P_N) - \mathcal{L}(f, P_N)] \\
&\leq \frac{\epsilon}{4} + \frac{\epsilon}{4} = \frac{\epsilon}{2}.
\end{aligned}$$

The case when $\mathcal{R}(f, Q) < \mathcal{R}(f, Q^*)$ is fairly similar:

$$\begin{aligned}
|\mathcal{R}(f, Q) - \mathcal{R}(f, Q^*)| &= \mathcal{R}(f, Q^*) - \mathcal{R}(f, Q) \\
&\leq \mathcal{U}(f, Q^*) - \mathcal{L}(f, Q) \\
&= [\mathcal{U}(f, Q^*) - \mathcal{L}(f, Q^*)] + [\mathcal{L}(f, Q^*) - \mathcal{L}(f, Q)] \\
&< [\mathcal{U}(f, P_N) - \mathcal{L}(f, P_N)] + \frac{\epsilon}{4} \\
&\leq \frac{\epsilon}{4} + \frac{\epsilon}{4} = \frac{\epsilon}{2}.
\end{aligned}$$

$\square$

The following alternative Cauchy criterion involving only upper and lower sums follows easily from Theorem 11.4.9.

Corollary 11.4.10 Let $K \subseteq \mathbb{R}$. Suppose that $a < b$ and that $[a, b] \subseteq K$. Let $f : K \to \mathbb{R}$ be a function. Then f is integrable on $[a, b]$ if and only if the following statement holds:

Upper Sum–Lower Sum Criterion for the Existence of the Integral: For every $\epsilon > 0$, there exists $\delta > 0$ such that if P is a partition of $[a, b]$ with $\|P\| < \delta$, then

$$|\mathcal{U}(f, P) - \mathcal{L}(f, P)| < \epsilon$$

Proof: The proof is Problem 7 at the end of this section. $\square$

The Cauchy criterion for the existence of the integral allows us to prove that a function is Riemann integrable without knowing ahead of time what the integral will be. It is immediately helpful in proving a large class of integrability theorems.

Problems 11.4

1. Prove that given a partition P of $[a, b]$ and a function f that is bounded on $[a, b]$, there exist Riemann sums of f on P that are arbitrarily close to the upper and lower sums (Lemma 11.4.4).

2. Suppose that P and Q are partitions of the interval $[a, b]$ and that Q is a refinement of P. In this problem you will show that, contrary to intuition, a Riemann sum of a function f on P can be a better estimate for the area under the graph of f than a Riemann sum of f on Q. Give an example with the following components:

   - A real-valued, integrable function f defined on some interval $[a, b]$
   - Two partitions P and Q of $[a, b]$ such that Q is a refinement of P
   - A Riemann sum $\mathcal{R}(f, P)$ on P and a Riemann sum $\mathcal{R}(f, Q)$ on Q such that $\mathcal{R}(f, P)$ is a better approximation for the integral of f than is $\mathcal{R}(f, Q)$

   The goal of this problem is for you to understand how this situation can occur. A proof is not necessary, but a clear, appropriately labeled diagram is.

3. Suppose that f is bounded on $[a, b]$ and that P and Q are partitions of $[a, b]$. Prove that if Q is a refinement of P, then the family $\mathcal{R}(f, Q)$ provides a "tighter" set of values than the family $\mathcal{R}(f, P)$ (Theorem 11.4.6).

4. Let a and b be real numbers with $a < b$. Let $K \subseteq \mathbb{R}$, and suppose that $[a, b] \subseteq K$. Let $f : K \to \mathbb{R}$ be bounded on $[a, b]$. Prove that if P and Q are any two partitions of $[a, b]$ then $\mathcal{L}(f, P) \leq \mathcal{U}(f, Q)$.

5. Prove Lemma 11.4.7. (*Hint*: For the first provision, see the first couple of parts of Problem 7 at the end of Section 11.2.)

6. Prove the first provision of Theorem 11.4.9: If f is Riemann integrable, then its Riemann sums satisfy the Cauchy criterion for the existence of the integral.

7. Prove the upper sum–lower sum criterion for the existence of the integral (Corollary 11.4.10).

11.5 Existence of the Integral

In Example 11.2.6, we saw that there are some subsets of the plane to which the Riemann integral does not assign an area. Given this limitation, it is reasonable to wonder about conditions on a plane region that will guarantee our ability to compute its area. An important subquestion might be phrased something like this: Suppose that f is a real-valued function defined at every point of $[a,b]$. What condition or conditions on a function will guarantee its integrability? The Cauchy criterion allows us to vigorously attack this problem and fairly quickly gives us a number of useful theorems.

Theorem 11.5.1 Let a and b be real numbers with $a < b$. Suppose that $[a,b] \subseteq K \subseteq \mathbb{R}$. Suppose also that $f : K \to \mathbb{R}$ is continuous on $[a,b]$. Then f is Riemann integrable on $[a,b]$.

Proof: The proof is Problem 1 at the end of this section. $\square$

Although you have seen many examples of discontinuous functions that are integrable, most of them haven't strayed too far from continuous functions. For instance, most have had a few isolated jump discontinuities or something similar. You may be getting the impression that integrable functions have to be fairly well behaved in this sense. Not necessarily, as Theorem 11.5.2 and the example that follows it show.

Theorem 11.5.2 Let a and b be real numbers with $a < b$. Suppose that $[a,b] \subseteq K \subseteq \mathbb{R}$. Suppose also that $f : K \to \mathbb{R}$ is monotonic on $[a,b]$. Then f is Riemann integrable on $[a,b]$.

Proof: The proof is Problem 2 at the end of this section. $\square$

Example 11.5.3 Take every rational number in $[0,1]$ and write it as a quotient of integers in lowest terms. Define a function f as follows:

$$f(x) = \begin{cases} 0 & \text{if } x \text{ is irrational} \\ \frac{1}{q} & \text{if } x \text{ is rational and } x = \frac{p}{q} \end{cases}.$$

This function is discontinuous at every rational number and continuous at every irrational number. In Problem 5 at the end of this section, you will be asked to show this and to show that this function is Riemann integrable on $[0,1]$ with integral zero.

Theorem 11.5.4 Let $a, b, c \in \mathbb{R}$. Let I be the smallest closed interval containing a, b, and c, and suppose that $I \subseteq K \subseteq \mathbb{R}$. If $f : K \to \mathbb{R}$ is a function, then the integral $\int_a^b f$ exists if and only if the integrals $\int_a^c f$ and $\int_c^b f$ exist.

In this case,
$$\int_a^b f = \int_a^c f + \int_c^b f.$$

Proof: The proof of this theorem is outlined for you in Problem 4. □

It is not, in general, true that the composition of two Riemann-integrable functions is Riemann integrable. (See Problem 6.) However, we have a slightly less strong theorem that is still very useful.

Theorem 11.5.5 [Composition Theorem for Integrals] Let a and b be real numbers with $a < b$. Let f be a real-valued function that is Riemann integrable on $[a, b]$. Suppose $f([a, b]) \subseteq S$, and let $h : S \to \mathbb{R}$ be uniformly continuous and bounded on $f([a, b])$. Then $h \circ f$ is Riemann integrable on $[a, b]$.

Proof: Let $\epsilon > 0$. Using the uniform continuity of h on $f([a, b])$, choose $\delta^* > 0$ such that

$$\text{if } |f(x) - f(y)| < \delta^*, \text{ then } |h(f(x)) - h(f(y))| < \frac{\epsilon}{2(b-a)}.$$

Also choose $K > 0$ such that $|h(z)| < K$ for all $z \in f([a, b])$. Now use the Riemann integrability of f on $[a, b]$ to choose δ such that given any partition P of $[a, b]$ with mesh less than δ,

$$\mathcal{U}(f, P) - \mathcal{L}(f, P) < \frac{\delta^* \epsilon}{4K}.$$

That said, we will fix an arbitrary partition $P = \{x_0, x_1, \ldots, x_n\}$ with mesh less than δ. For each $i \in \mathbb{N}$ such that $1 \leq i \leq n$, define

$$M_i = \sup\{f(x) : x \in [x_{i-1}, x_i]\} \quad \text{and} \quad m_i = \inf\{f(x) : x \in [x_{i-1}, x_i]\}.$$

For use later, we define two disjoint subsets A and B of $T = \{1, 2, \ldots, n\}$ as follows:

$$A = \{i \in T : M_i - m_i < \delta^*\} \quad \text{and} \quad B = \{i \in T : M_i - m_i \geq \delta^*\}.$$

Note that for $i \in A$, if $x, y \in [x_{i-1}, x_i]$, then $|f(x) - f(y)| < \delta^*$. It then follows that

$$|h(f(x)) - h(f(y))| < \frac{\epsilon}{2(b-a)}.$$

Now that the necessary notation is in place, consider the difference between Riemann sums for $h \circ f$ on P:

$$\left| \sum_{i=1}^{n} h \circ f(x_i^*)(x_i - x_{i-1}) - \sum_{i=1}^{n} h \circ f(\tilde{x}_i)(x_i - x_{i-1}) \right|$$

$$= \left| \sum_{i=1}^{n} [h(f(x_i^*)) - h(f(\tilde{x}_i))](x_i - x_{i-1}) \right|$$

$$= \left| \sum_{i \in A} [h(f(x_i^*)) - h(f(\tilde{x}_i))](x_i - x_{i-1}) + \sum_{i \in B} [h(f(x_i^*)) - h(f(\tilde{x}_i))](x_i - x_{i-1}) \right|$$

$$\leq \sum_{i \in A} |h(f(x_i^*)) - h(f(\tilde{x}_i))| |x_i - x_{i-1}| + \sum_{i \in B} |h(f(x_i^*)) - h(f(\tilde{x}_i))| |x_i - x_{i-1}|$$

$$< \frac{\epsilon}{2(b-a)} \sum_{i \in A} (x_i - x_{i-1}) + 2K \sum_{i \in B} (x_i - x_{i-1})$$

$$\leq \frac{\epsilon}{2} + 2K \sum_{i \in B} (x_i - x_{i-1}).$$

We will be done if we can show that

$$\sum_{i \in B} (x_i - x_{i-1}) < \frac{\epsilon}{4K}.$$

Here we go:

$$\sum_{i \in B} (x_i - x_{i-1}) = \frac{1}{\delta^*} \sum_{i \in B} \delta^*(x_i - x_{i-1})$$

$$\leq \frac{1}{\delta^*} \sum_{i \in B} (M_i - m_i)(x_i - x_{i-1})$$

$$\leq \frac{1}{\delta^*} \sum_{i=1}^{n} (M_i - m_i)(x_i - x_{i-1})$$

$$\leq \frac{1}{\delta^*} (\mathcal{U}(f, P) - \mathcal{L}(f, P))$$

$$< \frac{1}{\delta^*} \left(\frac{\delta^* \epsilon}{4K} \right) = \frac{\epsilon}{4K}.$$

□

Theorem 11.5.5 is sometimes stated in a slightly less general form which is often applied in practice. Because continuous functions on compact domains are both uniformly continuous and bounded, Corollary 11.5.6 follows immediately from Theorem 11.5.5. The two theorems are so closely related that we can refer to them interchangeably as "the composition theorem for integrals"; the context should make it clear to which version we are referring.

Corollary 11.5.6 [Composition Theorem for Integrals] Let a and b be real numbers with $a < b$. Let f be a real-valued function that is Riemann integrable on $[a, b]$. Suppose that $f([a, b]) \subseteq S$ and that $f([a, b])$ is compact. Let $h : S \to \mathbb{R}$ be continuous on $f([a, b])$. Then $h \circ f$ is Riemann integrable on $[a, b]$.

The composition theorem immediately allows us to expand our knowledge of integrable functions.

Theorem 11.5.7 Let a and b be real numbers with $a < b$. Let f and g be real-valued functions that are Riemann integrable on $[a, b]$. Then fg is Riemann integrable on $[a, b]$.

Proof: The proof is Problem 7a at the end of this section. □

Theorem 11.5.8 Let a and b be real numbers with $a < b$. Let f be a real-valued function that is Riemann integrable on $[a, b]$. Then $|f|$ is also Riemann integrable on $[a, b]$ and

$$\left| \int_a^b f \right| \leq \int_a^b |f|$$

Proof: The proof is Problem 8 at the end of this section. □

Problems 11.5

1. Use the Cauchy criterion for integrability to prove that continuous functions are integrable (Theorem 11.5.1).

2. Use the Cauchy criterion for integrability to prove that monotonic functions are integrable (Theorem 11.5.2).

3. Let a, b, c, and d be real numbers with $a < c < d < b$, and let r be any real number. Suppose that f is a real-valued function that is Riemann integrable on $[a, b]$. Define a function g as follows:

$$g(x) = \begin{cases} f(x) & \text{if } x \in [c, d] \\ r & \text{if } x \notin [c, d] \end{cases}$$

Prove that g is Riemann integrable on $[a, b]$. (*Hint*: You do *not* want to try to do this by directly estimating Riemann sums.)

4. This problem helps you to prove Theorem 11.5.4.

   (a) Let a, b, c, K, and f be as in the statement of the theorem with the added assumption that $a < c < b$. Suppose that f is integrable on $[a, b]$. You wish to prove that f is integrable on $[a, c]$ and $[c, b]$. Fill in the details of the following outline.

   Step 1. Let $\epsilon > 0$. Use the fact that f is integrable on $[a, b]$ and the Cauchy criterion for integrals to get a δ corresponding to ϵ. (*Looking ahead*: In the end you will show that, with this value of δ, f satisfies the Cauchy criterion on $[a, c]$ and $[c, b]$.)

   Step 2. Let P_1 and P_2 be partitions of $[a, c]$ and $[c, b]$, respectively, each with mesh less than δ. Note that $P = P_1 \cup P_2$ is a partition of $[a, b]$ with mesh less than δ.

   Step 3. Start by considering
   $$|\mathcal{R}_1(f, P_1) - \mathcal{R}_2(f, P_1)|.$$
   Then add and subtract the right quantity to turn this expression into a comparison of two Riemann sums of f on P.

   Step 4. Repeat the procedure for $|\mathcal{R}_1(f, P_2) - \mathcal{R}_2(f, P_2)|$.

   (b) Let a, b, c, K, and f be as in the statement of the theorem with the added assumption that $a < c < b$. Suppose that f is integrable on $[a, c]$ and on $[c, b]$. You are going to use the definition of the integral directly to show that if P is a partition of $[a, b]$ of sufficiently small mesh, then any Riemann sum of f on P is within ϵ of $\int_a^c f + \int_c^b f$. Fill in the details in the following outline.

   Step 1. Because f is integrable on $[a, c]$ and on $[c, b]$, f is bounded on each of those intervals and therefore on $[a, b]$. Let M be an upper bound for $|f|$ on $[a, b]$.

   Step 2. Let $\epsilon > 0$. Use the integrability of f on $[a, c]$ and $[c, b]$ to find δ_1 corresponding to $\frac{\epsilon}{3}$ on $[a, c]$ and δ_2 corresponding to $\frac{\epsilon}{3}$ on $[c, b]$.

   Step 3. Choose $\delta < \min(\delta_1, \delta_2, \eta)$, where η is a quantity depending on ϵ and M. I will leave you to discover the value of η in the course of your calculations. (You will need to write your final proof including that value for η, of course!)

   Step 4. Let $P = \{x_0, x_1, x_2, \ldots, x_n\}$ be a partition of $[a, b]$ of mesh less than δ.

   Step 5. You will want to consider the size of
   $$\left| \mathcal{R}(f, P) - \left(\int_a^c f + \int_c^b f \right) \right|$$

If $c \notin P$, then you will need to add and subtract some carefully chosen terms to insert c (as a partition point) in the Riemann sum. This will allow you to break it up into two Riemann sums for f: one on $[a, c]$ and one on $[c, b]$. You should be able to do this in such a way that any remaining "scraps" are as small as you need them to be. (You may be tempted to break the proof into two cases—one when $c \in P$ and one when $c \notin P$—but this isn't really necessary. A judicious use of $\leq$ rather than $<$ will allow you to subsume the first case into the second.)

(c) Now use the results from parts (a) and (b) to remove the restriction that $a < b < c$. (You may need to break this part of the proof into several cases.)

5. This problem concerns the function f defined in Example 11.5.3.

$$f(x) = \begin{cases} 0 & \text{if } x \text{ is irrational} \\ \frac{1}{q} & \text{if } x \text{ is rational and } x = \frac{p}{q} \text{ in "lowest terms"} \end{cases}$$

(a) Show that f is discontinuous at every rational number.

(b) Fix $t \in \mathbb{R}$ and $n \in \mathbb{N}$. Show that f maps only finitely many elements of $(t - \frac{1}{2}, t + \frac{1}{2})$ to $\frac{1}{n}$. (What is the maximum number?)

(c) Use the result from part (b) to prove that f is continuous at every irrational number.

(d) Let $\epsilon > 0$. Prove that f maps only finitely many elements of $[0, 1]$ to a value greater than ϵ. (What is the maximum number?)

(e) Prove that f is Riemann integrable on $[0, 1]$ and that

$$\int_0^1 f = 0.$$

6. Let f be the function defined in Example 11.5.3 (and again in Problem 5). Let $g : [0, 1] \to \mathbb{R}$ be given by

$$g(x) = \begin{cases} 0 & \text{if } x = 0 \\ 1 & \text{if } x \in (0, 1] \end{cases}.$$

Show that g and f are both integrable on $[0, 1]$ but that the composition $g \circ f$ is not.

7. Products of Riemann-integrable functions revisited:

    (a) Prove Theorem 11.5.7. (*Hint*: You will want to use Theorem 11.5.5 and think about the (algebraic) relationship between the functions $(f+g)^2$, $(f-g)^2$, and fg.)

    (b) Let a and b be real numbers with $a < b$. Let f be a real-valued function that is Riemann integrable on $[a,b]$. Prove that for all $n \in \mathbb{N}$, f^n is integrable on $[a,b]$ (where f^n is the nth power of f).

8. Prove that the absolute value of an integrable function is integrable (Theorem 11.5.8). (*Hint*: For the inequality, note that $f \leq |f|$ and $-f \leq |f|$.)

9. Consider the restriction that $a < b$ in Theorem 11.5.8. What happens if we remove this restriction? Formulate and prove a generalization that doesn't include this restriction.

10. Let a and b be real numbers with $a < b$. Let f be a real-valued function that is Riemann integrable on $[a,b]$. Suppose that there exists $\delta > 0$ such that $f(x) \geq \delta$ for all $x \in [a,b]$. Prove that $1/f$ is also Riemann integrable on $[a,b]$.

11.6 The Fundamental Theorem of Calculus

Theorem 11.5.1 tells us that we can define an "area accumulation function" for a function that is continuous on an interval.

Definition 11.6.1 Let $a \in \mathbb{R}$, let I be an interval containing a, and let $f : I \to \mathbb{R}$ be a function. Suppose that $\int_a^x f$ exists for all $x \in I$. Define the function $F_a : I \to \mathbb{R}$ by

$$F_a(x) = \int_a^x f(t)\, dt.$$

F_a is called the **area accumulation function of f on I based at a**.

Theorem 11.6.2 [Fundamental Theorem of Calculus] Let $a \in \mathbb{R}$, and let I be an interval containing a. Let $I \subseteq K \subseteq \mathbb{R}$. Suppose that the function $f : K \to \mathbb{R}$ is integrable on $[a, x]$ for all $x \in I$ with $x > a$ and on $[x, a]$ for all $x \in I$ with $x < a$. Let F_a be the area accumulation function for f on I based at a. Then F_a is uniformly continuous on I.

If f happens to be continuous on I, then F_a is differentiable on I and $F_a'(x) = f(x)$ for all $x \in I$.

Proof: The proof is Problem 2 at the end of this section. □

Corollary 11.6.3 [Integral Evaluation Theorem] Let a and b be real numbers with $a < b$. Let $[a,b] \subseteq K \subseteq \mathbb{R}$, and let $f : K \to \mathbb{R}$ be a continuous function. If $F : K \to \mathbb{R}$ is an antiderivative for f on $[a,b]$, then

$$\int_a^b f(t)\,dt = F(b) - F(a).$$

Proof: The proof is Problem 3 at the end of this section. □

Problems 11.6

1. Show by giving an example that it is possible for an area accumulation function to be continuous without being differentiable. Justify your answer.

2. In this problem you will prove the fundamental theorem of calculus (Theorem 11.6.2).

   (a) Prove that the area accumulation function F_a for f on I based at a is uniformly continuous on I. (In this proof you will, of course, assume only that f is integrable, not necessarily continuous.)

   (b) Suppose that f is continuous on I. To show that F_a is differentiable at the point c, you must consider the appropriate difference quotient:

   $$\frac{F_a(x) - F_a(c)}{x - c} = \frac{\int_a^x f(t)\,dt - \int_a^c f(t)\,dt}{x - c}$$
   $$= \frac{1}{x - c}\int_c^x f(t)\,dt.$$

   You will need to show that

   $$\lim_{x \to c} \frac{1}{x - c}\int_c^x f(t)\,dt = f(c).$$

   (*Hint:* $f(c) = \dfrac{1}{x - c}\int_c^x f(c)\,dt$.)

3. Prove the integral evaluation theorem (Corollary 11.6.3).

Chapter 12
Sequences of Functions

You may be familiar with the concept of a power series.[1] In the study of power series, we think about approximating a given function f more and more closely by means of a sequence of polynomials of increasing degree. If conditions are "sufficiently nice," the power series converges to the function f.

Thinking along these lines, we encounter an apparently different notion of sequence convergence. The terms of the sequence and its limit are all *functions*, rather than points.[2] What might we mean when we say that a sequence of functions converges to another function? Suppose we start with the most naive of definitions.

12.1 Pointwise Convergence

Definition 12.1.1 Let X and Y be metric spaces. For $n = 1, 2, 3, \ldots$, let $f_n : X \to Y$ be a sequence of functions. If $x \in X$, we say that the sequence (f_n) **converges at** x if the sequence of points $(f_n(x))$ converges in Y.

We say that the sequence (f_n) **converges pointwise** if it converges at each point of X.

Suppose that (f_n) converges pointwise and that $f : X \to Y$ satisfies

$$f(x) = \lim_{n \to \infty} f_n(x) \text{ for all } x \in X.$$

Then we say that (f_n) converges pointwise to f, and we write $f = \lim_{n \to \infty} f_n$ or $f_n \to f$.

---

1. If not, don't worry about it for now. Power series are the subject of Excursion J.

2. It is possible to view much of what we consider in this chapter as ordinary sequence convergence in an abstract metric space called $C(K)$—the set of all continuous functions on a compact metric space K. However, that viewpoint obscures some of the intuition and is helpful only in retrospect, so we won't begin by thinking in those terms. The space $C(K)$ is the subject of Excursion N.

What does this newly defined limiting process represent intuitively? What, if anything, can we say about the relationship between the functions (f_n) and the limit function f?

Example 12.1.2 Figures 12.1 and 12.2 depict two sequences of real-valued functions $f_n : [0, 1] \to \mathbb{R}$.

1. Figure 12.1 shows the sequence given by

$$f_n(x) = x\left(1 - \frac{1}{n}\right).$$

Because $\frac{1}{n}$ goes to zero as n goes to infinity, $\lim_{n \to \infty} f_n(x) = x$ for all $x \in [0, 1]$; $f(x) = x$ is the pointwise limit of the sequence (f_n).

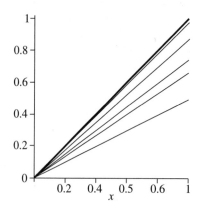

Figure 12.1 $f_n(x) = x\left(1 - \frac{1}{n}\right)$ converges pointwise to $f(x) = x$.

2. Now let $f_n(x) = x^n$ (see Figure 12.2). Then

$$f(x) = \lim_{n \to \infty} f_n(x) = \lim_{n \to \infty} x^n = \begin{cases} 0 & \text{when } 0 \leq x < 1 \\ 1 & \text{when } x = 1 \end{cases}.$$

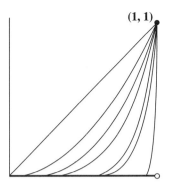

Figure 12.2 $f_n(x) = x^n$ converges pointwise to a discontinuous function!

For both examples, each function in the sequence is continuous. In the first example, so is the limiting function. Interestingly, the limiting function in the second example has a "jump" discontinuity. This immediately tells us something important: Even if we start with a sequence of continuous functions, we cannot guarantee that the limit function will share that property—a most inconvenient fact. How is the second example different from the first?

Let's go back to our original question. What is really going on in the process of pointwise convergence? One way to envision what it means for a sequence of real functions to converge pointwise is to picture a bunch of vertical lines superimposed on a graph showing the functions in the sequence (Figure 12.3). If we think of each vertical line as a copy of the real line, we can interpret the places where the line intersects the graphs of the f_n's as a sequence of points in $\mathbb{R}$. There is one such sequence for every vertical line. If all of these real number sequences converge, then the sequence (f_n) converges pointwise.

As you can see in Figure 12.3, the various sequences converge at different rates. The key difference between the first example and the second one is that in Figure 12.1, one of the sequences is converging the most slowly of all—the sequence on the line $x = 1$. In Figure 12.2, as we move to the right in the interval $(0, 1)$, the sequences converge more and more slowly—there is no sequence that is converging more slowly than all the others.

The fact that all of the functions in the second example go through the point $(1, 1)$ forces the limit to go through that same point. All points to the left of 1, however, yield sequences that go to zero. They just converge more and more slowly as the evaluation point moves to the right toward $x = 1$. This ultimately causes a "tear" in the limit function.

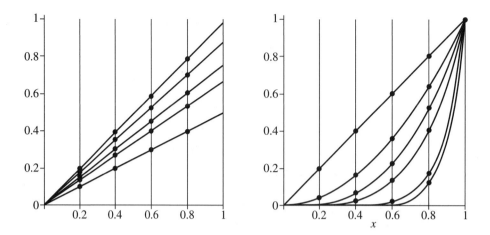

Figure 12.3 If all the vertical sequences converge, the sequence of functions converges pointwise.

12.2 Uniform Convergence

As noted when we first began our discussion of sequence convergence, for a fixed $\epsilon > 0$, some convergent sequences may be within ϵ of the limit after the 10th term, while others may not reach that level of tolerance for 10,000 or more terms. These differences don't affect the convergence of the sequence— they affect only the *rate of convergence*. That is, it makes sense to say that the sequences that require a larger N for a given ϵ are those that converge more slowly. The discussion at the end of Section 12.1 should lead you to conclude that the condition we are looking for will guarantee a minimum convergence rate for the pointwise sequences. The upshot is that for a given $\epsilon > 0$, there should be a single N that will work for all the pointwise sequences.

Definition 12.2.1 Let X and Y be metric spaces. Consider a sequence of functions $f_n : X \to Y$. We say that the sequence (f_n) **converges uniformly** to $f : X \to Y$ if for every $\epsilon > 0$, there exists $N \in \mathbb{N}$ such that if $n > N$ and $x \in X$, then $d(f(x), f_n(x)) < \epsilon$.

Clearly, uniform convergence implies pointwise convergence. The reverse is not true, however.

Example 12.2.2 The first sequence of functions given in Example 12.1.2 converges uniformly; the second does not. Problem 1 at the end of this section asks you to verify this explicitly.

Geometrically speaking, uniform convergence says that if we make a vertical "collar" of radius ϵ around the graph of the limit function f, the graphs of all functions in the sequence after the Nth one must lie within the "collar." This situation is illustrated in Figure 12.4.

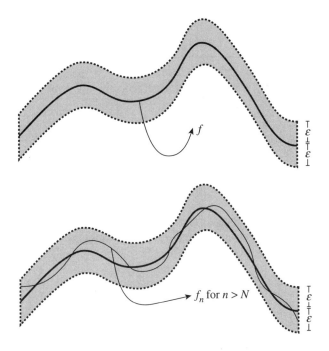

Figure 12.4 Uniform convergence. All functions after the Nth one must stay within the gray collar. (The collar appears to be narrower in some places than in others because our eyes want to measure the band moving out *perpendicularly*, rather than vertically, from the curve.)

Theorem 12.2.3 The uniform limit of continuous functions is continuous. That is, if X and Y are metric spaces, and $f_n : X \to Y$ is a sequence of continuous functions that converges uniformly to $f : X \to Y$, then f is also continuous.

Proof: The proof is Problem 6 at the end of this section. $\square$

Sometimes it is useful to have a criterion for determining whether a sequence of functions converges in the absence of a known limit. This point brings us to a Cauchy criterion for the uniform convergence of sequences of functions.

Definition 12.2.4 Let X and Y be metric spaces. Let (f_n) be a sequence of continuous functions from X to Y. We say that (f_n) is **uniformly Cauchy on X** if for every $\epsilon > 0$, there exists $N \in \mathbb{N}$ such that for all $m > N$, all $n > N$, and all $x \in X$,

$$d(f_n(x), f_m(x)) < \epsilon$$

Theorem 12.2.5 [Uniform Cauchy Criterion for Sequences of Functions] Let X be a metric space, and let Y be a complete metric space. If (f_n) is a sequence of functions from X to Y that is uniformly Cauchy on X, then (f_n) converges uniformly on X.

Proof: Because (f_n) is uniformly Cauchy, given $\epsilon > 0$, there exists $N \in \mathbb{N}$ such that for all $n > N$, $m > N$, and for all $x \in X$,

$$d(f_n(x), f_m(x)) < \epsilon$$

In particular, the sequence $(f_n(x))$ is Cauchy in Y for all $x \in X$. Because Y is complete, all such sequences converge. Thus we can define the pointwise limit of the sequence (f_n), $F: X \to Y$ as follows:

$$F(x) = \lim_{n \to \infty} f_n(x)$$

The claim is that (f_n) converges to F uniformly.

To establish this claim, let $\epsilon > 0$. Choose $N \in \mathbb{N}$ such that for all $n > N$, $m > N$, and $x \in X$,

$$d(f_n(x), f_m(x)) < \frac{\epsilon}{2}$$

Now fix any $x_0 \in X$. We know that $f_k(x_0) \to F(x_0)$, so we can find $m > N$ such that

$$d(f_m(x_0), F(x_0)) < \frac{\epsilon}{2}$$

Finally, fix any $n > N$.

$$d(f_n(x_0), F(x_0)) \le d(f_n(x_0), f_m(x_0)) + d(f_m(x_0), F(x_0))$$
$$< \frac{\epsilon}{2} + \frac{\epsilon}{2} = \epsilon$$

Given that the choice of N was independent of the choice of x_0, (f_n) converges to F uniformly. $\square$

Problems 12.2

In the following problems, all functions go from $\mathbb{R}$ to $\mathbb{R}$ unless otherwise specified. Feel free to use computer graphs to guide your intuition and illustrate your solution. However, your answers must ultimately be justified analytically.

1. This problem refers back to Example 12.1.2.

    (a) Show that the sequence of functions
    $$f_n(x) = x\left(1 - \frac{1}{n}\right)$$
    converges uniformly on $[0, 1]$.

    (b) Show that the sequence of functions
    $$f_n(x) = x^n$$
    does not converge uniformly on $[0, 1]$.

2. Say as much as you can about the convergence and the uniform convergence of the sequence of functions
$$f_n(x) = \frac{x^2}{1 + nx^4}.$$

3. Say as much as you can about the convergence and the uniform convergence of the sequence of functions
$$f_n(x) = \frac{x^{2n}}{1 + x^2}.$$

4. Discuss the convergence and uniform convergence of the sequence of functions
$$f_n(x) = \frac{x^n}{1 + x^n} \quad \text{on } [0, \infty).$$
In the case of non-uniform convergence on $[0, \infty)$, find subintervals on which the convergence is uniform. Also, discuss the continuity of the limit function.

5. Discuss the convergence and uniform convergence of the sequence of functions
$$f_n(x) = \frac{x^{2n}}{1 + x^{4n}}.$$
In the case of non-uniform convergence, find subintervals on which the convergence is uniform. Also, discuss the continuity of the limit function.

6. Prove that the uniform limit of continuous functions is continuous (Theorem 12.2.3).

7. Theorem 12.2.3 tells us that a sequence of continuous functions that converges uniformly has a continuous limit.

    (a) *It is possible for a sequence of discontinuous functions to converge uniformly to a continuous function.* Find a sequence of real functions that is discontinuous at every real number but that nevertheless converges uniformly to a continuous function.

    (b) *If a sequence of continuous functions converges pointwise but not uniformly, it may still be the case that the limit is continuous.* Consider the sequence
    $$f_n(x) = e^{-(x-n)^2}.$$
    Show that (f_n) converges to the function that is constantly zero but that it does so non-uniformly.

8. Prove the converse of the uniform Cauchy criterion. Let X and Y be metric spaces. Suppose that (f_n) is a sequence of functions from X to Y that is uniformly convergent to $F : X \to Y$. Prove that (f_n) is uniformly Cauchy on X.

9. Consider the sequence of functions
    $$f_n(x) = \frac{x}{1 + n^2 x^2}.$$

    (a) Show that f_n converges uniformly to a function f.

    (b) Say what you can about the correctness of the statement
    $$f'(x) = \lim_{n \to \infty} f'_n(x)?$$

10. **Uniform convergence and boundedness.** Let (f_n) be a sequence of real-valued continuous functions on a metric space X. Suppose that (f_n) converges uniformly to a function f.

    (a) Prove that if all terms of the sequence (f_n) are bounded, then f is bounded. Give an example to show that the pointwise limit of real-valued bounded functions need not be bounded.

    (b) Let $K \in \mathbb{R}$. Suppose that f is bounded with $|f(x)| \leq K$ for all $x \in X$. Show that for all $\epsilon > 0$ there exists $N \in \mathbb{N}$ such that
    $$|f_n(x)| \leq K + \epsilon \text{ for all } x \in X \text{ and for all } n > N.$$

    This result motivates the following **definition**: Let X be a metric space. Let (f_n) be a sequence of real-valued functions on X. The sequence (f_n)

is said to be **uniformly bounded** if there exists a positive real number M such that

$$|f_n(x)| \leq M \text{ for all } n \in \mathbb{N} \text{ and for all } x \in X.$$

Note that part (b) tells us that if (f_n) is a uniformly convergent sequence of real-valued functions on X converging to a bounded function, then a tail of the sequence is uniformly bounded.

  (c) Use the results from parts (a) and (b) to prove that if all terms of the sequence (f_n) are bounded, then (f_n) is uniformly bounded.

  (d) Suppose that f is bounded away from zero. That is, suppose that there exists $K \in \mathbb{R}^+$ such that $|f(x)| \geq K$ for all $x \in X$. Show that there exists $N \in \mathbb{N}$ such that

$$|f_n(x)| \geq \frac{K}{2} \text{ for all } x \in X \text{ and all } n > N.$$

When such a condition holds, we say that the sequence is **uniformly bounded away from zero**.

11. **Uniform convergence and arithmetic.** Let (f_n) and (g_n) be sequences of real-valued functions on a metric space X. Suppose that $f_n \to f$, $g_n \to g$, and that the convergence is uniform in each case.

  (a) Let $k \in \mathbb{R}$. Prove that (kf_n) converges uniformly to kf.

  (b) Prove that $(f_n + g_n)$ converges uniformly to $f + g$.

  (c) Prove that if f and g are bounded functions, then $(f_n g_n)$ converges uniformly to the function fg. (*Hint*: Problem 10b may be helpful to you.)

  (d) Suppose that g is bounded away from zero. Prove that $\dfrac{1}{g_n} \to \dfrac{1}{g}$ uniformly. (*Hint*: Problem 10d may be of use to you.)

  (e) Suppose that f is a bounded function and that g is bounded away from zero. Prove that $\dfrac{f_n}{g_n} \to \dfrac{f}{g}$ uniformly.

12. Let (f_n) be a sequence of continuous functions which converges uniformly to a function f on a subset S of a metric space X. Let x_n be a sequence of points in S that converges to some point $x \in S$. The for all $m \in \mathbb{N}, \lim_{n \to \infty} f_n(x_m) = f(x_m)$. Furthermore, the continuity of f implies that $\lim_{m \to \infty} = f(x)$. Can we somehow get "the best of both worlds"?

  (a) Prove that $\lim_{n \to \infty} f_n(x_n) = f(x)$.

(b) Is the converse of part (a) true? That is, suppose (f_n) is a sequence of continuous functions on a metric space X, and that for every sequence of points (x_n) in S that converges to some point $x \in S$, $\lim_{n \to \infty} f_n(x_n) = f(x)$. Is it true that (f_n) converges uniformly to f? (*Note*: It is trivial to see that (f_n) converges pointwise to f. Why?)

12.3 Series of Functions

When we are thinking about approximating a complicated real-valued function f with simpler functions, it is often convenient to start with a naive approximation and then to refine our approximation by adding on a "correction term." To improve our result, we add on yet another "correction term," getting a third approximation that is even better, and so on. In effect, our approach is to approximate some function f by a sequence of functions:

$$f_1, \quad f_1 + f_2, \quad f_1 + f_2 + f_3, \quad f_1 + f_2 + f_3 + f_4, \ldots$$

Each function is a more precise approximation than the last. (Taylor polynomials are an excellent example of this approach; see Section 9.7.[3]) Ultimately, we hope that the sequence of sums will converge to f.

Definition 12.3.1 Let X be a metric space. Let $f_n : X \to \mathbb{R}$ be a sequence of functions. For each $n \in \mathbb{N}$ and for each $x \in X$, define

$$S_n(x) = \sum_{i=1}^{n} f_i(x).$$

This new sequence of functions from X to $\mathbb{R}$ is denoted by $\sum_{n=1}^{\infty} f_n$ and is called a **series of functions**. The terms of the sequence (S_n) are called the **partial sums of the series**.[4] As you might expect, if the sequence (S_n) converges pointwise or uniformly to f, we say that the series $\sum_{n=1}^{\infty} f_n$ converges pointwise or uniformly to f.

---

[3]. The limiting Taylor *series* are examples of power series, which have very nice properties and are discussed in Excursion J.

[4]. We read $\sum_{n=1}^{\infty} f_n$ as "the sum from 1 to infinity of f_n" or as "the sum of the f_n from 1 to infinity."

Remark: When we have convergence of a series of functions to some function f, we use the notation
$$\sum_{n=1}^{\infty} f_n$$
to refer to both the sequence (S_n) and to its limit f. The context usually makes it clear which meaning is intended, so the ambiguity is not ordinarily a problem in practice. Nevertheless, it is important to keep the dual usage in mind so as not to become confused.

As you might expect, many theorems about the convergence of sequences of functions translate easily into theorems about the convergence of series of functions.

Exercise 12.3.2 Let (f_n) be a sequence of continuous functions. Suppose that
$$\sum_{n=1}^{\infty} f_n$$
converges uniformly to some function f. Show that f is continuous. ∎

In our discussion, we will refer tangentially to series of real numbers. If (a_n) is a sequence of real numbers, we can define the partial sums of the series $\sum_{n=1}^{\infty} a_n$ as we did for series of functions. The series converges to a real number S provided that
$$S = \lim_{n \to \infty} \sum_{i=1}^{n} a_i$$
More detailed information on series of real numbers can be found in Excursion H.

There is a very useful criterion due to Karl Weierstrass, for showing that a series of functions converges uniformly. First we need a lemma.

Lemma 12.3.3 [Uniform Cauchy Criterion for Series of Functions] Let X be a metric space, and let $f_n : X \to \mathbb{R}$ be a sequence of real-valued functions. Suppose that for all $\epsilon > 0$, there exists $N \in \mathbb{N}$ such that for all $m > n > N$ and for all $x \in X$,
$$\left| \sum_{i=n}^{m} f_i(x) \right| < \epsilon.$$
Then the series $\sum_{i=1}^{\infty} f_i$ converges uniformly.

Proof: The proof is Problem 1 at the end of this section. □

Theorem 12.3.4 [Weierstrass M-Test] Let X be a metric space, and let $f_n : X \to \mathbb{R}$ be a sequence of real-valued functions. Suppose there is a sequence (M_n) of positive real numbers such that for each $n \in \mathbb{N}$ and each $x \in X$,

$$|f_n(x)| \leq M_n.$$

If $\sum_{n=1}^{\infty} M_n < \infty$, then the series $\sum_{n=1}^{\infty} f_n$ converges uniformly.

It is enlightening to think about the Weierstrass test in terms of our original motivation. We began by picturing the successive terms in a series of functions as "correction" terms. Suppose we are trying to construct a uniformly convergent series of functions. We begin by fixing a convergent series of positive real numbers—for example, $\sum_{n=1}^{\infty} 2^{-n}$. The Weierstrass test then tells us that all we have to do is make sure that the absolute value of the nth correction term is uniformly smaller than 2^{-n}; our correction terms need to stay within a rapidly shrinking collar about the x-axis. If they do, then the convergence of the resulting series is guaranteed to be uniform.

Problems 12.3

1. Prove the Cauchy criterion for convergence of series of functions (Lemma 12.3.3).

2. Prove that the Weierstrass M-test works (Theorem 12.3.4).

12.4 Interchange of Limit Operations

In Section 12.1, we showed that it was possible for a sequence of continuous functions to converge pointwise to a discontinuous function. In Section 12.2, we showed that if a sequence of continuous functions converges *uniformly*, the limit will be continuous as well. It is useful to think about this issue in a more general context: that of the interchange of limit operations.

If we hope to have the continuity of the f_n's "carry over" to f, what exactly are we asking? Remember that f is continuous at a if and only if

$\lim_{x \to a} f(x) = f(a)$. Using the fact that f is the pointwise limit of the f_n's, this statement can be rewritten as follows:

$$\lim_{x \to a} \lim_{n \to \infty} f_n(x) = \lim_{n \to \infty} f_n(a).$$

Using the continuity of the f_n's, this expression in turn gives

$$\lim_{x \to a} \lim_{n \to \infty} f_n(x) = \lim_{n \to \infty} \lim_{x \to a} f_n(x).$$

The continuity of the limit function requires that we be able to interchange the order of two limit operations!

With this in mind, let us reexamine the second case in Example 12.1.2:

$$\lim_{x \to 1} \lim_{n \to \infty} f_n(x) = \lim_{x \to 1} f(x) = 0.$$

$$\lim_{n \to \infty} \lim_{x \to 1} f_n(x) = \lim_{n \to \infty} 1 = 1.$$

Even though both "double limits" exist, they are not equal. The order in which we take the limits cannot be interchanged!

Many formal manipulations, which may tempt us at one time or another, amount to interchanging the order in which two limiting operations are performed. Ordinarily, we have to be very careful in such situations because this step often leads to incorrect conclusions.

Here is another example that shows why interchanging the order of two limiting processes can be a dicey proposition.

Example 12.4.1 The limit of a sequence of integrals is not always the integral of the limit. Consider the following sequence of real-valued functions defined on $[0,1]$.

$$f_n(x) = \begin{cases} 4n^2 x & \text{if } 0 \leq x \leq \frac{1}{2n} \\ 4n - 4n^2 x & \text{if } \frac{1}{2n} < x \leq \frac{1}{n} \\ 0 & \text{if } \frac{1}{n} < x \leq 1 \end{cases}.$$

Some representative graphs are shown in Figure 12.5.

You will show in Problem 2 at the end of this section that (f_n) converges pointwise to the zero function and that

$$\lim_{n \to \infty} \int_0^1 f_n = 1.$$

In other words,

$$\lim_{n \to \infty} \int_0^1 f_n \neq \int_0^1 \lim_{n \to \infty} f_n.$$

250 Chapter 12 ■ Sequences of Functions

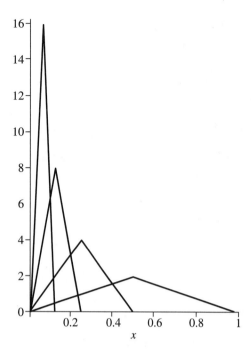

Figure 12.5 The limit of the integrals is not the integral of the limit.

We can vary this example a little to get a sequence of functions that goes to zero, yet whose integrals go to ∞ (see Problem 3 at the end of this section). In other words, one limit may exist without the other.

We cannot solve all possible mysteries associated with the interchange of limits in the next few pages. However, there are some things that we *can* say about this complicated issue. We know already that the uniform limit of continuous functions is continuous. Uniform convergence can give us a boost in other situations as well.

Theorem 12.4.2 Let $a, b \in \mathbb{R}$ with $a < b$, and let (f_n) be a uniformly convergent sequence of continuous functions on $[a, b]$. Then, if $f = \lim f_n$, f is integrable and

$$\int_a^b f(x)\, dx = \lim_{n \to \infty} \int_a^b f_n(x)\, dx$$

Proof: The proof is Problem 4a at the end of this section. □

This immediately yields a corresponding result about series of functions.

12.4 Interchange of Limit Operations

Corollary 12.4.3 Let $a, b \in \mathbb{R}$ with $a < b$, and let $\sum_{i=1}^{\infty} f_i$ be a uniformly convergent series of continuous functions on $[a, b]$. Then if $f(x) = \sum_{i=1}^{\infty} f_i(x)$, f is integrable and

$$\int_a^b f(x)\,dx = \int_a^b \sum_{i=1}^{\infty} f_i(x)\,dx = \sum_{i=1}^{\infty} \int_a^b f_i(x)\,dx.$$

Proof: The proof is Problem 4b at the end of this section. □

Similar questions can be asked about differentiation of sequences and series. If we have a function that is the limit of a series of functions, under what circumstances can we differentiate the series term by term and get a series whose limit is the derivative of the limit function for the original series? That is, if $f(x) = \sum_{i=1}^{\infty} f_i(x)$, under what circumstances can we say the following:

$$\frac{d}{dx} f(x) = \sum_{i=1}^{\infty} \frac{d}{dx} f_n(x)?$$

The answer is that sometimes we can (see Excursion J on power series) and sometimes we can't.

The question about the term-by-term differentiation of a series can be rephrased as a question about limits of sequences of functions. Suppose that (f_n) is a sequence of differentiable functions that converges pointwise to f. Under what circumstances can we guarantee that f is differentiable and (more importantly) that

$$\lim_{n \to \infty} f'_n(x) = f'(x)?$$

Unfortunately, Problem 9b on page 244 tells us that even uniform convergence of the sequence (f_n) does not allow us to reach this conclusion! Luckily, something similar does.

Theorem 12.4.4 Let (g_n) be a sequence of real-valued functions defined on an open interval (a, b) in $\mathbb{R}$. Suppose that each g_n has a continuous derivative. If the sequence (g'_n) converges uniformly on (a, b), and for some $z \in (a, b)$ the sequence $(g_n(z))$ converges, then (g_n) converges pointwise to some differentiable function g and

$$g'(x) = \lim_{n \to \infty} g'_n(x) \text{ for all } x \in (a, b)$$

Proof: Because g'_n is continuous, we can integrate it, and the fundamental theorem of calculus tells us that (for the z mentioned in the hypothesis and any $x \in (a,b)$)

$$\int_z^x g'_n(t)\, dt = g_n(x) - g_n(z).$$

Let $\mathcal{G}(x) = \lim_{n \to \infty} g'_n(x)$. Then by Theorem 12.4.2, we know that $\mathcal{G}$ is integrable and

$$\int_z^x \mathcal{G}(t)\, dt = \lim_{n \to \infty} \int_z^x g'_n(t)\, dt.$$

In particular, $\lim_{n \to \infty} (g_n(x) - g_n(z))$ exists. Now $\lim_{n \to \infty} g_n(z)$ exists by hypothesis, so $\lim_{n \to \infty} g_n(x)$ exists for all $x \in (a,b)$ because the sum of two convergent sequences is convergent (Theorem 5.3.1).

Define $g(x) = \lim_{n \to \infty} g_n(x)$. Thus

$$\int_z^x \mathcal{G}(t)\, dt = \lim_{n \to \infty} \int_z^x g'_n(t)\, dt$$
$$= \lim_{n \to \infty} g_n(x) - \lim_{n \to \infty} g_n(z)$$
$$= g(x) - g(z).$$

Finally, we know that $\mathcal{G}$ is continuous (because it is the uniform limit of a sequence of continuous functions), so the fundamental theorem of calculus tells us that $\int_z^x \mathcal{G}(t)\, dt$ is differentiable with derivative $\mathcal{G}(x)$. We therefore conclude that

$$\lim_{n \to \infty} g'_n(x) = \mathcal{G}(x) = \frac{d}{dx} \int_z^x \mathcal{G}(t)\, dt = \frac{d}{dx}(g(x) - g(z)) = g'(x).$$

And that is what we wanted to prove! □

Corollary 12.4.5 Suppose that (f_n) is a sequence of real functions that is continuously differentiable on the interval (a,b). Suppose that $\sum_{n=1}^{\infty} f'_n(x)$ converges uniformly on (a,b) and that $\sum_{n=1}^{\infty} f_n(t)$ converges for some $t \in (a,b)$. Then $\sum_{n=1}^{\infty} f_n(x)$ converges pointwise on (a,b) to a differentiable function f, and

$$f'(x) = \sum_{n=1}^{\infty} f'_n(x).$$

Proof: The proof is Problem 5 at the end of this section. □

You are now ready for several of the excursions.

Excursion F. *Doubly Indexed Sequences*

Synopsis: This excursion could, in principle, be studied immediately after considering the convergence of sequences (Sections 3.3 and 3.4). However, the relationship between the limit of a doubly indexed sequence and the associated iterated limits reinforces the concept of uniform convergence and its role in "allowing" the interchange of limit operations. Excursion F is written as a set of interconnected exercises that can be assigned as a mini-project, or individual parts can be assigned as additional exercises.

Excursion J. *Power Series*

Synopsis: Power series are perhaps the most familiar (and certainly one of the most important) applications of the convergence of sequences of functions. Their convergence behavior is very regular, so functions that are given by a power series have some extremely nice properties. The study of power series also relies on the facts about series of real numbers, as discussed in Excursion H.

Excursion K. *Everywhere Continuous, Nowhere Differentiable*

Synopsis: Students of calculus know that every differentiable function is continuous but that some continuous functions fail to be differentiable. They also know that a continuous function fails to be differentiable either at a "corner" or when the graph is vertical (for instance, the absolute value function and the cube root function). It is not too difficult for students to see that a continuous function can fail to be differentiable at infinitely many points (the saw-tooth curve). But could there be a function that is continuous everywhere and differentiable nowhere? Excursion K introduces such a function, leaving the proofs as exercises.

Excursion N. *Spaces of Continuous Functions*

Synopsis: It is natural to ask how the convergence of sequences of functions is related to "ordinary" convergence of sequences in a metric space. In Excursion N, we describe a metric space in which each point is a continuous function and where converging sequences *are* sequences of functions. This is a very powerful point of view. Function spaces play a large role in a branch of mathematics called functional analysis, which has applications in everything from electrical engineering to theoretical physics. The power of this point of view is glimpsed in Excursion O "Solutions to Differential Equations."

In addition to Chapter 12, Excursion N assumes familiarity with the ideas discussed in Chapters 5, 6, and 7.

Excursion O. *Solutions to Differential Equations*

Synopsis: This excursion discusses the Picard–Lindelöf theorem for the existence and uniqueness of solutions to differential equations. Chapter 10 and Excursion N are also assumed as background for this excursion.

Problems 12.4

1. Find a sequence of continuous functions $f_n : [0, 1] \to \mathbb{R}$ such that
$$\lim_{n \to \infty} \lim_{x \to 0} f_n(x) \text{ and } \lim_{x \to 0} \lim_{n \to \infty} f_n(x)$$
both exist but are unequal.

2. This problem refers to the sequence of functions defined in Example 12.4.1:
$$f_n(x) = \begin{cases} 4n^2 x & \text{if } 0 \leq x \leq \frac{1}{2n} \\ 4n - 4n^2 x & \text{if } \frac{1}{2n} < x \leq \frac{1}{n} \\ 0 & \text{if } \frac{1}{n} < x \leq 1 \end{cases}.$$

   (a) Verify that $f_n(x) \to 0$ as $n \to \infty$.

   (b) Verify that $\int_0^1 f_n(x)\,dx = 1$ for all $n \in \mathbb{N}$.

(c) Use the calculations you did in parts (a) and (b) to conclude that

$$\lim_{n\to\infty} f_n(x) = f(x) \text{ for all } x \not\Rightarrow \lim_{n\to\infty} \int_0^1 f_n = \int_0^1 f.$$

3. Construct a sequence of continuous real-valued functions that converge to zero, yet whose integrals go to infinity.

4. Limits and Integrals:

   (a) Prove Theorem 12.4.2, which discusses the integration of uniform limits of integrable functions.

   (b) Prove Corollary 12.4.3, which discusses the term-by-term integration of series of integrable functions.

5. Prove Corollary 12.4.5, which discusses the term-by-term differentiation of series of differentiable functions.

6. Prove the following variation of Theorem 12.4.4:

   Let (g_n) be a sequence of real-valued functions defined on an open interval (a, b) in $\mathbb{R}$. Suppose that each g_n has a continuous derivative. If the sequence (g'_n) converges uniformly on every closed subinterval of (a, b) and for some $z \in (a, b)$, the sequence $(g_n(z))$ converges, then (g_n) converges pointwise to some differentiable function g and

   $$g'(x) = \lim_{n\to\infty} g'_n(x) \text{ for all } x \in (a, b).$$

Chapter 13
Differentiating $\mathbf{f} : \mathbb{R}^n \to \mathbb{R}^m$

In this chapter, we will talk about differentiating functions in which the domain is a subset of $\mathbb{R}^n$ and the range is a subset of $\mathbb{R}^m$. It is assumed that the reader is familiar with the basic facts of linear algebra on n-dimensional real spaces.

13.1 What Are We Studying?

Language and Interpretation: Stemming from their different usages, three special categories of functions $\mathbf{f} : \mathbb{R}^n \to \mathbb{R}^m$ are especially important.

- **Scalar fields or scalar-valued functions**: Let $E \subseteq \mathbb{R}^n$. A function $f : E \to \mathbb{R}$ is sometimes called a scalar field because it assigns a scalar to each point in its domain.

- **Vector fields or vector-valued functions**: Let $E \subseteq \mathbb{R}^n$. A function $\mathbf{f} : E \to \mathbb{R}^m$ with $m > 1$ is sometimes called a **vector field** because it assigns a vector to each point in its domain.[1]

- **Parametric curves in space**: Let $I \subseteq \mathbb{R}$ be an interval, and let $m > 1$. A continuous function $\mathbf{f} : I \to \mathbb{R}^m$ is often viewed as a (parametrically defined) curve in m-dimensional space; the set of outputs of $\mathbf{f}$ is a curve in $\mathbb{R}^m$. In this context, the input, which is a real number, is called a *parameter*. If we picture a particle moving along the curve described by the outputs of $\mathbf{f}$, the input parameter can be interpreted as the time at which the particle occupies the corresponding point on the curve.

For some useful pictures, see Figure 13.1.

---

1. Many authors restrict the phrase "vector field" to the case when $n = m$, but we will use the more general definition because it is useful to have a name for these functions.

258 Chapter 13 ■ Differentiating $\mathbf{f} : \mathbb{R}^n \to \mathbb{R}^m$

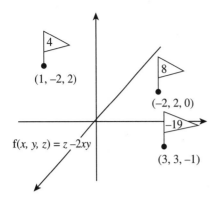

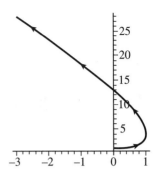

Scalar Fields. Think of the domain as a "field" in which each point is "tagged" with a number. For example, each point in a room can be associated with a temperature in degrees Celsius.

Parametric Curves. The curve shown consists of the outputs of the function $f(t) = (2t - t^2,\ 3t^2 + 1)$ on the interval $[0, 3]$. Notice that t does not appear in the picture. But we can picture a particle moving along the curve. For each t in $[0, 3]$, we can imagine that the particle is at the point $(2t - t^2,\ 3t^2 + 1)$ at time t. The arrows indicate the "direction of motion."

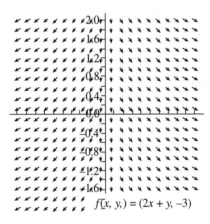

Vector Fields. Think of the domain as a "field" in which each point is "tagged" with a vector. For example, if the domain is the surface of a river, we can associate each point with a current, which has both magnitude and direction and is therefore a vector.

Figure 13.1 When we get beyond one or two variables, picturing functions becomes more difficult. These are useful images to keep in mind.

Let E be a subset of $\mathbb{R}^n$. Note that a vector field $\mathbf{f} : E \to \mathbb{R}^m$ can be viewed as an m-vector of scalar fields: $\mathbf{f}(\mathbf{x}) = (f_1(\mathbf{x}), f_2(\mathbf{x}), \ldots, f_m(\mathbf{x}))$. The scalar-valued functions $f_1 : E \to \mathbb{R}$, $f_2 : E \to \mathbb{R}$, $\ldots$, $f_m : E \to \mathbb{R}$ are called the **coordinate functions** of the vector field $\mathbf{f}$.

Example 13.1.1 Suppose $\mathbf{f} : \mathbb{R}^3 \to \mathbb{R}^2$ is given by

$$\mathbf{f}(x, y, z) = (2x - yz,\ x^2 + y^2 + z^2).$$

Then the coordinate functions of $\mathbf{f} = (f_1, f_2)$ are $f_1(x, y, z) = 2x - yz$ and $f_2(x, y, z) = x^2 + y^2 + z^2$.

In Chapter 5, you learned quite a bit about real-valued functions or scalar fields. Theorem 13.1.2 begins to explore the connections between a vector field and its scalar coordinate functions.

Theorem 13.1.2 Let E be an open subset of $\mathbb{R}^n$. Let $\mathbf{f} : E \to \mathbb{R}^m$ be a vector-valued function with scalar coordinates $\mathbf{f} = (f_1, f_2, f_3, \ldots, f_m)$. Then the following statements are true:

1. Let $(\mathbf{y}_k)$ be a sequence in E, and let $\mathbf{b} = (b_1, b_2, \ldots, b_m) \in \mathbb{R}^m$. Then $\lim_{k \to \infty} \mathbf{f}(\mathbf{y}_n) = \mathbf{b}$ if and only if for each $i = 1, 2, \ldots, m$, $\lim_{k \to \infty} f_i(\mathbf{y}_n) = b_i$.

2. Let $\mathbf{a} = (a_1, a_2, \ldots, a_n) \in E$, and let $\mathbf{b} = (b_1, b_2, \ldots, b_m) \in \mathbb{R}^m$. Then $\lim_{\mathbf{y} \to \mathbf{a}} \mathbf{f}(\mathbf{y}) = \mathbf{b}$ if and only if for each $i = 1, 2, \ldots, m$, $\lim_{\mathbf{y} \to \mathbf{a}} f_i(\mathbf{y}) = b_i$.

3. Let $\mathbf{a} = (a_1, a_2, \ldots, a_n) \in E$. The vector field $\mathbf{f}$ is continuous at $\mathbf{a}$ if and only if each of its scalar coordinate functions $f_1, f_2, \ldots, f_m$ is continuous at $\mathbf{a}$.

Proof: The proof is Problem 2 at the end of this section. $\square$

Problems 13.1

1. Let $\mathbf{f} : \mathbb{R}^n \to \mathbb{R}^m$ be a vector field. Prove that the set of coordinate functions representing $\mathbf{f}$ is unique. That is, if $\mathbf{f} = (f_1, f_2, \ldots, f_n)$ and $\mathbf{f} = (g_1, g_2, \ldots, g_n)$, then $f_i = g_i$ for $i = 1, 2, \ldots, n$.

2. Prove Theorem 13.1.2.

13.2 Thinking Intuitively

In Chapter 9, we learned that given a differentiable function $f : \mathbb{R} \to \mathbb{R}$, we can interpret the derivative in two different ways. First, the number $f'(x)$ given by the limit

$$\lim_{y \to x} \frac{f(y) - f(x)}{y - x}$$

is the rate of change of the function f at the point x.

Second, it is the slope of the best local linear approximation to the graph of f at the point x. This is given by

$$L(y) = f'(x)(y - x) + f(x)$$

where $f'(x)$ is chosen so that

$$r(y) = f(y) - L(y) \text{ satisfies } \frac{r(y)}{y - x} \to 0 \text{ as } y \to x.[2]$$

It is not difficult to show that these formulations are equivalent and that each yields the same number $f'(x)$, which we call the derivative of f at x.

For functions of several variables, rates of change and local linear approximations are also closely related ideas, and both will play an important role in the theory. But for $n > 1$ and/or $m > 1$, the geometry of the function $\mathbf{f} : \mathbb{R}^n \to \mathbb{R}^m$ is more complicated than that of its close cousin $f : \mathbb{R} \to \mathbb{R}$. Thus the generalizations require a bit of care. In particular, it is easy to see that the expression

$$\frac{\mathbf{f(y)} - \mathbf{f(x)}}{\mathbf{y} - \mathbf{x}}$$

will not get us anywhere, because the numerator is a vector in $\mathbb{R}^m$ and the denominator is a vector in $\mathbb{R}^n$. The quotient is not even defined! The formulation that involves local linear approximation is more amenable to full generalization. It is to this idea that we now turn.

Tangent Planes

If we "zoom in" on the graph of a one-variable differentiable function, we expect to see a straight line. To see how this idea generalizes to higher dimensions, we turn—for intuition and inspiration—to functions of two variables.

As before, we can imagine "zooming in" on the surface that is the graph of our function. Figure 13.2 suggests that if we zoom in on the graph of a "nice" function of two variables, we should expect that our snapshot of the graph will come to look more and more like a plane. Thus, for functions from $\mathbb{R}^2$ to $\mathbb{R}$, the "best local linear approximation" will be a plane. The plane that we see is called the *tangent plane to the graph at the point*. The derivative of $f : \mathbb{R}^n \to \mathbb{R}^n$ at the point $\mathbf{x}$ will have to specify the tangent plane.

To elaborate on this last observation, it is useful to recall the relationship between the derivative of a one-variable function f and the straight line approximation to the graph of f at the point $(x, f(x))$. The former is a number, while the latter is a function. Because we already know that our straight-line

---

2. Remember that $r(y) \to 0$ is weaker and is insufficient to guarantee the differentiability of f at x. See Section 9.2 for details.

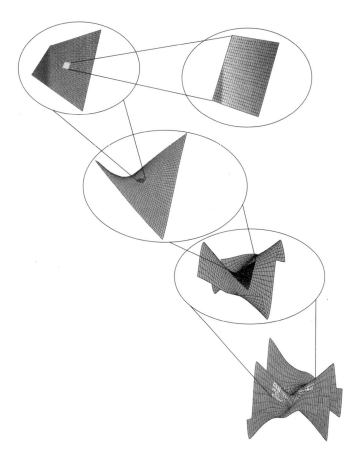

Figure 13.2 Zooming in on the surface. The gray section at each stage becomes the black section at the next stage.

approximation must go through the point $(x, f(x))$, we need only specify the slope of the line to determine the tangent line. That's what the derivative does—or is. But to specify a plane going through a point in 3-space, we need two numbers, not just one. Hence our derivative is no longer just a number.

Recall that a "generic" plane going through the point $(a, b, f(a, b))$ has an equation of the form

$$z = r(x - a) + s(y - b) + f(a, b),$$

where r is the slope of the plane in the x-direction and s is the slope of the plane in the y-direction.

To see where the *big generalization* comes from, it is useful to think about this matter in a slightly different way. When we find the derivative of a function $f : \mathbb{R} \to \mathbb{R}$, we identify the unique straight line through the origin with the same slope as the one we want and then we shift that line into the right place. This last twist may seem unnecessarily cumbersome, but it really helps us to expand our thinking into many dimensions. The benefit of thinking about "generic" lines or "generic" planes through the *origin* is that these functions are truly linear functions, in the linear algebraic sense, whereas their shifted affine "twins" are not.[3] (Remember that a linear transformation from one vector space to another must take the zero vector to the zero vector!)

The main thrust of differentiation is to approximate a "nice" function by the shift of an appropriate linear transformation. Thus, the derivative of a function $\mathbf{f} : \mathbb{R}^n \to \mathbb{R}^m$ will be a linear transformation from $\mathbb{R}^n$ to $\mathbb{R}^m$! With this point in mind, we must take a bit of a detour through linear algebra and its connections to analysis.

13.3 Analysis in Linear Spaces

This section has two important goals. First, we will review the linear algebraic ideas that we need to discuss linear transformations from $\mathbb{R}^n$ to $\mathbb{R}^m$. Second, we will see how algebra and analysis fit together in these finite-dimensional real spaces.

We start by recalling some familiar definitions and theorems.

Definition 13.3.1 Let $\mathbf{x} = (x_1, x_2, \ldots, x_n)$ and $\mathbf{y} = (y_1, y_2, \ldots, y_n)$ be vectors in $\mathbb{R}^n$, and let $t \in \mathbb{R}$ be any scalar.

We define **scalar multiplication** as follows. For the scalar t and the vector $\mathbf{x}$, we define the vector
$$t\mathbf{x} = (tx_1, tx_2, \ldots, tx_n).$$

We define **vector addition** as follows. For the two vectors $\mathbf{x}$ and $\mathbf{y}$, we define the vector
$$\mathbf{x} + \mathbf{y} = (x_1 + y_1, x_2 + y_2, \ldots, x_n + y_n).$$

It is easy to show that these operations make $\mathbb{R}^n$ into a **vector space**. In other words, $\mathbb{R}^n$, together with the scalar multiplication and vector addition given in Definition 13.3.1, satisfies the following properties:

1. Vector addition is commutative and associative.

2. The zero vector $\mathbf{0} = (0, 0, \ldots, 0)$ satisfies $\mathbf{0} + \mathbf{x} = \mathbf{x}$ for all $\mathbf{x} \in \mathbb{R}^n$.

---

3. A shift of a linear transformation is called an **affine transformation**.

3. For all $\mathbf{x} \in \mathbb{R}^n$, $1\mathbf{x} = \mathbf{x}$.

4. For each $\mathbf{x} \in \mathbb{R}^n$, $(-1)\mathbf{x} = -\mathbf{x}$ satisfies $-\mathbf{x} + \mathbf{x} = \mathbf{0}$.

5. For all scalars s and $t \in \mathbb{R}$ and all $\mathbf{x} \in \mathbb{R}^n$, $r(s\mathbf{x}) = (rs)\mathbf{x}$.

6. For all scalars s and $t \in \mathbb{R}$ and all $\mathbf{x} \in \mathbb{R}^n$, $(r+s)\mathbf{x} = r\mathbf{x} + s\mathbf{x}$.

7. For all scalars $s \in \mathbb{R}$ and vectors $\mathbf{x}$ and $\mathbf{y} \in \mathbb{R}^n$, $s(\mathbf{x} + \mathbf{y}) = s\mathbf{x} + s\mathbf{y}$.

All of these statements follow easily from related properties in $\mathbb{R}$. (Try a few to see how it goes.)

Definition 13.3.2 Consider a collection $\mathbf{b}_1, \mathbf{b}_2, \ldots, \mathbf{b}_k$ in $\mathbb{R}^n$.

- Any expression of the form $r_1\mathbf{b}_1 + r_2\mathbf{b}_2 + \cdots + r_k\mathbf{b}_k$ (with $r_1, r_2, \ldots, r_k$ scalars) is called a **linear combination** of the vectors $\mathbf{b}_1, \mathbf{b}_2, \ldots, \mathbf{b}_k$.

- The collection $\{\mathbf{b}_i\}_{i=1}^k$ is said to **span** all of $\mathbb{R}^n$ provided that every vector in $\mathbb{R}^n$ can be written as a linear combination of the vectors $\mathbf{b}_1, \mathbf{b}_2, \ldots, \mathbf{b}_k$.

- $\{\mathbf{b}_i\}_{i=1}^k$ is said to be **linearly independent** provided that

$$r_1\mathbf{b}_1 + r_2\mathbf{b}_2 + \cdots + r_k\mathbf{b}_k = \mathbf{0}$$

if and only if $r_1 = r_2 = \cdots = r_k$.

- The collection $\{\mathbf{b}_i\}_{i=1}^k$ is said to be a **basis** for $\mathbb{R}^n$ provided that $\{\mathbf{b}_i\}_{i=1}^k$ is both linearly independent and spans $\mathbb{R}^n$.

Exercise 13.3.3 Suppose that $\{\mathbf{b}_i\}_{i=1}^k$ is a basis for $\mathbb{R}^n$. Prove that given any $\mathbf{v} \in \mathbb{R}^n$, there exist *unique* real numbers $a_1, a_2, \ldots, a_n$ such that

$$\mathbf{v} = \sum_{i=1}^n a_i \mathbf{b}_i.$$

∎

It is relatively easy to show that if $\{\mathbf{b}_i\}_{i=1}^k$ is any linearly independent set in $\mathbb{R}^n$, then $k \leq n$. Moreover, if $\{\mathbf{b}_i\}_{i=1}^k$ is a basis for $\mathbb{R}^n$, then $k = n$. We will assume these facts from now on. (They are proved in most linear algebra texts. See, for instance, [AN], Chapter 5.)

Definition 13.3.4 Consider the collection $\mathbf{e}_1, \mathbf{e}_2, \mathbf{e}_3, \ldots, \mathbf{e}_n$ of vectors in $\mathbb{R}^n$, where $\mathbf{e}_i = (e_{i1}, e_{i2}, e_{i3}, \ldots, e_{in})$ is given by

$$e_{ij} = \begin{cases} 1 & \text{if } i = j \\ 0 & \text{if } i \neq j \end{cases}.$$

This collection of vectors is called the **unit vector basis** or the **standard basis** for $\mathbb{R}^n$. (It is straightforward to show that the standard basis is a basis in the sense of Definition 13.3.2.)

Definition 13.3.5 Let $\mathbf{x} = (x_1, x_2, \ldots, x_n)$ and $\mathbf{y} = (y_1, y_2, \ldots, y_n)$ be vectors in $\mathbb{R}^n$. The **dot product** of $\mathbf{x}$ and $\mathbf{y}$ is a real number denoted by $\mathbf{x} \cdot \mathbf{y}$. It is given by

$$\mathbf{x} \cdot \mathbf{y} = x_1 y_1 + x_2 y_2 + x_3 y_3 + \ldots + x_n y_n.$$

Exercise 13.3.6 Let $\mathbf{x} = (x_1, x_2, \ldots, x_n) \in \mathbb{R}^n$, and let $\{\mathbf{e}_i\}_{i=1}^n$ be the unit vector basis for $\mathbb{R}^n$. Prove the following easy facts:

1. For each $i \leq n$, $\mathbf{e}_i \cdot \mathbf{x} = x_i$.

   Remark: In practical terms, if we have a vector, taking its dot product with $\mathbf{e}_i$ allows us to "extract" the ith coordinate.

2. $\mathbf{x} = \sum_{i=1}^n x_i \mathbf{e}_i = \sum_{i=1}^n (\mathbf{e}_i \cdot \mathbf{x}) \mathbf{e}_i.$

■

The following properties of the dot product are easy to check.

Theorem 13.3.7 Let $\mathbf{x} = (x_1, x_2, x_3, \ldots, x_n)$, $\mathbf{y} = (y_1, y_2, y_3, \ldots, y_n)$, and $\mathbf{z} = (z_1, z_2, z_3, \ldots, z_n)$ be vectors in $\mathbb{R}^n$. Let $t \in \mathbb{R}$ be a scalar. Then the following identities hold:

1. $\mathbf{x} \cdot \mathbf{0} = 0.$

2. $\mathbf{x} \cdot \mathbf{x} \geq 0$; equality holds if and only if $\mathbf{x} = \mathbf{0}.$

3. $\mathbf{x} \cdot \mathbf{y} = \mathbf{y} \cdot \mathbf{x}.$

4. $(t\mathbf{x}) \cdot \mathbf{y} = t(\mathbf{x} \cdot \mathbf{y}).$

5. $(\mathbf{x} + \mathbf{y}) \cdot \mathbf{z} = \mathbf{x} \cdot \mathbf{z} + \mathbf{y} \cdot \mathbf{z}.$

Proof: The proof is Problem 1 at the end of this section. □

Linear Transformations

Definition 13.3.8 Let $\mathbf{L}: \mathbb{R}^n \to \mathbb{R}^m$ be a function. We say that L is a **linear transformation** provided that the following conditions hold:

1. For all $\mathbf{x}$ and $\mathbf{y} \in \mathbb{R}^n$, $\mathbf{L}(\mathbf{x}+\mathbf{y}) = \mathbf{L}(\mathbf{x}) + \mathbf{L}(\mathbf{y})$.

2. For all $r \in \mathbb{R}$ and $\mathbf{x} \in \mathbb{R}^n$, $\mathbf{L}(r\mathbf{x}) = r\mathbf{L}(\mathbf{x})$.

Theorem 13.3.9 Let $\mathbf{L}: \mathbb{R}^n \to \mathbb{R}^m$ be a linear transformation. Then the following facts hold:

1. $\mathbf{L}(\mathbf{0}) = \mathbf{0}$. Linear transformations always take the zero vector to the zero vector.[4]

2. Suppose that $\mathbf{L}(\mathbf{x}) = \mathbf{0}$ if and only if $\mathbf{x} = \mathbf{0}$. Then $\mathbf{L}$ is one-to-one.

3. The set $\{L(\mathbf{e}_i)\}_{i=1}^n$, the image of the standard basis, is linearly independent in $\mathbb{R}^m$.

4. If $\mathbf{L}$ is onto, then $m \leq n$.

5. If $\mathbf{L}$ is both one-to-one and onto, $m = n$.

Proof: The proof is Problem 3 at the end of this section. □

Exercise 13.3.10 Let A be an $m \times n$ matrix. Define the function $\mathbf{T}: \mathbb{R}^n \to \mathbb{R}^m$ by the product $\mathbf{T}(\mathbf{x}) = A\mathbf{x}$. Show that $\mathbf{T}$ is a linear transformation. ■

Theorem 13.3.11 Let $\mathbf{L}: \mathbb{R}^n \to \mathbb{R}^m$ be a linear transformation. There exists a unique $m \times n$ matrix A such that for all $\mathbf{x} \in \mathbb{R}^n$, $\mathbf{L}(\mathbf{x}) = A\mathbf{x}$. The *columns* of A are given by the vectors $\{L(\mathbf{e}_i)\}_{i=1}^n$.

Proof: The proof is Problem 4 at the end of this section. □

Theorem 13.3.11 tells us that *every* linear transformation is given by left-multiplication by an appropriate matrix. Furthermore, the uniqueness clause in the theorem tells us that there is a one-to-one correspondence between matrices

---

4. Note that the symbol **0** is used ambiguously here. On the left side of the expression, it refers to the zero vector in $\mathbb{R}^n$; on the right side, it refers to the zero vector in $\mathbb{R}^m$. Context tells us how to interpret the symbol.

and linear transformations: Different matrices yield different linear transformations, and vice versa. This is a very useful result, because it gives us a concrete procedure for describing and working with linear transformations.

Definition 13.3.12 Let $\mathbf{L} : \mathbb{R}^n \to \mathbb{R}^m$ be a linear transformation. If A is the matrix that satisfies $\mathbf{L}(\mathbf{x}) = A\mathbf{x}$ for all $\mathbf{x} \in \mathbb{R}^n$, we say that **A represents L**.

Exercise 13.3.13 It is easy to see that the identity function $\mathbf{I}_n : \mathbb{R}^n \to \mathbb{R}^n$ given by $\mathbf{I}_n(\mathbf{x}) = \mathbf{x}$ is a linear transformation. Let $\mathbf{I}_n$ be the $n \times n$ identity matrix:

$$\mathbf{I}_n = \begin{pmatrix} 1 & 0 & 0 & \cdots & 0 \\ 0 & 1 & 0 & \cdots & 0 \\ 0 & 0 & 1 & \cdots & 0 \\ \vdots & \vdots & \vdots & \ddots & \vdots \\ 0 & 0 & 0 & \cdots & 1 \end{pmatrix}$$

Show that $\mathbf{I}_n$ is the (unique) matrix that represents the identity transformation. ∎

Example 13.3.14 [Important Special Case] Let $\mathbf{L} : \mathbb{R}^n \to \mathbb{R}$ be a (scalar-valued) linear transformation. Then the matrix representing $\mathbf{L}$ has n columns but only one row, so we can think of it as a vector rather than a matrix. In the representation of $\mathbf{L}$, matrix multiplication becomes a dot product. Indeed, if $\mathbf{a}$ is the vector representing $\mathbf{L}$, then

$$\mathbf{L}(\mathbf{x}) = \mathbf{a} \cdot \mathbf{x} \quad \text{for all} \quad \mathbf{x} \in \mathbb{R}^n$$

Theorem 13.3.15 Let $\mathbf{L} : \mathbb{R}^n \to \mathbb{R}^m$ and $\mathbf{T} : \mathbb{R}^n \to \mathbb{R}^m$ be linear transformations represented by matrices A and B, respectively. Suppose that $t \in \mathbb{R}$ is any scalar. Then the functions $\mathbf{L} + \mathbf{T}$ and $t\mathbf{L}$ are linear transformations and are represented by the matrices $A + B$ and tA, respectively.

Proof: The proof is Problem 6 at the end of this section. □

Theorem 13.3.16 Let $\mathbf{L} : \mathbb{R}^n \to \mathbb{R}^m$ and $\mathbf{T} : \mathbb{R}^m \to \mathbb{R}^k$ be linear transformations represented by matrices A and B, respectively. Then the function $\mathbf{T} \circ \mathbf{L} : \mathbb{R}^n \to \mathbb{R}^k$ is a linear transformation and is represented by the matrix product BA.

Proof: The proof is Problem 7 at the end of this section. □

Theorem 13.3.17 Let $\mathbf{L} : \mathbb{R}^n \to \mathbb{R}^n$ be a linear transformation represented by the matrix A. Then $\mathbf{L}$ is invertible if and only if A is invertible.

Proof: The proof is Problem 8 at the end of this section. □

Linear Algebra and Analysis

So far, all of our discussion has been purely algebraic. We have not yet referred in any way to the analytic properties of n-dimensional real spaces. The key to bringing linear algebra and analysis together is to introduce a norm.

Definition 13.3.18 The **norm**[5] of $\mathbf{x} = (x_1, x_2, \ldots, x_n)$ is given by

$$\|\mathbf{x}\| = \sqrt{x_1^2 + x_2^2 + x_3^2 + \cdots + x_n^2}$$

A vector of norm one is called a **unit vector**.

Exercise 13.3.19 Establish the following simple facts about the norm function:

1. Let $\mathbf{x} \in \mathbb{R}^n$. Then $\|\mathbf{x}\| = 0$ if and only if $\mathbf{x} = \mathbf{0}$.

2. Let $s \in \mathbb{R}$ be any scalar, and let $\mathbf{x} = (x_1, x_2, \ldots, x_n)$ be any vector in $\mathbb{R}^n$. Then $\|s\mathbf{x}\| = |s|\,\|\mathbf{x}\|$.

3. Let $\mathbf{x}$ and $\mathbf{y} \in \mathbb{R}^n$. Then $\|\mathbf{x} + \mathbf{y}\| \leq \|\mathbf{x}\| + \|\mathbf{y}\|$.

■

It is easy to see that if $\mathbf{x} = (x_1, x_2, \ldots, x_n)$ and $\mathbf{y} = (y_1, y_2, \ldots, y_n)$ are vectors in $\mathbb{R}^n$, then

$$\|\mathbf{x} - \mathbf{y}\| = \sqrt{(x_1 - y_1)^2 + (x_2 - y_2)^2 + \cdots + (x_n - y_n)^2}$$

is the usual distance in $\mathbb{R}^n$ between the two vectors. We can now bring all of our previous theorems about metric spaces into the conversation and prove some familiar analogues.

Theorem 13.3.20 [Limits and Vector Operations in $\mathbb{R}^n$] Let $(\mathbf{x}_n)$ and $(\mathbf{y}_n)$ be sequences in $\mathbb{R}^n$ converging to $\mathbf{x}$ and $\mathbf{y}$, respectively. Let (t_n) be a sequence in $\mathbb{R}$ that converges to a scalar t. Let k be an arbitrary scalar. Then the following facts hold:

1. $(k\mathbf{x}_n)$ converges to $k\mathbf{x}$.

2. $(t_n\mathbf{x}_n)$ converges to $t\mathbf{x}$.

---

5. You may have heard the norm be called the *length* or *magnitude* of the vector.

3. $(\mathbf{x}_n + \mathbf{y}_n)$ converges to $\mathbf{x} + \mathbf{y}$.

4. $(\mathbf{x}_n \cdot \mathbf{y}_n)$ converges to $\mathbf{x} \cdot \mathbf{y}$.

5. $\|\mathbf{x}_n\| \to \|\mathbf{x}\|$.

Proof: The proof is Problem 9 at the end of this section. □

Notice that $B_1(\mathbf{0}) = \{\mathbf{x} \in \mathbb{R}^n : \|\mathbf{x}\| < 1\}$, and that $C_1(\mathbf{0}) = \{\mathbf{x} \in \mathbb{R}^n : \|\mathbf{x}\| \leq 1\}$ (general open balls are described in Problem 15 at the end of this section).

Lemma 13.3.21 Let $\mathbf{L} : \mathbb{R}^n \to \mathbb{R}^m$ be a linear transformation. If $\mathbf{L}$ is represented by the matrix $A = [a_{ij}]$ and $\mathbf{u} = (u_1, u_2, \ldots, u_n)$ satisfies $\|\mathbf{u}\| \leq 1$, then

$$\|\mathbf{L}(\mathbf{u})\| \leq \left(\sum_{i=1}^{m} \sum_{j=1}^{n} a_{ij}^2 \right)^{\frac{1}{2}}.$$

Proof: The proof is Problem 11a at the end of this section. □

Lemma 13.3.21 tells us that $\mathbf{L}$ is bounded on the closed unit ball of $\mathbb{R}^n$. This fact justifies the following definition.

Definition 13.3.22 Let $\mathbf{L} : \mathbb{R}^n \to \mathbb{R}^m$ be a linear transformation. We define the **norm of the linear transformation** $\mathbf{T}$ to be

$$\|\mathbf{L}\| = \sup\{\|\mathbf{L}(\mathbf{v})\| : \|\mathbf{v}\| \leq 1\}.$$

In keeping with this definition, we define the **norm of a matrix** to be the norm of the linear transformation that it represents.

Corollary 13.3.23 Let $\mathbf{T} : \mathbb{R}^n \to \mathbb{R}^m$ be a linear transformation represented by the matrix $A = [a_{ij}]$. Then

$$\|\mathbf{T}\| \leq \left(\sum_{i=1}^{m} \sum_{j=1}^{n} a_{ij}^2 \right)^{\frac{1}{2}}.$$

Proof: The proof is Problem 11b at the end of this section. □

Theorem 13.3.24 Let $\mathbf{L}: \mathbb{R}^n \to \mathbb{R}^m$ and $\mathbf{T}: \mathbb{R}^n \to \mathbb{R}^m$ be linear transformations, and let $s \in \mathbb{R}$ be a scalar. Then the following facts hold:

1. $\|\mathbf{T}\| = 0$ if and only if $\mathbf{T}$ is the linear transformation that takes every vector to the zero vector.

2. $\|s\mathbf{L}\| = |s|\,\|\mathbf{L}\|$.

3. $\|\mathbf{L} + \mathbf{T}\| \leq \|\mathbf{L}\| + \|\mathbf{T}\|$.

4. For all $\mathbf{v} \in \mathbb{R}^n$, $\|\mathbf{L}(\mathbf{v})\| \leq \|\mathbf{L}\|\,\|\mathbf{v}\|$.

Proof: The proof is Problem 13 at the end of this section. □

Theorem 13.3.25 Let $\mathbf{L}: \mathbb{R}^n \to \mathbb{R}^m$ be a linear transformation. Then $\mathbf{L}$ is uniformly continuous on $\mathbb{R}^n$.

Proof: The proof is Problem 14 at the end of this section. □

We end with a technical lemma that will play a key role in establishing the connection between local linear approximation and rates of change for functions of several variables.

Lemma 13.3.26 Let n and m be in $\mathbb{N}$. Suppose that for $i = 1, 2, \ldots, m$ and $j = 1, 2, \ldots, n$, the functions

$$f_{ij}: \mathbb{R}^n \to \mathbb{R}$$

are continuous at the point $\mathbf{x} \in \mathbb{R}^n$. For $\mathbf{z} \in \mathbb{R}^n$, let $\mathbf{T}(\mathbf{z})$ be the linear transformation represented by the matrix

$$A(\mathbf{z}) = \begin{bmatrix} f_{11}(\mathbf{z}) & f_{12}(\mathbf{z}) & \cdots & f_{1n}(\mathbf{z}) \\ f_{21}(\mathbf{z}) & f_{22}(\mathbf{z}) & \cdots & f_{2n}(\mathbf{z}) \\ \vdots & \vdots & \ddots & \vdots \\ f_{m1}(\mathbf{z}) & f_{m2}(\mathbf{z}) & \cdots & f_{mn}(\mathbf{z}) \end{bmatrix}$$

Let $\epsilon > 0$. Then there exists $\delta > 0$ such that if $\|\mathbf{z} - \mathbf{x}\| < \delta$, then

$$\|\mathbf{T}(\mathbf{z}) - \mathbf{T}(\mathbf{x})\| < \epsilon.$$

Proof: The proof is Problem 20 at the end of this section. □

Problems 13.3

1. Prove Theorem 13.3.7, which establishes the standard properties of the dot product.

2. Let $\mathbf{L} : \mathbb{R}^n \to \mathbb{R}^m$ be a vector field with scalar coordinate functions $L_1 : \mathbb{R}^n \to \mathbb{R}$, $L_2 : \mathbb{R}^n \to \mathbb{R}$, ..., $L_m : \mathbb{R}^n \to \mathbb{R}$. Prove $\mathbf{L}$ is a linear transformation if and only if $L_1, L_2, \ldots, L_n$ are all linear transformations.

3. Prove Theorem 13.3.9, which establishes some general properties of linear transformations from $\mathbb{R}^n$ to $\mathbb{R}^m$.

4. Prove that linear transformations are represented by matrices (Theorem 13.3.11).

5. Let $\mathbf{L} : \mathbb{R}^n \to \mathbb{R}^m$ be a linear transformation. Theorem 13.3.11 tells us that $\mathbf{L}$ is represented by an $m \times n$ matrix $\mathbf{A} = [a_{ji}]$, $1 \leq j \leq m$ and $1 \leq i \leq n$.

    (a) Prove that if $\mathbf{L} = (L_1, L_2, \ldots, L_m)$, then the entries of the matrix $\mathbf{A}$ are $a_{ji} = L_j(\mathbf{e}_i)$.

    (b) Suppose that $\mathbf{L} = (L_1, L_2, \ldots, L_m)$. Fix j such that $1 \leq j \leq m$. Set $\mathbf{a}_j = (a_{j1}, a_{j2}, \ldots, a_{jn})$—the jth row of the matrix $\mathbf{A}$. Prove that for all $\mathbf{x} \in \mathbb{R}^n$, $L_j(\mathbf{x}) = \mathbf{a}_j \cdot \mathbf{x}$. This result shows that the vectors representing the m-coordinate functions of $\mathbf{L}$ are the m rows of the matrix that represents $\mathbf{L}$.

6. Prove that the sum of two linear transformations is represented by the sum of the matrices that represent them, and similarly for the scalar multiple of a linear transformation (Theorem 13.3.15).

7. Prove that the composition of two linear transformations is represented by the *product* of the matrices that represent them (Theorem 13.3.16).

8. Prove that a linear transformation is invertible if and only if the matrix that represents it is invertible (Theorem 13.3.17).

9. Establish the result about the relationship between vector operations and sequence convergence in $\mathbb{R}^n$ (Theorem 13.3.20).

10. Let $(\mathbf{x}_n)$ be a sequence in $\mathbb{R}^n$, and let (t_n) be a sequence of scalars.

    (a) Suppose that $(\mathbf{x}_n)$ converges to $\mathbf{0}$ and that (t_n) is bounded in $\mathbb{R}$. Prove that $(t_n \mathbf{x}_n)$ converges to $\mathbf{0}$.

    (b) Suppose that (t_n) is a sequence in $\mathbb{R}$ that converges to 0, and that $(\mathbf{x}_n)$ is a bounded sequence in $\mathbb{R}^n$. Prove that $(t_n \mathbf{x}_n)$ converges to $\mathbf{0}$.

(c) Give examples showing that the boundedness conditions are required for the conclusions in parts (a) and (b).

11. (a) Prove Lemma 13.3.21. (*Hint*: The proof uses the Cauchy–Schwarz inequality on $\mathbf{a}_i \cdot \mathbf{u}$, where $\mathbf{a}_i$ is the ith row of A. Otherwise, just compute!)

    (b) Use Lemma 13.3.21 to establish Corollary 13.3.23.

12. Show that the inequality established in Lemma 13.3.21 may be a strict inequality unless the range of $\mathbf{T}$ is a subset of $\mathbb{R}$.

13. Prove the theorem that establishes the standard properties of the norm for linear transformations (Theorem 13.3.24).

14. Prove that linear transformations are uniformly continuous (Theorem 13.3.25).

15. Let $\mathbf{x} \in \mathbb{R}^n$ and $t \in \mathbb{R}$. Consider $B_r(\mathbf{x})$.

    (a) Prove that $B_r(\mathbf{x}) = \{\mathbf{x} + s\mathbf{u} : 0 \leq s \leq r \text{ and } \|\mathbf{u}\| = 1\}$.

    (b) Prove that $B_r(\mathbf{x})$ is convex. That is, prove that if $\mathbf{y}$ and $\mathbf{z} \in B_r(\mathbf{x})$ and $t \in [0, 1]$, then $\mathbf{w} = t\mathbf{y} + (1-t)\mathbf{z}$ is also in $B_r(\mathbf{x})$.

16. Let $\mathbf{T} : \mathbb{R}^n \to \mathbb{R}^m$ and $\mathbf{S} : \mathbb{R}^n \to \mathbb{R}^m$ be linear transformations. Prove that $\mathbf{T} = \mathbf{S}$ if and only if $\mathbf{L}(\mathbf{e}_i) = \mathbf{S}(\mathbf{e}_i)$ for $i = 1, 2, \ldots, n$.

17. Let $\mathbf{K} : \mathbb{R}^n \to \mathbb{R}^m$ and $\mathbf{L} : \mathbb{R}^m \to \mathbb{R}^k$ be linear transformations. Prove that
$$\|\mathbf{L} \circ \mathbf{K}\| \leq \|\mathbf{L}\|\,\|\mathbf{K}\| .$$

18. Let $\mathbf{T} : \mathbb{R}^n \to \mathbb{R}^m$ be a linear transformation. Prove that there exists $\mathbf{v} \in \mathbb{R}^n$ with $\|\mathbf{v}\| = 1$, such that
$$\|\mathbf{T}\| = \|\mathbf{T}(\mathbf{v})\| .$$

19. Let $\mathbf{L} : \mathbb{R}^n \to \mathbb{R}^m$ be a linear transformation and let $s \in \mathbb{R}$. Prove that, if $\|\mathbf{L}(\mathbf{x})\| \leq s\,\|\mathbf{x}\|$ for all $\mathbf{x} \in \mathbb{R}^n$, then $\|\mathbf{L}\| \leq s$.

20. Prove Lemma 13.3.26.

13.4 Local Linear Approximation for Functions of Several Variables

We can now return to the question of differentiation for functions of several variables.

Definition 13.4.1 Let E be an open subset of $\mathbb{R}^n$. Let $\mathbf{f} : E \to \mathbb{R}^m$ be a function, and let $\mathbf{a} \in E$. The function $\mathbf{f}$ is **differentiable** at $\mathbf{a}$ if there exists a linear transformation $\mathbf{f}'(\mathbf{a}) : \mathbb{R}^n \to \mathbb{R}^m$ and a function $\mathbf{r} : E \to \mathbb{R}^m$ such that for all $\mathbf{x} \in E$

$$\mathbf{f}(\mathbf{x}) = \mathbf{f}'(\mathbf{a})(\mathbf{x} - \mathbf{a}) + \mathbf{f}(\mathbf{a}) + \mathbf{r}(\mathbf{x}) \text{ and } \lim_{\mathbf{x} \to \mathbf{a}} \frac{\|\mathbf{r}(\mathbf{x})\|}{\|\mathbf{x} - \mathbf{a}\|} = 0$$

The linear transformation $\mathbf{f}'(\mathbf{a})$ is called the **total derivative of f at a** or just the **derivative of f at a**. As usual, the function $\mathbf{f}$ is itself said to be differentiable if it is differentiable at every point of its domain.[6]

Note that the expression, $\mathbf{f}'(\mathbf{a})(\mathbf{x} - \mathbf{a})$ is not a product. It is the vector in $\mathbb{R}^m$ obtained by the *action* of the linear transformation $\mathbf{f}'(\mathbf{a})$ on the n-coordinate vector $(\mathbf{x} - \mathbf{a})$.

If we think of $\mathbf{f}'(a)$ as an $m \times n$ matrix, the vectors $\mathbf{f}(\mathbf{x})$, $(\mathbf{x} - \mathbf{a})$, $\mathbf{f}(\mathbf{a})$, and $\mathbf{r}(\mathbf{x})$ in Definition 13.4.1 must all be interpreted as column vectors for the vector equation to make sense. In the discussion that follows, we will assume that vectors are interpreted as row or column vectors, based on the context. Furthermore, in some cases it is convenient to interpret a $1 \times n$ matrix as an n-vector or an $m \times 1$ matrix as an m-vector. Matrix multiplication will need to be reinterpreted accordingly. These are natural associations, and we will feel free to make them whenever it is useful to do so.

Exercise 13.4.2 [Another Formulation] Let E be an open subset of $\mathbb{R}^n$. Let $\mathbf{f} : E \to \mathbb{R}^m$ be a function, and let $\mathbf{a} \in E$. Let $E_\mathbf{a} = \{\mathbf{v} \in \mathbb{R}^n : \mathbf{v} = \mathbf{x} - \mathbf{a} \text{ for some } \mathbf{x} \in E\}$. Prove that the function $\mathbf{f}$ is differentiable at $\mathbf{a}$ if and only if there

---

6. Observe that

$$\lim_{\mathbf{x} \to \mathbf{a}} \frac{\|\mathbf{r}(\mathbf{x})\|}{\|\mathbf{x} - \mathbf{a}\|} = 0 \quad \text{and} \quad \lim_{\mathbf{x} \to \mathbf{a}} \frac{\mathbf{r}(\mathbf{x})}{\|\mathbf{x} - \mathbf{a}\|} = \mathbf{0}$$

are equivalent. You may use whichever formulation is most convenient.

13.4 Local Linear Approximation for Functions of Several Variables

exists a linear transformation $\mathbf{f}'(\mathbf{a}) : \mathbb{R}^n \to \mathbb{R}^m$ and a function $\mathbf{r} : E_{\mathbf{a}} \to \mathbb{R}^m$ such that for all $\mathbf{h} \in E_{\mathbf{a}}$

$$\mathbf{f}(\mathbf{a}+\mathbf{h}) = \mathbf{f}'(\mathbf{a})(\mathbf{h}) + \mathbf{f}(\mathbf{a}) + \mathbf{r}(\mathbf{h}) \text{ and } \lim_{\mathbf{h} \to 0} \frac{\|\mathbf{r}(\mathbf{h})\|}{\|\mathbf{h}\|} = 0.$$

(This is not too difficult, but there is a small wrinkle. Why is $E_{\mathbf{a}}$ needed? That is, why isn't E sufficient?) ∎

Exercise 13.4.3 Here are some examples. The first two are very easy; the third may be a bit of a puzzle, but you should think about it carefully.

1. Let $\mathbf{v} \in \mathbb{R}^m$, and let $\mathbf{f} : \mathbb{R}^n \to \mathbb{R}^m$ be the constant function $\mathbf{f}(\mathbf{x}) = \mathbf{v}$. Find the derivative of $\mathbf{f}$ at an arbitrary $\mathbf{x} \in \mathbb{R}^n$. (*Hint:* Guess and then show that your guess "works." That is, compute $\mathbf{r}$ and show that it satisfies Definition 13.4.1.)

2. Let $\mathbf{L} : \mathbb{R}^n \to \mathbb{R}^m$ be a linear transformation. Prove that $\mathbf{L}$ is differentiable, and that at every $\mathbf{x} \in \mathbb{R}^n$, $\mathbf{L}'(\mathbf{x}) = \mathbf{L}$. Give an intuitive explanation of this result.

3. Suppose that $f : \mathbb{R} \to \mathbb{R}$ is the linear transformation $f(x) = ax$. The result in part (2) says that for all $x \in \mathbb{R}$, $f'(x) = f$. This may appear to be saying that $f'(x) = ax$, which should clash with your previous notion that $f'(x) = a$. Keeping in mind that, in moving to functions of several variables, we have changed our point of view about what a derivative is, resolve this conundrum. ∎

The definition of the derivative tells us immediately that differentiable functions are also continuous.

Theorem 13.4.4 Let E be an open subset of $\mathbb{R}^n$. Let $\mathbf{f} : E \to \mathbb{R}^m$ be a function, and let $\mathbf{x} \in E$. Suppose that $\mathbf{f}$ is **differentiable** at $\mathbf{x}$. Then $\mathbf{f}$ is continuous at $\mathbf{x}$.

Proof: The proof is Problem 5 at the end of this section. □

Definition 13.4.1 says that a function is differentiable at $\mathbf{x}$ if there exist a linear transformation $\mathbf{f}'(\mathbf{x})$ and a function $\mathbf{r} : E \to \mathbb{R}^m$ that satisfy certain conditions. Given the definition, it seems plausible that two such pairs could exist, but that turns out not to be the case.

Theorem 13.4.5 Let E be an open subset of $\mathbb{R}^n$. Let $\mathbf{f} : E \to \mathbb{R}^m$ be a function, and let $\mathbf{x} \in E$. Suppose that $\mathbf{f}$ is differentiable at the point $\mathbf{x}$. Then the derivative $\mathbf{f}'(\mathbf{x})$ is unique.

Proof: The proof is Problem 6 at the end of this section. □

Connections: Total and Partial Derivatives

So far, our discussion has dealt exclusively with the local linearity of a function of several variables. Our task now is to bring rates of change into the mix. Before we can discern the relationship between rates of change and the derivative of a function $\mathbf{f} : \mathbb{R}^n \to \mathbb{R}^m$, we have to think a bit more about this derivative.

Let E be an open subset of $\mathbb{R}^n$, and let the function $f : E \to \mathbb{R}^m$ be differentiable at $\mathbf{a} \in E$. At present, we know two things about $\mathbf{f}'(\mathbf{a})$. First, it is a linear transformation from $\mathbb{R}^n$ to $\mathbb{R}^m$. And, second, the function

$$\mathbf{T}(\mathbf{x}) = \mathbf{f}'(\mathbf{a})(\mathbf{x} - \mathbf{a}) + \mathbf{f}(\mathbf{a})$$

is the "best" local linear approximation for the function $\mathbf{f}$ at the point $\mathbf{a}$. At this juncture, you may feel like your intuition is a bit thin. In particular, aside from a few trivial examples where you can make a lucky guess, you are probably at a loss as to how you would actually *compute* a total derivative.

Recall that a linear transformation $\mathbf{L} : \mathbb{R}^n \to \mathbb{R}^m$ corresponds to an $m \times n$ matrix of real numbers. Therefore, computing the derivative of a function of several variables amounts to finding the right set of mn real numbers. What are these numbers? Interestingly (but perhaps not surprisingly), the answer involves slopes. Once again, we turn to scalar functions of one and two variables for intuition.

Suppose we have a function $f : \mathbb{R} \to \mathbb{R}$ that is differentiable at $x = a$. The derivative of f at a is the number[7] required to specify the tangent line to the graph of the function through the point $(a, f(a))$—that is, the slope (in the positive x-direction) of the graph of f at the point $(a, f(a))$.

If $f : \mathbb{R}^2 \to \mathbb{R}$ is differentiable at $\mathbf{a}$, $f'(\mathbf{a})$ will be given by a 1×2 matrix; we need two numbers to specify it. Furthermore, the fact that f is differentiable at $\mathbf{a}$ means the graph of f is "locally planar" at $(\mathbf{a}, f(\mathbf{a}))$. The derivative of f at $\mathbf{a}$ will need to specify the tangent plane to the graph of $\mathbf{f}$ at the point $(\mathbf{a}, f(\mathbf{a}))$. We have already said that, in addition to a point, the two numbers we need to specify a plane are its slope in the (positive) x-direction and its slope in the (positive) y-direction. The slopes of f at the point $\mathbf{a} = (a_1, a_2)$ in the x and y directions will be the derivatives of the one-variable functions

$$x \to f(x, a_2) \qquad \text{and} \qquad y \to f(a_1, y)$$

that we get by fixing one variable of f and letting the other vary. (See Figure 13.3.) As you probably know, the slopes of f in the x- and y-directions are called the partial derivatives of f with respect to x and y, respectively.

When we have more variables, we will have more partial derivatives but the basic idea is the same. We now give a precise definition for the partial derivatives of a scalar field at a point.

---

7. A 1×1 matrix!

13.4 Local Linear Approximation for Functions of Several Variables

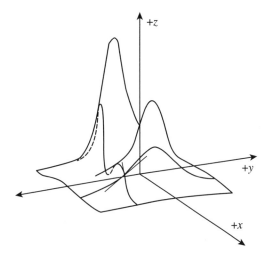

Figure 13.3 The perpendicular paths along the surface show the functions obtained by setting one variable to a constant and letting the other one vary. The tangents at the intersection have slopes that are the partial derivatives with respect to x and y at the point of intersection.

Definition 13.4.6 [Partial Derivatives in Scalar Fields] Let E be an open subset of $\mathbb{R}^n$. Let $f : E \to \mathbb{R}$ be a scalar field in the variables x_1, x_2, x_3, ..., x_n. Fix i such that $1 \leq i \leq n$. Provided that the limit exists, the **partial derivative of f with respect to x_i at the point a** or the **partial derivative of f with respect to the ith variable at the point a** is given by

$$\lim_{h \to 0} \frac{f(\mathbf{a} + h\mathbf{e}_i) - f(\mathbf{a})}{h},$$

or equivalently by

$$\lim_{h \to 0} \frac{f(a_1, \ldots, a_{i-1}, a_i + h, a_{i+1}, \ldots, a_n) - f(a_1, \ldots, a_n)}{h},$$

or

$$\lim_{x_i \to a_i} \frac{f(a_1, \ldots, a_{i-1}, x_i, a_{i+1}, \ldots, a_n) - f(a_1, \ldots, a_n)}{x_i - a_i}.$$

A number of standard notations for the partial derivatives of f exist, but we will (interchangeably) use only two. Fix i with $1 \leq i \leq n$. Then the partial derivative of f with respect to x_i is denoted by

$$\frac{\partial f}{\partial x_i} \quad \text{or} \quad D_i f$$

When each of these is evaluated at the point **a**, we have

$$\frac{\partial f}{\partial x_i}(\mathbf{a}) \quad \text{or} \quad D_i f(\mathbf{a})$$

> **Note:** The partial derivative of f with respect to x_i at the point $\mathbf{a} = (a_1, a_2, \ldots, a_n)$ is just the derivative of the function $g_i : \mathbb{R} \to \mathbb{R}$ given by
>
> $$g_i(x) = f(a_1, a_2, \ldots, a_{i-1}, x, a_{i+1}, \ldots, a_n)$$
>
> evaluated at the point a_i: We fix the values of all but the ith variables of f and let the ith one vary. Then we take the derivative and evaluate it at $x = a_i$.
>
> $$\frac{\partial f}{\partial x_i}(\mathbf{a}) = g_i'(a_i)$$

The definition for partial derivatives of scalar-valued functions generalizes easily to the case of vector-valued functions.

Definition 13.4.7 [Partial Derivatives in Vector Fields] Let E be an open subset of $\mathbb{R}^n$. Let $\mathbf{f}(\mathbf{x}) = (f_1(\mathbf{x}), f_2(\mathbf{x}), \ldots, f_m(\mathbf{x}))$ be a vector-valued function from E to $\mathbb{R}^m$. Then the **partial derivatives of f** are just the partial derivatives of the scalar-valued components $f_1, f_2, \ldots, f_m$ of $\mathbf{f}$.

We can easily adapt the notation for the partial derivatives of scalar functions to the vector-valued case. Note that each of the m component functions of $\mathbf{f}$ has a partial derivative for each of the n variables. Thus f has partial derivatives:

$$\begin{array}{cccc} D_1 f_1 & D_2 f_1 & \cdots & D_n f_1 \\ D_1 f_2 & D_2 f_2 & \cdots & D_n f_2 \\ \vdots & \vdots & \ddots & \vdots \\ D_1 f_m & D_2 f_m & \cdots & D_n f_m \end{array}$$

Notice that we have an $m \times n$ array of partial derivatives. At this point, Theorem 13.4.8 will probably not surprise you. It says that for a function $\mathbf{f} : \mathbb{R}^n \to \mathbb{R}^m$, if $\mathbf{f}'(\mathbf{a})$ exists, then all the partial derivatives of $\mathbf{f}$ exist at $\mathbf{a}$, and the mn partial derivatives of $\mathbf{f}$ at $\mathbf{a}$ specify the entries in the necessary derivative matrix.

13.4 Local Linear Approximation for Functions of Several Variables

Theorem 13.4.8 Let E be an open subset of $\mathbb{R}^n$. Let $\mathbf{f} : E \to \mathbb{R}^m$ be a vector field. Let $\mathbf{a} \in \mathbb{R}^n$. Suppose that f is differentiable at $\mathbf{a}$. Then all the partial derivatives of f exist at $\mathbf{a}$.

Furthermore, if $\{\mathbf{e}_i\}_{i=1}^n$ and $\{\mathbf{u}_j\}_{j=1}^m$ are the standard unit vector bases for $\mathbb{R}^n$ and $\mathbb{R}^m$, respectively, it follows that for $i = 1, 2, \ldots, n$, and $j = 1, 2, \ldots, m$,

$$[\mathbf{f}'(\mathbf{a})](\mathbf{e}_i) = \sum_{j=1}^m \frac{\partial f_j}{\partial x_i}(\mathbf{a})\mathbf{u}_j.$$

Proof: The proof is Problem 7 at the end of this section. □

Corollary 13.4.9 Let E be an open subset of $\mathbb{R}^n$. Let $\mathbf{f} : E \to \mathbb{R}^m$ be a vector field in the variables $x_1, x_2, x_3, \ldots, x_n$. Let $\mathbf{a} \in \mathbb{R}^n$. Suppose that f is differentiable at $\mathbf{a}$. Then all the partial derivatives of $\mathbf{f}$ exist at $\mathbf{a}$, and the matrix representing the linear transformation $\mathbf{f}'(\mathbf{a})$ is the matrix $[J_\mathbf{f}](\mathbf{a})$ given by

$$[J_\mathbf{f}](\mathbf{a}) = \begin{bmatrix} \frac{\partial f_1}{\partial x_1}(\mathbf{a}) & \frac{\partial f_1}{\partial x_2}(\mathbf{a}) & \cdots & \frac{\partial f_1}{\partial x_n}(\mathbf{a}) \\ \frac{\partial f_2}{\partial x_1}(\mathbf{a}) & \frac{\partial f_2}{\partial x_2}(\mathbf{a}) & \cdots & \frac{\partial f_2}{\partial x_n}(\mathbf{a}) \\ \vdots & \vdots & \ddots & \vdots \\ \frac{\partial f_m}{\partial x_1}(\mathbf{a}) & \frac{\partial f_m}{\partial x_2}(\mathbf{a}) & \cdots & \frac{\partial f_m}{\partial x_n}(\mathbf{a}) \end{bmatrix}$$

This matrix is called the **Jacobian matrix of f at a**.

Proof: Exercise. Do it now! [8] □

---

[8]. Corollary 13.4.9 is just a rephrasing of Theorem 13.4.8. The latter formulation is more easily understood and simpler to use. In fact, you may ask yourself why we didn't just go straight to the corollary. The answer, as is often true in such cases, is that the formulation given in Theorem 13.4.8 gives us a clue about how to prove the result. After we've proved it, we then restate things in a more usable form. Conversely, having worked out the relationship between the two statements before proving the main result, you will find it easier to prove the theorem because you will understand it better.

> **Important Special Case: The Derivative of a Scalar Field.** Let E be an open subset of $\mathbb{R}^n$, and let $f : E \to \mathbb{R}$ be a scalar field in the variables $x_1, x_2, x_3, \ldots, x_n$. Suppose that f is differentiable at $\mathbf{a} \in E$. Then Corollary 13.4.9 implies that all of the partial derivatives of f exist at $\mathbf{a}$; the derivative is given by the Jacobian matrix, which will have a single row. We can interpret it as an n-vector given by
>
> $$\nabla f(\mathbf{a}) = \left(\frac{\partial f}{\partial x_1}(\mathbf{a}), \frac{\partial f}{\partial x_2}(\mathbf{a}), \frac{\partial f}{\partial x_3}(\mathbf{a}), \ldots, \frac{\partial f}{\partial x_n}(\mathbf{a}) \right)$$
>
> This vector is called the **gradient** of f at $\mathbf{a}$. Matrix multiplication becomes a dot product; the vector $\nabla f(\mathbf{a})$ represents the linear transformation $\mathbf{f}'(\mathbf{a})$ in the sense that for all $\mathbf{v} \in \mathbb{R}^n$,
>
> $$\mathbf{f}'(\mathbf{a})(\mathbf{v}) = \nabla f(\mathbf{a}) \cdot \mathbf{v}$$

Theorem 13.4.10 Let E be an open subset of $\mathbb{R}^n$. Let $\mathbf{f} : E \to \mathbb{R}^m$ be a vector field with scalar component functions $f_1, f_2, \ldots, f_m$. Let $\mathbf{a} \in E$. Then $\mathbf{f}$ is differentiable at $\mathbf{a}$ if and only if for each $i = 1, 2, \ldots, m$, f_i is differentiable at $\mathbf{a}$.

Proof: The proof is Problem 16 at the end of this section. $\square$

Theorem 13.4.8 says that if a function $\mathbf{f} : \mathbb{R}^n \to \mathbb{R}^m$ is differentiable at $\mathbf{a}$, then the partial derivatives of f at $\mathbf{a}$ all exist. The converse is not true. In fact, the existence of the partial derivatives of f at a point $\mathbf{a}$ does not guarantee the continuity of $\mathbf{f}$ at $\mathbf{a}$ even for scalar fields!

Example 13.4.11 Consider the function $f : \mathbb{R}^2 \to \mathbb{R}$ given by

$$f(x,y) = \begin{cases} 1 & \text{if } x \neq 1 \text{ and } y \neq 1 \\ 2 & \text{if } x = 1 \text{ or } y = 1 \end{cases}$$

(See Figure 13.4.) It is clear that f is not continuous at $(1,1)$, but the partial derivatives of f exist and are zero there.

However, if we insist that the partial derivatives exist at and near $\mathbf{a}$, and also that they be continuous at $\mathbf{a}$, we can deduce the differentiability of f at $\mathbf{a}$. We begin by proving the result for scalar fields. The more general result will then follow.

13.4 Local Linear Approximation for Functions of Several Variables 279

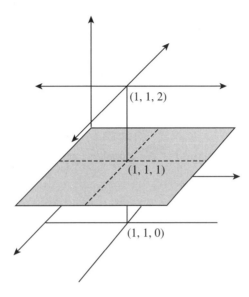

Figure 13.4 The graph of the function f given in Example 13.4.11.

Theorem 13.4.12 Let E be an open subset of $\mathbb{R}^n$. Let $f : E \to \mathbb{R}$ be a function, and let $\mathbf{a} = (a_1, a_2, \ldots, a_n) \in E$. Suppose that there exists $r > 0$ such that the partial derivatives $D_1 f, D_2 f, \ldots, D_\mathbf{n} f$ of f exist at every point of $B_r(\mathbf{a})$ and that each is continuous at $\mathbf{a}$. Then f is differentiable at $\mathbf{a}$.

Proof Outline:
Here is a detailed outline of the proof of this theorem. You should read the proof with paper and pencil, filling in the details as you go along.

Step 1. Define $r : E \to \mathbb{R}$ such that for all $\mathbf{y} \in E$,
$$f(\mathbf{y}) = \nabla f(\mathbf{a}) \cdot (\mathbf{y} - \mathbf{a}) + f(\mathbf{a}) + e(\mathbf{y}).$$
If we can show that
$$\frac{|e(\mathbf{y})|}{\|\mathbf{y} - \mathbf{a}\|} \to 0 \text{ as } \mathbf{y} \to \mathbf{a}$$
we will be done. Note that
$$\frac{|e(\mathbf{y})|}{\|\mathbf{y} - \mathbf{a}\|} = \frac{1}{\|\mathbf{y} - \mathbf{a}\|} |f(\mathbf{y}) - \nabla f(\mathbf{a}) \cdot (\mathbf{y} - \mathbf{a}) - f(\mathbf{a})|.$$

Step 2. Let $\epsilon > 0$. Given that for all $i \leq n$, $D_i(f)$ is continuous at $\mathbf{a}$, we can choose $\delta^* > 0$ such that if $\|\mathbf{w} - \mathbf{a}\| < \delta^*$ and $i = 1, 2, \ldots, n$, then
$$|D_i f(\mathbf{w}) - D_i f(\mathbf{a})| < \frac{\epsilon}{n}.$$

Step 3. To exploit the hypothesis that the partial derivatives of f exist at every point in some open ball $B_r(\mathbf{a})$, we wish to restrict our attention to points in $B_r(\mathbf{a})$. So we choose our value of $\delta = \min(r, \delta^*)$. Now we choose $\mathbf{y} \in E$ such that $\|\mathbf{y} - \mathbf{a}\| < \delta$, say, $\mathbf{y} = (y_1, y_2, \ldots, y_n)$.

Notation: For $i = 1, 2, \ldots, n$, designate the closed interval between y_i and a_i by K_i.

Useful Fact: Fix $i \leq n$ and $t \in K_i$. Consider the element

$$\mathbf{w} = (y_1, y_2, \ldots, y_{i-1}, t, a_{i+1}, a_{i+2}, \ldots, a_n)$$

(If $i = 1$, $\mathbf{w} = (t, a_2, a_3, \ldots, a_n)$, and if $i = n$, $\mathbf{w} = (y_1, y_2, \ldots, y_{n-1}, t)$.) It is easy to see that

$$\|\mathbf{w} - \mathbf{a}\| \leq \|\mathbf{y} - \mathbf{a}\| < \delta. \ (\textit{Work it out!})$$

See Figure 13.5 for a geometric interpretation of this fact.

Step 4. Set $\mathbf{w}_0 = \mathbf{a}$, $\mathbf{w}_i = (y_1, y_2, \ldots, y_i, a_{i+1}, a_{i+2}, \ldots, a_n)$, and $\mathbf{w}_n = \mathbf{y}$.

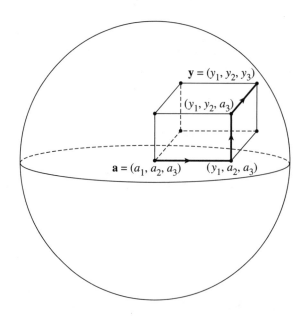

Figure 13.5 If $\mathbf{y}$ is in $B_r(\mathbf{a})$, then so are the edges of the parallelepiped determined by $\mathbf{a}$ and $\mathbf{y}$. The diagram illustrates this idea in three dimensions; the "useful fact" shows that an analogous fact holds in higher dimensions.

13.4 Local Linear Approximation for Functions of Several Variables

Consider the quantity $f(\mathbf{y}) - f(\mathbf{a})$. Using the vectors $\mathbf{w}_i$, we can write it as a telescoping sum:

$$f(\mathbf{y}) - f(\mathbf{a}) = \sum_{i=1}^{n} f(\mathbf{w}_i) - f(\mathbf{w}_{i-1})$$

Step 5. Our useful fact tells us that for $i = 1, 2, \ldots, n$, $\mathbf{w}_{i-1}$, $\mathbf{w}_i$, and the entire line segment joining the two are all contained in $\mathrm{B}_r(\mathbf{a})$. Thus the partial derivatives of f exist at every point on each of these intervals. It follows that for a fixed $i \leq n$, the function $g_i : K_i \to \mathbb{R}$ given by

$$g_i(t) = f(y_1, y_2, \ldots, y_{i-1}, t, a_{i+1}, a_{i+2}, \ldots, a_n)$$

is differentiable on its domain, and for $t \in K_i$,

$$g_i'(t) = D_i f(y_1, y_2, \ldots, y_{i-1}, t, a_{i+1}, a_{i+2}, \ldots, a_n).$$

Thus the (one-variable) mean value theorem holds for g_i; therefore, for $i = 1, 2, \ldots, n$, we can find $c_i \in K_i$ such that

$$g_i'(c_i)(y_i - a_i) = g(y_i) - g(a_i)$$

But then

$$D_i f(y_1, y_2, \ldots, y_{i-1}, c_i, a_{i+1}, a_{i+2}, \ldots, a_n)(y_i - a_i)$$
$$= f(\mathbf{w}_i) - f(\mathbf{w}_{i-1})$$

For $i = 1, 2, \ldots, n$, set $\mathbf{t}_i = (y_1, y_2, \ldots, y_{i-1}, c_i, a_{i+1}, a_{i+2}, \ldots, a_n)$.

Step 6. Steps 4 and 5 tell us that

$$f(\mathbf{y}) - f(\mathbf{a}) = \sum_{i=1}^{n} f(\mathbf{w}_i) - f(\mathbf{w}_{i-1}) = \sum_{i=1}^{n} D_i f(\mathbf{t}_i)(y_i - a_i)$$

Last Step. You can now complete the proof. Show that

$$\frac{|e(\mathbf{y})|}{\|\mathbf{y} - \mathbf{a}\|} = \frac{1}{\|\mathbf{y} - \mathbf{a}\|} |f(\mathbf{y}) - f(\mathbf{a}) - \nabla f(\mathbf{a}) \cdot (\mathbf{y} - \mathbf{a})|$$

is less than ϵ.

$\square$

Corollary 13.4.13 Let E be an open subset of $\mathbb{R}^n$. Let $f : E \to \mathbb{R}^m$ be a function, and let $\mathbf{a} \in E$. Let $U \subseteq E$ be an open subset of $\mathbb{R}^n$ containing $\mathbf{a} = (a_1, a_2, \ldots, a_n)$. Suppose that the partial derivatives

$$\begin{array}{cccc} D_1 f_1 & D_2 f_1 & \cdots & D_n f_1 \\ D_1 f_2 & D_2 f_2 & \cdots & D_n f_2 \\ \vdots & \vdots & \ddots & \vdots \\ D_1 f_m & D_2 f_m & \cdots & D_n f_m \end{array}$$

of f exist at every point of U and that each is continuous at $\mathbf{a}$. Then f is differentiable at $\mathbf{a}$.

Proof: The proof is Problem 17 at the end of this section. □

Problems 13.4

1. Let E be an open subset of $\mathbb{R}^n$. Let $\mathbf{f} : E \to \mathbb{R}^m$ and $\mathbf{g} : E \to \mathbb{R}^m$ be functions, and let α be a scalar. Suppose that $\mathbf{f}$ and $\mathbf{g}$ are differentiable at $\mathbf{x} \in E$. Prove that the functions $\alpha \mathbf{f}$ and $\mathbf{f} + \mathbf{g}$ are differentiable at $\mathbf{x}$ with

$$(\alpha \mathbf{f})'(\mathbf{x}) = \alpha(\mathbf{f}'(\mathbf{x})) \quad \text{and} \quad [\mathbf{f} + \mathbf{g}]'(\mathbf{x}) = \mathbf{f}'(\mathbf{x}) + \mathbf{g}'(\mathbf{x})$$

2. Let $\mathbf{f} : \mathbb{R}^n \to \mathbb{R}^n$ be the function $\mathbf{f}(\mathbf{y}) = \mathbf{y}$. Using only Definition 13.4.1, prove that $\mathbf{f}$ is differentiable at every point of $\mathbb{R}^n$ and that for all $\mathbf{a} \in \mathbb{R}^n$, $\mathbf{f}'(\mathbf{a})$ is represented by I_n, the $n \times n$ identity matrix. Compare this result to Exercise 13.4.3, part 2, and to Corollary 13.4.9.

3. Let E be an open subset of $\mathbb{R}^n$. Let $f : E \to \mathbb{R}^m$ be differentiable at $\mathbf{a} \in E$. Let $\mathbf{L}$ be any linear transformation from $\mathbb{R}^m$ to $\mathbb{R}^k$. Prove that the function $\mathbf{g} = \mathbf{L} \circ \mathbf{f}$ is differentiable at $\mathbf{a}$ and that $\mathbf{g}'(\mathbf{a}) = \mathbf{L} \circ [\mathbf{f}'(\mathbf{a})]$.

> Consider Problems 2 and 3. What results in single variable calculus do they generalize?

4. **The chain rule:** Let E be an open subset of $\mathbb{R}^n$, and let V be an open subset of $\mathbb{R}^m$. Let $\mathbf{f} : E \to \mathbb{R}^m$ and $\mathbf{g} : V \to \mathbb{R}^k$ be functions with $\mathbf{f}(E) \subseteq V$, such that $\mathbf{g} \circ \mathbf{f}$ is defined on E. Suppose that $\mathbf{f}$ is differentiable at

13.4 Local Linear Approximation for Functions of Several Variables

$\mathbf{a} \in E$ and that $\mathbf{g}$ is differentiable at $\mathbf{f}(\mathbf{a})$. Prove that $\mathbf{g} \circ \mathbf{f}$ is differentiable at $\mathbf{a}$, and $[\mathbf{g} \circ \mathbf{f}]'(\mathbf{a}) = \mathbf{g}'(\mathbf{f}(\mathbf{a})) \circ \mathbf{f}'(\mathbf{a})$.

[*Hint*: Mimic the proof of the related one-variable result (Theorem 9.2.7). Don't forget that the derivative is a *linear* function and that linear functions are continuous. These facts will be important in the proof.]

5. Prove that if $\mathbf{f} : \mathbb{R}^n \to \mathbb{R}^m$ is differentiable at $\mathbf{x}$, then it is continuous at $\mathbf{x}$ (Theorem 13.4.4).

6. Prove that if $\mathbf{f}$ is differentiable at $\mathbf{x}$, then the derivative of $\mathbf{f}$ at $\mathbf{x}$ is unique (Theorem 13.4.5).

7. Prove Theorem 13.4.8. (*Hint*: Start with the expression guaranteed by the differentiability of f at $\mathbf{a}$. Let t_n be a decreasing sequence of positive real numbers that goes to zero. Consider the sequence $\mathbf{y}_n = \mathbf{a} + t_n \mathbf{e}_i$.)

8. Let E be an open subset of $\mathbb{R}^n$, let $f : E \to \mathbb{R}$ be a scalar field, and let $\mathbf{a} \in E$. Suppose that all of the partial derivatives of f exist at $\mathbf{a}$. Prove that f is differentiable at $\mathbf{a}$ if and only if there exists a function $r : E \to \mathbb{R}$ such that for all $\mathbf{x} \in E$,

$$f(\mathbf{x}) = \nabla f(\mathbf{a}) \cdot (\mathbf{x} - \mathbf{a}) + f(\mathbf{a}) + r(\mathbf{x}) \quad \text{and} \quad \lim_{\mathbf{x} \to \mathbf{a}} \frac{|r(\mathbf{x})|}{\|\mathbf{x} - \mathbf{a}\|} \to 0 \text{ as } \mathbf{x} \to \mathbf{a}.$$

9. Consider the three-variable scalar field $f(x, y, z) = x - 2xy + z^2$. Let $\mathbf{x} = (x, y, z)$ and $\mathbf{h} = (\Delta x, \Delta y, \Delta z)$.

    (a) Compute $f(\mathbf{x} + \mathbf{h})$ explicitly in terms of x, y, z, Δx, Δy, and Δz.

    (b) Write the expression you got in part (a) in the form $\mathbf{f}(\mathbf{x} + \mathbf{h}) = \mathbf{f}'(\mathbf{x})(\mathbf{h}) + \mathbf{f}(\mathbf{x}) + \mathbf{r}(\mathbf{h})$ or, in more expansive terms,

    $$f(x + \Delta x, y + \Delta y, z + \Delta z)$$
    $$= \nabla f(\mathbf{x}) \cdot (\Delta x, \Delta y, \Delta z) + f(x, y, z) + r(\Delta x, \Delta y, \Delta z)$$

    See the box on page 278 for a definition of the gradient vector $\nabla f(\mathbf{x})$.

    (c) Using the expression that you derived in part (b) for $r(\mathbf{h})$, Show that $\dfrac{|r(\mathbf{h})|}{\|\mathbf{h}\|} \to 0$ as $\mathbf{h} \to \mathbf{0}$.

    You can now conclude that f is differentiable everywhere.

10. Let E be an open subset of $\mathbb{R}^n$. Let f and g be scalar-valued functions on E that are differentiable at $\mathbf{x} \in E$. Prove that the product f and g given by $fg(\mathbf{x}) = f(\mathbf{x})g(\mathbf{x})$ is differentiable at $\mathbf{x}$ and that $\nabla (fg)(\mathbf{x}) = g(\mathbf{x})\nabla f(\mathbf{x}) + f(\mathbf{x})\nabla g(\mathbf{x})$.

> What about a product rule for vector fields? Obviously $\mathbf{f}(\mathbf{x})\mathbf{g}(\mathbf{x})$ makes no sense for general vector fields, but what about $\mathbf{f}(\mathbf{x}) \cdot \mathbf{g}(\mathbf{x})$? Could there be a product rule associated with the *dot product* of two such vectors? Or would some other vector product give us a reasonable generalization? If so, what would such a generalization look like? I leave it to you to consider the question.

11. Differentiating parametric curves in space. Let $\mathbf{f} : \mathbb{R} \to \mathbb{R}^m$ be given by $\mathbf{f}(x) = (f_1(x), f_2(x), \ldots, f_m(x))$. Let $x \in \mathbb{R}$. Prove that $\mathbf{f}$ is differentiable at x in the sense of Definition 13.4.1 if and only if $f_1, f_2, \ldots, f_m$ are all differentiable at x and $\mathbf{f}'(x) = (f_1'(x), f_2'(x), \ldots, f_m'(x))$.

    Notice that we are associating the $1 \times m$ column matrix $\mathbf{f}'(x)$ with its natural m-vector cousin.[9]

12. Suppose that E is an open subset of $\mathbb{R}^n$ and that $f : E \to \mathbb{R}$ is a differentiable scalar field. Let $\mathbf{a} = (a_1, a_2, \ldots, a_n) \in E$. Suppose that f has a local maximum at $\mathbf{a}$. Prove that $\nabla f(\mathbf{a}) = \mathbf{0}$.

13. Consider the function $f : \mathbb{R}^2 \to \mathbb{R}$ given by

    $$f(x) = \begin{cases} \dfrac{x^2 + 2y^2}{x^2 + y^2} & \text{if } (x,y) \neq (0,0) \\ 0 & \text{if } (x,y) = (0,0) \end{cases}.$$

    It is clear that the partial derivatives of f exist everywhere except possibly at $\mathbf{0}$.

    (a) Show that the partial derivatives of f both exist at $\mathbf{0}$.

    (b) Show that f is not continuous at $\mathbf{0}$.

    A computer-generated graph of this function around zero may help you see what is happening.

14. Let E be an open subset of $\mathbb{R}^n$. Let $f : E \to \mathbb{R}$ be a function, and let $\mathbf{a} = (a_1, a_2, \ldots, a_n) \in E$. Suppose that there exists $r > 0$ such that the partial derivatives $D_1 f, D_2 f, \ldots, D_n f$ of f exist at every point of $B_r(\mathbf{a})$ and are

---

[9]. Many sources *define* a parametric curve f to be differentiable at x if each of its component functions is differentiable at x. In Problem 11, you show that component-wise differentiation of parametric functions is consistent with the more general definition we have been discussing.

uniformly bounded there. That is, suppose that there exists $M \in \mathbb{R}$ such that for all $\mathbf{x} \in B_r(\mathbf{a})$ and all $i \leq n$, $|D_i f(\mathbf{x})| \leq M$. Prove that f is continuous at $\mathbf{a}$.

15. Consider the function $f : \mathbb{R}^2 \to \mathbb{R}$ given by

$$f(x) = \begin{cases} \dfrac{2xy^2}{x^2 + y^2} & \text{if } (x, y) \neq (0, 0) \\ 0 & \text{if } (x, y) = (0, 0) \end{cases}.$$

It is clear that the partial derivatives of f exist everywhere except possibly at $\mathbf{0}$.

   (a) Show that the partial derivatives of f exist at $\mathbf{0}$.
   (b) Show that the partial derivatives of f are bounded in a region around $\mathbf{0}$ (and thus that f is continuous at $\mathbf{0}$).
   (c) Show that f is not differentiable at $\mathbf{0}$.

16. In this problem you will prove Theorem 13.4.10. But first, here are some preliminary things to consider by way of a hint.

    Let E be an open subset of $\mathbb{R}^n$. Let $\mathbf{f} : E \to \mathbb{R}^m$ be a vector field with component functions $f_1, f_2, \ldots, f_m$. Let $\mathbf{a} \in E$.

    (a) Suppose that $\mathbf{f}$ is differentiable at $\mathbf{a}$, as is each of its component functions. Consider the vector equation

    $$\mathbf{f}(\mathbf{y}) = \mathbf{f}'(\mathbf{a})(\mathbf{y} - \mathbf{a}) + \mathbf{f}(\mathbf{a}) + \mathbf{r}(\mathbf{y})$$

    and kindred scalar equations corresponding to $f_1, f_2, \ldots, f_m$. What is the relationship between the vector equation and the set of scalar equations?

    (b) Suppose that $\mathbf{f}$ is differentiable at $\mathbf{a}$, as is each of its component functions. What is the relationship between $\mathbf{f}'(\mathbf{a})$ and $f_1'(\mathbf{a})$, $f_2'(\mathbf{a})$, $\ldots$, $f_m'(\mathbf{a})$?

    (c) Prove that $\mathbf{f}$ is differentiable at $\mathbf{a}$ if and only if for each $i = 1, 2, \ldots, m$, f_i is differentiable at $\mathbf{a}$.

17. Prove Corollary 13.4.13. (*Hint*: Consider Theorems 13.4.10 and 13.4.12.)

18. **Equality of mixed partials.** We begin with a definition and some notation. Let E be an open subset of $\mathbb{R}^n$, and let $f : E \to \mathbb{R}$ be a function in the variables $x_1, x_2, \ldots, x_n$. Suppose that all of the partial derivatives of

f exist and that each is itself differentiable. Then the **second-order partial derivatives** of f are the partial derivatives of the partial derivatives. We denote these as follows:

$$\frac{\partial}{\partial x_j}\left(\frac{\partial f}{\partial x_i}\right) = \frac{\partial^2 f}{\partial x_j \partial x_i} \quad \text{or} \quad D_j D_i f.$$

If $j = i$, we contract this notation:

$$\frac{\partial^2 f}{\partial x_i^2} \quad \text{or} \quad D_i^2 \mathbf{f}.$$

When each of these is evaluated at the point $\mathbf{a}$, we have

$$\frac{\partial^2 f}{\partial x_j \partial x_i}(\mathbf{a}) \quad \text{or} \quad D_j D_i f(\mathbf{a}).$$

When $i \neq j$, we call the second partials **"mixed" partials**.

The goal of this problem is to help you prove the following

> **Mixed Partials theorem:** Let E be an open subset of $\mathbb{R}^n$, and let $f : E \to \mathbb{R}$ be a function in the variables $x_1, x_2, \ldots, x_n$. Let $\mathbf{a} \in E$. Suppose that f is such that $D_i f$, $D_j f$, and $D_i D_j f$ all exist at every point of E. Suppose also that $D_i D_j f$ is continuous at $\mathbf{a}$. Then $D_j D_i f$ exists at $\mathbf{a}$ and $D_i D_j f(\mathbf{a}) = D_j D_i f(\mathbf{a})$.

(a) Explain why there is no loss of generality in proving this result only for functions of two variables. (Later in this problem, you will be asked to account for this result in a rigorous way, but a "loose" argument is fine for now.)

For the remainder of the problem, E is an open subset of $\mathbb{R}^2$, $f : E \to \mathbb{R}$, and $\mathbf{a} = (a_1, a_2) \in E$.

(b) *Building intuition:* Consider the part of the surface (given by the graph of f) that sits above the rectangle shown in Figure 13.6. Roughly speaking, the second-order mixed partial $D_1 D_2 f(\mathbf{a})$ measures the rate of change in the "north–south" slope between the left end of the rectangle and the right end of the rectangle.

$$D_1 D_2 f(\mathbf{a}) \approx \frac{\frac{f(a_1, a_2+k) - f(a_1, a_2)}{k} - \frac{f(a_1+h, a_2+k) - f(a_1+h, a_2)}{k}}{h}.$$

Compute a similar approximation for $D_2 D_1 f(\mathbf{a})$.

Use these expressions to argue (heuristically) that $D_1 D_2 f(\mathbf{a})$ and $D_2 D_1 f(\mathbf{a})$ should be approximately equal to each other. (Ignore issues such as continuity conditions, which will obviously come into play when an actual proof is given.)

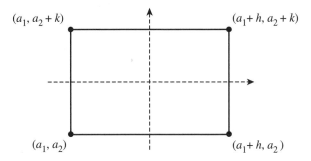

Figure 13.6 Diagram for Problem 18.

Previous experience may tell you that the move from

$$D_1 D_2 f(\mathbf{a}) \approx D_2 D_1 f(\mathbf{a})$$

to

$$D_1 D_2 f(\mathbf{a}) = D_2 D_1 f(\mathbf{a})$$

will require some form of the mean value theorem. And, indeed, this is so.

(c) Use the one-variable mean value theorem to prove the following lemma:

Suppose that $D_1 f$ and $D_1 D_2 f$ exist at every point of E. Suppose that $\mathcal{R} \subseteq E$ is the rectangle shown in Figure 13.6. Then there exists a point $\mathbf{c}$ in the interior of $\mathcal{R}$ such that

$$(f(a_1 + h, a_2 + k) - f(a_1 + h, a_2)) - (f(a_1, a_2 + k) - f(a_1, a_2))$$
$$= hk D_1 D_2 f(\mathbf{c}).$$

(d) Prove the mixed partials theorem for a function of two variables. Conclude your proof with a careful account of why this result is sufficient to allow us to deduce the more general n-variable theorem.

13.5 The Mean Value Theorem for Functions of Several Variables

The mean value theorem was a crucial piece of the theoretical framework for the theory of differentiation of functions from $\mathbb{R}$ to $\mathbb{R}$. It stands to reason that something similar will be useful for functions of several variables.

As always, we take our cue from the one-variable case. If $f : \mathbb{R} \to \mathbb{R}$ is differentiable on (a, b) and continuous on $[a, b]$, then there exists $c \in (a, b)$ such that
$$f(b) - f(a) = f'(c)(b - a).^{10}$$
Suppose that $\mathbf{f} : \mathbb{R}^n \to \mathbb{R}^m$ is differentiable. Ideally, we would like to say that given $\mathbf{a}$ and $\mathbf{b} \in \mathbb{R}^n$, there exists $\mathbf{c}$ lying on the line segment joining $\mathbf{a}$ and $\mathbf{b}$ such that
$$\mathbf{f}(\mathbf{b}) - \mathbf{f}(\mathbf{a}) = [\mathbf{f}'(\mathbf{c})](\mathbf{b} - \mathbf{a})$$
where, as usual, we must interpret the expression on the right as the linear transformation $\mathbf{f}'(\mathbf{c})$ *acting* on the vector $\mathbf{b} - \mathbf{a}$.

Unfortunately, things are not quite so simple. In fact, this supposition is false for vector fields, even when the domain is $\mathbb{R}$ and the codomain is $\mathbb{R}^2$.

Example 13.5.1 Let $\mathbf{f} = (x, y) : \mathbb{R} \to \mathbb{R}^2$ be given by $x(t) = \cos(t)$ and $y(t) = \sin(t)$. Problem 11 at the end of Section 13.4 tells us that $\mathbf{f}$ is differentiable and that
$$\mathbf{f}'(t) = (x'(t), y'(t)) = (-\sin(t), \cos(t))$$
Note that $(x(0), y(0)) = (x(2\pi), y(2\pi))$ but that there is no place between 0 and 2π where $\sin(t)$ and $\cos(t)$ are both zero. Thus we never have a value $c \in (0, 2\pi)$ such that
$$\mathbf{0} = \mathbf{f}(2\pi) - \mathbf{f}(0) = \mathbf{f}'(c)(2\pi - 0) = (-2\pi \sin(c), 2\pi \cos(c))$$

Interpreting Example 13.5.1:
You may recall from your multivariable calculus course that if we interpret a set of parametric equations $(x(t), y(t))$ as a particle moving along the curve, where t is time, the derivative at a given "time" t can be interpreted as the velocity of the particle at that time. Our proposed generalization of the mean value theorem would imply that if the particle starts at some point, moves around in the plane for a while, and then comes back to the starting position, then it must have had a velocity of zero at some time. This is clearly not true.

You may recognize the set of parametric equations in the example as describing uniform motion around the unit circle. The velocity vectors all have length 1; the particle never has a velocity of zero. (See Figure 13.7.)

---

10. Note that the more familiar
$$f'(c) = \frac{f(b) - f(a)}{b - a}$$
is not a good way to think about generalizing the mean value theorem, because the "corresponding" vector equation
$$\mathbf{f}'(\mathbf{c}) = \frac{\mathbf{f}(\mathbf{b}) - \mathbf{f}(\mathbf{a})}{\mathbf{b} - \mathbf{a}}$$
makes no sense. (*Be sure you understand why.*)

13.5 The Mean Value Theorem for Functions of Several Variables

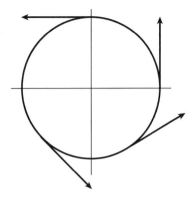

Figure 13.7 Uniform circular motion.

We have just seen that the "obvious" generalization of mean value theorem does not hold for vector fields, but it is true for scalar fields (with a few caveats, of course).

Theorem 13.5.2 Let E be an open subset of $\mathbb{R}^n$. Let $f : E \to \mathbb{R}$ be a scalar field. Suppose that $\mathbf{a}$ and $\mathbf{b}$ and the entire line segment joining them are contained in E. If f is differentiable at every point of the segment joining $\mathbf{a}$ and $\mathbf{b}$ (including the endpoints), then there exists $\mathbf{c}$ lying on the segment joining $\mathbf{a}$ and $\mathbf{b}$ such that $f(\mathbf{b}) - f(\mathbf{a}) = [f'(\mathbf{c})](\mathbf{b} - \mathbf{a})$.

Proof: The proof is Problem 1 at the end of this section. □

Exercise 13.5.3 will help you unravel the statement of Theorem 13.5.2 and (the second part) will help you rewrite it in a form that should be reminiscent of ideas you encountered in a multivariable calculus course.

Exercise 13.5.3 Different looks at the mean value theorem for scalar fields:

1. Theorem 13.5.2 is often applied to differentiable functions whose domain is convex. The theorem can be considerably simplified for this context. Restate the theorem for this special case.

2. Carefully formulate a version of Theorem 13.5.2 that avoids the language of linear transformations and uses partial derivatives instead.

■

We now turn to vector fields. As you saw in Example 13.5.1, the mean value theorem cannot be fully generalized, even in the simplest of vector fields. However, a related condition, involving an inequality instead of an equality, does hold for vector fields.

Theorem 13.5.4 Let E be an open subset of $\mathbb{R}^n$. Let $f : E \to \mathbb{R}^m$ be a vector field. Suppose that $\mathbf{a}$ and $\mathbf{b}$ and the entire line segment joining them are contained in E. If f is differentiable at every point of the segment joining $\mathbf{a}$ and $\mathbf{b}$ (including the endpoints), then there exists $\mathbf{c}$ lying on the segment joining $\mathbf{a}$ and $\mathbf{b}$ such that $\|\mathbf{f}(\mathbf{b}) - \mathbf{f}(\mathbf{a})\| \leq \|\mathbf{f}'(\mathbf{c})(\mathbf{b} - \mathbf{a})\|$.

Proof: The proof is Problem 2 at the end of this section. □

Theorem 13.5.4 allows us to conclude that if the domain of $\mathbf{f}$ is convex and the derivative of $\mathbf{f}$ is uniformly bounded, then $\mathbf{f}$ must be a Lipschitz function.

Corollary 13.5.5 Let E be an open, convex subset of $\mathbb{R}^n$, and let $\mathbf{f} : E \to \mathbb{R}^m$ be a differentiable function. Suppose that there exists $K \in \mathbb{R}$ such that $\|\mathbf{f}'(\mathbf{x})\| \leq K$ for all $\mathbf{x} \in E$. Then

$$\|\mathbf{f}(\mathbf{b}) - \mathbf{f}(\mathbf{a})\| \leq K \|\mathbf{b} - \mathbf{a}\| \quad \text{for all} \quad \mathbf{b}, \mathbf{a} \in E$$

You are now ready for Excursion M.

Excursion M. *The Implicit Function Theorem*

Synopsis: Excursion M provides an in-depth discussion of this very important theorem. In addition to the proof, the reader is treated to a discussion of the intuition behind the theorem and the intuition that underlies the proof. Excursion M requires a good working knowledge of Chapters 10 and 13. A general knowledge of Newton's method for finding roots, such as might be gotten from any calculus text, is also helpful for understanding the discussion of the relationship between quasi-Newton's methods and the proof of the implicit function theorem.

Problems 13.5

1. Prove the scalar field generalization of the mean value theorem (Theorem 13.5.2). (*Hint*: The key is to parametrize the line segment joining $\mathbf{a}$ and $\mathbf{b}$ and compose $\mathbf{f}$ with the parametrization. The resulting function is a one-variable, real-valued function, to which the one-variable mean value theorem can be applied.)

2. Prove the vector field generalization of the mean value theorem (Theorem 13.5.4).

3. Let E be an open, convex subset of $\mathbb{R}^n$, and let $\mathbf{f} : E \to \mathbb{R}^m$ be a differentiable function. Suppose that $\mathbf{f}'(\mathbf{x}) = \mathbf{0}$ for all $\mathbf{x} \in E$. Prove that $\mathbf{f}$ is constant in E.

PART II
Excursions

Excursion A

Truth and Provability

> *Required Background*
> The co-requisite for this excursion is Section 1.2.

At the end of Section 1.2, I make the claim that the statement "$\mathbb{R}$ is infinite" is *unprovable* within the formal system given by set theory plus the field axioms. This does not, of course, mean that the statement is false. It just means that we will never be able to produce a proof.

To see why, suppose that from the field axioms we have deduced the truth of theorems $T_1, T_2, \ldots, T_n$. Will the field axioms together with the theorems we have proved so far allow us to deduce that $\mathbb{R}$ is infinite?

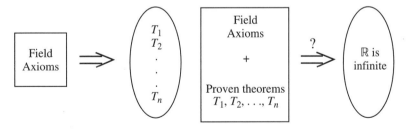

Because every field axiom holds true if $(\mathbb{R}, +, \cdot)$ is replaced by $(A, \oplus, \otimes)$ (see Exercise 1.2.9), the logic that gets us theorems $T_1, T_2, \ldots, T_n$ for $\mathbb{R}$ will also get us theorems $T_1, T_2, \ldots, T_n$ for $(A, \oplus, \otimes)$. That is, $T_1, T_2, \ldots, T_n$ are also true for $(A, \oplus, \otimes)$. In fact, every theorem that can be proved from the field axioms must hold true for both $\mathbb{R}$ and A! Given that A is not infinite and $\mathbb{R}$ is, the fact that $\mathbb{R}$ is infinite must be unprovable in the logical system given only by the field axioms. (Of course, the same logic tells us that we will not be able to *disprove* it either!) This means that the field axioms are not sufficient to give a complete mathematical description of the object we call the real numbers. At least one more axiom is needed to prove that $\mathbb{R}$ is infinite. Similar reasoning shows us that the field axioms and the order axiom are not sufficient to describe $\mathbb{R}$. We need the least upper bound axiom. Establishing that no more axioms are needed is a more complicated issue and is beyond the scope of this book.

Excursion B
Number Properties

> *Required Background*
> The prerequisite for this excursion is Chapter 1.

This excursion comprises a set of exercises that ask you to prove some miscellaneous useful number properties.

1. **Properties of $\mathbb{N}$ and $\mathbb{Z}$:** Given the definitions of $\mathbb{N}$ and $\mathbb{Z}$ on page 48 in Chapter 1, prove the following properties:

   (a) $\mathbb{N}$ and $\mathbb{Z}$ are closed under $+$ and $\cdot$.

   (b) If $k \in \mathbb{N}$ and $a \in \mathbb{R}$, then
   $$k \cdot a = \underbrace{a + a + a + \cdots + a}_{k \text{ times}}.$$

2. The following useful property of $\mathbb{N}$ and its corollary in $\mathbb{Z}$ also appeared as Theorem 0.3.4 in the discussion of mathematical induction in Chapter 0.3. If you didn't prove them at that time, correct that omission and do so now.

   (a) **Well-Ordering Principle:** Prove that every non-empty subset of $\mathbb{N}$ has a least element. (*Hint:* You will need to use the principle of mathematical induction. Assume that $K \subseteq \mathbb{N}$ has no least element. Prove by induction that $K = \emptyset$.)

   (b) Prove that every non-empty subset of $\mathbb{Z}$ that is bounded below has a least element.

3. **Arithmetic of Rational and Irrational Numbers.**

    (a) Prove that the sum and the product of two rational numbers are rational numbers.

    (b) Prove that the sum of a rational number and an irrational number is an irrational number. Prove that the product of a non-zero rational number and an irrational number is an irrational number.

    (c) Prove that the additive inverse and the multiplicative inverse of rational numbers are rational numbers.

    (d) Prove by giving examples that the sum and the product of two irrational numbers may be irrational or rational.

    (e) Let $n \in \mathbb{N}$ with n not a perfect square. Prove that $\sqrt{n}$ is irrational.

4. Outline the necessary steps for establishing rigorously that the rational numbers satisfy the field and order axioms from Chapter 1. (You may find yourself making assumptions about the additive and multiplicative structures in the real numbers. This is okay—go ahead and do it—but think about what is being assumed. Pay special attention to places where you use "facts" that are "common knowledge" but for which you don't have good mathematical grounding. As you go, carefully state the facts that are being assumed without proof.)

5. Let k and S be positive real numbers. Prove that there exists $n \in \mathbb{N}$ such that $n \cdot k > S$.

6. Suppose you have finitely many positive numbers that are all less than 1. Prove that their product is less than 1. Likewise, prove that the product of finitely many numbers that are greater than 1 is greater than 1.

7. Let k be a positive real number. The geometric sequence $k, k^2, k^3, k^4, \ldots$ is very important because it crops up in many situations. The sequence is, of course, constant if $k = 1$. Otherwise, the properties of the sequence are determined by whether $k < 1$ or $k > 1$.

    (a) Prove that if $k > 1$, the sequence $k, k^2, k^3, \ldots$ is strictly increasing. Prove that if $k < 1$, the sequence is strictly decreasing.

    (b) Suppose $k > 1$. Let S be any real number. Prove that there exists $N \in \mathbb{N}$ such that for all $n > N$, $k^n > S$.

    (c) Suppose that $k < 1$. Let ϵ be any positive real number. Prove that there exists $N \in \mathbb{N}$ such that for all $n > N$, $k^n < \epsilon$.

8. Use mathematical induction to prove the following identities:

    (a) $\sum_{i=1}^{n} i = \dfrac{n(n+1)}{2}$

(b) $\displaystyle\sum_{i=1}^{n} i^2 = \frac{n(n+1)(2n+1)}{6}$

(c) If $r \neq 1$, $\displaystyle\sum_{i=0}^{n} ar^i = a\left(\frac{r^{n+1}-1}{r-1}\right)$ (What happens if $r = 1$?)

9. **Bernoulli's Inequality:** Let x be any real number with $x > -1$. Let $n \in \mathbb{N}$. Prove the following inequality, which was discovered by the famous mathematician Johann Bernoulli:

$$(1+x)^n \geq 1 + nx.$$

(*Hint*: Be sure that you see why it is important that $x > -1$. (Without it, the statement is false when n is odd.) Make a note of it when you use it in the proof.)

Excursion C

Exponents

> *Required Background*
> Section C.1 of this excursion, which covers integer and rational powers, requires only the background provided by Chapter 1. Further background on convergence and continuity that is provided by Chapters 4 and 5 are required for Section C.2, which deals with irrational powers.

In this excursion, we will discuss the process of exponentiating positive real numbers. The excursion is structured as a sequence of interconnected problems: The theorems build on one another and, in the end, show us how to raise a positive real number a to any real power. The proofs of these theorems are left to you.

C.1 Integer and Rational Powers

Positive Integer Powers

Let $a \in \mathbb{R}$, and let $n \in \mathbb{N}$. Then, following the usual convention, we interpret the expression a^n to mean a multiplied by itself n times:

$$a^n = \underbrace{a \cdot a \cdot a \cdots a}_{n \text{ times}}$$

An easy application of mathematical induction to the field axioms shows that a^n exists and is equal to a real number.

Theorem C.1.1 Let $a \in \mathbb{R}$, and let $n \in \mathbb{N}$.

1. Suppose that a is positive. Then a^n is also positive.

2. Suppose that a is negative. Then a^n is positive if n is even and negative if n is odd. □

Theorem C.1.2 Let $a, b \in \mathbb{R}$. Let $m, n \in \mathbb{N}$. Then the following identities hold:

1. $a^{n+m} = a^n a^m = a^m a^n$
2. $a^{nm} = (a^n)^m = (a^m)^n$
3. $(ab)^n = a^n b^n$
4. $\left(\dfrac{a}{b}\right)^n = \dfrac{a^n}{b^n}$ □

Theorem C.1.3 Let $n \in \mathbb{N}$. Let $f : \mathbb{R} \to \mathbb{R}$ be given by $f(x) = x^n$. Then f is strictly increasing for $x > 0$. That is, if $0 < a < b$, then $a^n < b^n$. □

Exercise C.1.4 Use Theorem C.1.3 to show that for $x < 0$, $f(x) = x^n$ is decreasing if n is even and increasing if n is odd. ■

Theorem C.1.5 Let a be a positive real number, and suppose that $n > 1$ is a natural number. Then

$$\text{if } a > 1, \text{ then } a^n > a > 1,$$

and

$$\text{if } 0 < a < 1, \text{ then } 0 < a^n < a < 1.$$

□

Definition C.1.6 Let $n \in \mathbb{N}$. Define **n-factorial** to be

$$n! = 1 \cdot 2 \cdot 3 \cdot \ldots \cdot n.$$

For any natural number $k \leq n$, define **n choose k** to be

$$\binom{n}{k} = \frac{n!}{k!(n-k)!}.$$

For $k > n$ and for negative integers, $\binom{n}{k}$ is defined to be 0.

It is an interesting (and not entirely obvious) fact that for $0 < k < n$, $\binom{n}{k}$ is a positive integer. This follows easily from the following technical lemma (which we will need for other reasons).

Lemma C.1.7 Let $n, k \in \mathbb{N}$ with $k < n$. Then
$$\binom{n+1}{k} = \binom{n}{k-1} + \binom{n}{k}.$$
□

> The number $\binom{n}{k}$ is called "n choose k" because it counts the number of different ways that one can pick k unordered outcomes from n possibilities. This relationship is not difficult to show; it is frequently used in probability calculations and elsewhere. Our main interest in the number $\binom{n}{k}$ is in its role as a coefficient in the binomial expansion, as described in the binomial theorem.

Theorem C.1.8 [Binomial Theorem] Let a and $b \in \mathbb{R}$. Then for all $n \in \mathbb{N}$,
$$(a+b)^n = \sum_{i=0}^{n} \binom{n}{i} a^{n-i} b^i.$$
□

Proof: Prove this by induction on n. You will need to make use of Lemma C.1.7.

Roots

The fact that every non-negative real number has a non-negative square root was proved in Chapter 1 (Theorem 1.4.8). The proof can be generalized to show that every non-negative real number has a non-negative n^{th} root.

Theorem C.1.9 Let $n \in \mathbb{N}$. Let a be a non-negative real number. Then there exists a unique non-negative real number b such that $b^n = a$. □

Exercise C.1.10 Let $n \in \mathbb{N}$. Let a be a negative real number.

1. Suppose that n is odd. Prove that there exists a unique negative real number b such that $b^n = a$.

2. Suppose that n is even. Prove that there does not exist a real number b such that $b^n = a$.

■

Definition C.1.11 Let a be a non-negative real number. Then if $b \in \mathbb{R}$ is the unique non-negative real number such that $b^n = a$, we denote b by $\mathbf{a^{\frac{1}{n}}}$ and call it the $\mathbf{n^{th}}$ **root of a**.[1]

Theorem C.1.12 Let $n, m \in \mathbb{N}$, and let a and b be positive real numbers. Then the following identities hold:

1. $\left(a^{\frac{1}{n}}\right)^n = a = (a^n)^{\frac{1}{n}}$

2. $\left(a^{\frac{1}{m}}\right)^{\frac{1}{n}} = \left(a^{\frac{1}{nm}}\right) = \left(a^{\frac{1}{n}}\right)^{\frac{1}{m}}$

3. $(ab)^{\frac{1}{n}} = (a)^{\frac{1}{n}} (b)^{\frac{1}{n}}$

4. $\left(\dfrac{a}{b}\right)^{\frac{1}{n}} = \dfrac{a^{\frac{1}{n}}}{b^{\frac{1}{n}}}$

$\square$

Theorem C.1.13 Suppose that $n > 1$ is a natural number.

$$\text{If } a > 1, \text{ then } a > a^{\frac{1}{n}} > 1,$$

and

$$\text{if } 0 < a < 1, \text{ then } 0 < a < a^{\frac{1}{n}} < 1.$$

$\square$

Definition C.1.14 [**Positive rational powers**] Fix $n, m \in \mathbb{N}$. Let a be a positive real number. In Theorem C.1.9 you showed that $a^{\frac{1}{n}}$ is a well-defined positive real number. We also know that we can raise any real number to any positive integer power. Thus we can say what it means to raise a to a positive rational power. Define

$$a^{\frac{m}{n}} = \left(a^{\frac{1}{n}}\right)^m.$$

Definition C.1.14 raises the issue of well-definedness. (The following exercise might, therefore, be thought of as a technical lemma that justifies Definition C.1.14.)

---

[1]. If a is a negative real number and n is odd, the unique negative real number b such that $b^n = a$, is also denoted by $\mathbf{a^{\frac{1}{n}}}$ and called the $\mathbf{n^{th}}$ **root of a**. However, we will be considering only positive bases as we progress with our study of exponents, so we leave this case here.

Exercise C.1.15 Let a be a positive real number. Prove that the operation of raising a to a rational power is well defined. That is, prove that

$$\text{if } m, n, s, \text{ and } t \in \mathbb{N} \text{ and } \frac{m}{n} = \frac{s}{t}, \text{ then } a^{\frac{m}{n}} = a^{\frac{s}{t}}.$$

It may be useful to assume that $\frac{m}{n}$ is in "lowest terms." What justifies this simplification? ∎

Theorem C.1.16 Let a be a positive real number. Let $m, n \in \mathbb{N}$. Then

$$a^{\frac{m}{n}} = (a^m)^{\frac{1}{n}}$$

and

$$\left(a^{\frac{m}{n}}\right)^{\frac{n}{m}} = a.$$

□

Definition C.1.17 [Negative rational powers] Let a be a positive real number, and let r be a positive rational number. Set $q = -r$. Then we define

$$a^q = a^{-r} = \frac{1}{a^r}.$$

Exercise C.1.18 Explain how you know that the definition given for raising a positive real number to a negative rational power is well defined. ∎

To complete our definitions for raising every positive real number to rational powers, we need only one more detail.

Definition C.1.19 [0th power] Let a be a positive real number. Then we define $a^0 = 1$.

Theorem C.1.20 Let a and b be positive real numbers. Let p and q be rational numbers. Then the following facts hold:

1. $a^p > 0$
2. $1^p = 1$
3. $a^{p+q} = a^p a^q = a^q a^p$
4. $a^{pq} = (a^p)^q = (a^q)^p$
5. $(ab)^p = a^p b^p$

Hint: When you prove these statements you may need to handle the cases for positive and negative exponents separately. □

Theorem C.1.21 Let $f : \mathbb{Q} \to \mathbb{R}$ be given by $f(x) = a^x$. If $a > 1$, this function is strictly increasing; if $a = 1$, the function is constant; and if $a < 1$, it is strictly decreasing. □

> In conclusion: Positive real numbers can be raised to rational powers, and the operation of rational exponentiation behaves in "expected" ways. You have probably heard it asserted that positive real numbers can also be raised to irrational powers.
>
> For a positive number $a > 1$ and an irrational number t, we might define
> $$a^t = \sup\{a^r : r \in \mathbb{Q} \text{ and } r < t\}.$$
> (How would you modify this definition for $a < 1$?)
>
> It is easy to see that this set is non-empty and bounded above; thus the definition makes sense. And we can readily believe that this will yield the "right" answer. Using this definition, however, it is not easy to show that the standard properties of exponents hold for irrational exponents. For this reason, we will appeal to a definition with more theoretical power. The "answer" we get for a^t will be the same, but the definition will be more useful as a tool for proving theorems (which is, after all, the purpose of a mathematical definition!).

C.2 Irrational Powers

We would like to make sense of the expression x^r, where x is a positive real number and r is any real number. The general idea, as suggested by math teachers everywhere, is to find a sequence (r_i) of rational numbers converging to r and then to define x^r to be the limit of the sequence (x^{r_i}). This, of course, begs a number of important questions. (*How many questions can you list?*) The theorems in this section will outline a path by which you may answer them.

Lemma C.2.1 appeared as Problem 7 at the end of Section 4.4. That problem was inspired by the following question: If we have a "nice" function defined on the rational numbers, under what conditions can we extend the function to all of the real numbers in such a way that the extension shares the "nice" properties of its progenitor? Here we need to extend the operation of exponentiation from the rational numbers to all of the real numbers, preserving the usual rules of exponentiation. The easiest way to do this is to find a continuous extension.

Lemma C.2.1 Let X be a metric space. Let D be a dense subset of X. Suppose that $f : D \to \mathbb{R}$ is bounded and uniformly continuous on each bounded subset of D. Then f can be extended to a continuous function on all of X.

Proof Outline:

Step 1. Define the function $F : X \to \mathbb{R}$ as follows. Let $x \in X$. Choose any sequence (t_i) in D such that $t_i \to x$. Note that the sequence $(f(t_i))$ has a convergent subsequence. (*Why?*) Extract such a subsequence and define $F(x)$ to be its limit.

Step 2. Prove that the function F is well defined. That is, prove that it doesn't matter which sequence (t_i) we choose or how we extract the convergent subsequence from $(f(t_i))$. The limit is always the same.

Step 3. Prove that $F|_D = f$. (F is an *extension* of f.)

Step 4. Prove that the function F is uniformly continuous on each bounded subset of X and, therefore, is continuous on X.

$\square$

Question to Ponder: Now that we have the function F, it is reasonable to ask whether it is unique. That is, is F the *only* continuous extension for f? Or do others exist? (Problem 9a at the end of Section 4.3, is relevant to this question.) Why is this issue important in the context of defining irrational exponents?

Clearly, we are starting with a positive real number a and the function $f : \mathbb{Q} \to \mathbb{R}$ given by $f(r) = a^r$. Lemma C.2.1 tells us that if we can show that f is uniformly continuous and bounded on bounded subsets of $\mathbb{Q}$, then we can find a continuous extension of f to all of $\mathbb{R}$.

Exercise C.2.2 Let a be a positive real number, and let $f : \mathbb{Q} \to \mathbb{R}$ be the function $f(x) = a^x$. Let K be a positive rational number. Prove that f is bounded on $[-K, K] \cap \mathbb{Q}$ and, therefore, on all bounded subsets of $\mathbb{Q}$. ∎

Lemma C.2.3 Let a be a positive real number. Then the sequence $\left(a^{\frac{1}{n}}\right)$ converges to 1. $\square$

Lemma C.2.4 Let a be a positive real number, and let $f : \mathbb{Q} \to \mathbb{R}$ be the function $f(x) = a^x$. Then f is continuous at $x = 0$. (*Hint*: Use Lemma C.2.3 to help you get the bound you need.) $\square$

Theorem C.2.5 Let a be a positive real number, and let $f : \mathbb{Q} \to \mathbb{R}$ be the function $f(x) = a^x$. Let K be a positive rational number. Then f is uniformly continuous on $[-K, K] \cap \mathbb{Q}$, and, therefore, on every bounded subset of $\mathbb{Q}$. $\square$

Corollary C.2.6 Let a be a positive real number, and let $f : \mathbb{Q} \to \mathbb{R}$ be the function $f(x) = a^x$. Then f can be extended to a continuous function on all of $\mathbb{R}$.

Definition C.2.7 [Exponential Function (*Finally!*)] Let a be a positive real number. Let x be any real number. Then we define a^x to be $F(x)$ where F is the continuous extension of the function $f : \mathbb{Q} \to \mathbb{R}$ given by $f(x) = a^x$.

Theorem C.2.8 Let a and b be positive real numbers. Let $r, s \in \mathbb{R}$. Then the following conditions hold:

1. $a^r > 0$.

2. $1^r = 1$.

3. $a^{r+s} = a^r a^s = a^s a^r$.

4. $(ab)^r = a^r b^r$.

5. $\left(\dfrac{a}{b}\right)^r = \dfrac{a^r}{b^r}$.

6. $x^{-r} = \dfrac{1}{x^r}$. $\square$

The last remaining rules for exponents turn out to be a bit trickier. To make them work, we need the following lemma.

Lemma C.2.9 Let $p \in \mathbb{Q}$. Then the function $f : \mathbb{R}^+ \to \mathbb{R}$ given by $f(x) = x^p$ is continuous.

Theorem C.2.10 Let a and b be positive real numbers. Let r and s be real numbers. Then $a^{rs} = (a^r)^s = (a^r)^s$. (*Hint*: For the proof you will need to employ the result of Problem 3 at the end of Section 4.3. Continuity of $f(x) = x^p$ ($p \in \mathbb{Q}$) and $f(x) = a^x$ both play a role.) $\square$

Theorem C.2.11 Let a be a positive real number, and let $F : \mathbb{R} \to \mathbb{R}$ be given by $F(x) = a^x$. Then F is strictly increasing if $a > 1$, constant if $a = 1$, and strictly decreasing if $a < 1$. $\square$

Theorem C.2.12 Let a be a positive real number. Consider the function $F : \mathbb{R} \to \mathbb{R}$ given by $F(x) = a^x$.

1. Suppose that $a > 1$.

$$\text{If } x > 1, \text{ then } 1 < a < a^x,$$

and

$$\text{if } 0 < x < 1, \text{ then } 1 < a^x < a.$$

2. Suppose that $a < 1$.

$$\text{If } x > 1, \text{ then } a^x < a < 1.$$

and

$$\text{if } 0 < x < 1, \text{ then } a < a^x < 1.$$

$\square$

Exercise C.2.13 Considering the inequalities you established in Theorem C.2.12, what can you say if $x = 0$, $x = 1$, or $x < 0$? (Consider both $a > 1$ and $a < 1$.) ∎

Excursion D

Sequences in $\mathbb{R}$ and $\mathbb{R}^n$

> *Required Background*
> The prerequisite for this excursion is the definition of sequence convergence given in Section 3.2.

D.1 Sequence Convergence in $\mathbb{R}$ and $\mathbb{R}^n$

Example D.1.1 Consider the sequence

$$\frac{1}{2}, \frac{2}{3}, \frac{3}{4}, \frac{4}{5}, \frac{5}{6}, \ldots, \frac{n}{n+1}, \ldots$$

that was discussed in Section 3.2. Our intuition tells us that this sequence converges to 1. Here is a proof of this fact.

Proof: Let $\epsilon > 0$. Choose $N \in \mathbb{N}$ large enough so that $\dfrac{1}{N} < \epsilon$. Now we consider $n > N$. For all $n \geq 1$, $a_n = \dfrac{n}{n+1}$. Thus

$$\begin{aligned} d(a_n, 1) &= \left| \frac{n}{n+1} - 1 \right| \\ &= \left| \frac{n - (n+1)}{n+1} \right| \\ &= \left| -\frac{1}{n+1} \right| \\ &= \frac{1}{n+1} < \frac{1}{N} < \epsilon. \end{aligned}$$

We conclude that (a_n) converges to 1, as we guessed. □

Exercise D.1.2 Say what it means for a sequence (a_n) to fail to converge to a given limit L. (That is, negate Definition 3.3.1.) How would you have to modify your statement to express the fact that (a_n) fails to converge *at all*?

Hint: Strictly speaking, Example D.1.3 relies on a proper understanding of this exercise. If you get stuck, think about the example for a bit, then come back and try again. Going back and forth between the two should make *both* easier, as each will help in understanding the other. ■

Example D.1.3 We now prove that

$$1, 0, 0, 1, 1, 0, 0, 1, 1, 0, 0, 1, \ldots$$

does not converge at all.

Proof: Let L be any real number. We have two cases.

Case i. $L = 0$. Let $\epsilon = \frac{1}{2}$. Then no matter how large we pick N, there will always be an $n > N$ such that $a_n = 1$, in which case $d(a_n, L) = d(0, 1) = 1 > \frac{1}{2}$. The sequence cannot converge to L.

Case ii. $L \neq 0$. Choose $\epsilon < |L|$. Then no matter how large we pick N, there will always be an $n > N$ such that $a_n = 0$. In this case, $d(a_n, L) = d(0, L) = |0 - L| = |L| > \epsilon$. Once again, we see that (a_n) cannot converge to L.

□

So far, we have confined our discussion to sequences of real numbers. To broaden our perspective, let us consider sequences in $\mathbb{R}^2$. We will use this point as a springboard from which to consider sequence convergence in $\mathbb{R}^n$.

Example D.1.4 Consider the sequence

$$a_n = \left(\frac{10n - 1}{2n + 1}, \frac{(-1)^n}{n} \right).$$

We want to show that this sequence converges. (See Figure D.1.)

Proof: We conjecture that the limit of this sequence is the point $L = (5, 0)$. To prove this, we fix any $\epsilon > 0$, and then choose N large enough so that $\dfrac{1}{N} < \dfrac{\epsilon}{\sqrt{37}}$.

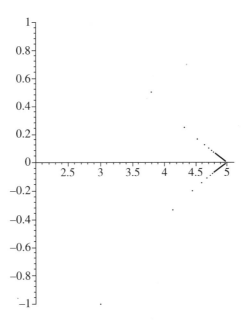

Figure D.1 The sequence $\left(\dfrac{10n-1}{2n+1}, \dfrac{(-1)^n}{n}\right)$

Suppose now that $n > N$. Then

$$d(a_n, L) = \sqrt{\left(\frac{10n-1}{2n+1} - 5\right)^2 + \left(\frac{(-1)^n}{n}\right)^2}$$

$$= \sqrt{\left(\frac{(10n-1) - 5(2n+1)}{2n+1}\right)^2 + \frac{1}{n^2}}$$

$$= \sqrt{\left(\frac{-6}{2n+1}\right)^2 + \frac{1}{n^2}}$$

$$= \sqrt{\frac{36}{(2n+1)^2} + \frac{1}{n^2}}.$$

Before we continue, notice that $\dfrac{1}{2n+1} < \dfrac{1}{n}$ and thus that

$$\left(\frac{1}{2n+1}\right)^2 < \left(\frac{1}{n}\right)^2.$$

Therefore,
$$\frac{36}{(2n+1)^2} + \frac{1}{n^2} < \frac{37}{n^2}.$$

This, in turn, implies that
$$\sqrt{\frac{36}{(2n+1)^2} + \frac{1}{n^2}} < \sqrt{\frac{37}{n^2}} = \frac{\sqrt{37}}{n}.$$

Finally, if $n > N$,
$$d(a_n, L) < \frac{\sqrt{37}}{n} < \frac{\sqrt{37}}{N} < \sqrt{37}\frac{\epsilon}{\sqrt{37}} = \epsilon.$$

We conclude that (a_n) converges to $(5, 0)$, as we asserted earlier. □

In Example D.1.4, we proved that the sequence
$$a_n = \left(\frac{10n-1}{2n+1}, \frac{(-1)^n}{n}\right)$$

of points in $\mathbb{R}^2$ converges to the point $(5, 0)$. We conjectured that this was true by observing that
$$\frac{10n-1}{2n+1} \longrightarrow 5 \text{ and that } \frac{(-1)^n}{n} \longrightarrow 0.$$

This suggests that there is a connection between the convergence of a sequence of points in $\mathbb{R}^2$ and the convergence of the sequences of real numbers formed by the individual coordinates. This is, indeed, the case in $\mathbb{R}^2$. In fact, the obvious generalization is true in $\mathbb{R}^n$ for all $n \in \mathbb{N}$.

Theorem D.1.5 Consider the sequence $(\mathbf{a}_k)$ in $\mathbb{R}^n$ where, for each k,
$$\mathbf{a}_k = (a_{1k}, a_{2k}, a_{3k}, \ldots, a_{nk})$$

Then the following statements are equivalent:

1. $(\mathbf{a}_k)$ converges to $\mathbf{b} = (b_1, b_2, \ldots, b_n)$ in $\mathbb{R}^n$.

2. For each j ($1 \leq j \leq n$), the sequence $(a_{jk})_{k=1}^{\infty}$ converges to b_j in $\mathbb{R}$.

Proof: The proof is Problem 4 at the end of Section D.2. □

D.2 Epsilonics: Playing the Game

When you read a proof, two questions should come to mind:

1. Is this correct? Does it prove what it purports to prove?

2. What is the intuition here? How did someone think this up?

In the case of the theorems proved in Section D.1, even if you were able to answer the first question affirmatively, it is entirely possible that your answer to the question "How did someone think this up?" was "*VOODOO!*," or at least "*I have no idea!*"

Voodoo Mathematics?

Naturally, voodoo played no part in the construction of the proofs given in Section D.1. The arguments given represent one half of a game that mathematicians informally, and for obvious reasons, call *epsilonics*. The intuition is unclear because you are seeing the end result of a reasoning process that took place "off screen."

The "worst enemy" metaphor alluded to in Section 3.2, is a good model for explaining what this means. Your worst enemy challenges you to a game of epsilonics. In this game it is your job to prove that a particular sequence converges; your worst enemy will try to foil you in the attempt. You have the first move—you must specify a supposed value for the limit of the sequence. In view of this choice, your worst enemy then gets to give you *any* positive value for ϵ, as fiendishly small as he or she likes. You have to counter that choice by producing a large enough value for N so that from that point on, all terms of the sequence are within ϵ of your choice for the limit. When that is accomplished, your worst enemy gets to specify another—very likely smaller—value for ϵ, and you must once again counter with a choice of N. This exchange continues back and forth. If it is ever impossible for you to counter your worst enemy's choice of ϵ, then your worst enemy wins. If you can always counter it, then you win.

From a mathematical point of view, the problem with the game of epsilonics is that your worst enemy gets to choose values of ϵ that get ever closer to zero. To be successful you must counter those choices over and over again, *ad infinitum*, making the game potentially infinite in length. (Your worst enemy can win the game at any time by merely producing a value for ϵ to which you cannot respond.) To turn this metaphorical game into a strategy for proof, we have to figure out how to avoid this infinity. What is required is a strategy that will allow us to counter *any* value our enemy could give us at *any time*. If we can prove that our strategy works, then we will have defeated the enemy.

Herein lies the crux of the voodoo question. The reason that only half of our epsilonics game becomes part of the actual proof is that we don't show how we *came up* with our strategy. From a mathematical standpoint, it is immaterial whether we came up with our strategy for picking N by means of solid mathematical reasoning or by reading the entrails of birds. It is only necessary that we *have* a strategy and that we can prove that it will always work. That is, the thinking that went into devising the strategy in the first place does not become part of the written proof because it is mathematically irrelevant!

Exercise D.2.1 Describe a game of epsilonics in which your worst enemy challenges you to prove that a particular sequence *does not* converge. Who moves first in this game—you or your worst enemy? Describe your moves and those of your worst enemy. What will it take for you to win? What will it take for your worst enemy to win? Under what circumstances does the game go on infinitely? What sort of strategy must you exhibit to circumvent this infinity so that you can make the game into an actual mathematical proof? ∎

Scratch Work: Devising a Strategy

Although peering into crystal balls or reading tea leaves as a means of foiling your worst enemy is not intrinsically objectionable from a mathematical standpoint, I do not recommend it as a likely avenue for success. You would do much better to look ahead and pick the value for N so as to ensure that the eventual outcome favors you. This will require a fair amount of scratch work before you can decide how to pick N. The good news is that much of this scratch work will eventually show up in your proof.

Let us return to the proof that

$$a_n = \left(\frac{10n-1}{2n+1}, \frac{(-1)^n}{n} \right)$$

converges. This is the way that my thinking goes—*ad nauseum*!

Our first move is to decide what the limit has to be. Well, we can tell that for very large values of n, $\frac{(-1)^n}{n}$ must be very close to 0. Thus the second coordinate of the limit of the sequence should be 0. Likewise, $\frac{10n-1}{2n+1}$ tends toward $\frac{10n}{2n} = 5$ as n gets larger without bound, so we make the educated guess that the limit ought to be $(5,0)$.

We hope to show that for "large" values of n, the distance between an arbitrary term a_n and the point $(5,0)$ will be small. But how large does n have to be to accomplish this? Here is where the scratch work comes in: We need to know the distance between a_n and $(5,0)$, so we set to calculating.

$$d(a_n, L) = \sqrt{\left(\frac{10n-1}{2n+1} - 5\right)^2 + \left(\frac{(-1)^n}{n} - 0\right)^2}$$

$$= \sqrt{\left(\frac{(10n-1) - 5(2n+1)}{2n+1}\right)^2 + \frac{1}{n^2}}$$

$$= \sqrt{\left(\frac{-6}{2n+1}\right)^2 + \frac{1}{n^2}}$$

$$= \sqrt{\frac{36}{(2n+1)^2} + \frac{1}{n^2}}$$

We need to choose N large enough so that this quantity is smaller than some arbitrary— but fixed—ϵ whenever $n > N$.

We could try to solve the inequality

$$\sqrt{\frac{36}{(2n+1)^2} + \frac{1}{n^2}} < \epsilon$$

for n. The result would tell us what we need to know, but the calculation would be, shall we say, ... unpleasant. Fortunately, this step isn't necessary. That calculation would give the best (i.e., smallest) possible value for N, but we are not required to produce such a precise value. Any choice of N that works will do just as well as any other.[2]

To make life more pleasant, we set about simplifying our expression to make it easier to analyze. The only rule is that each simplification must make the expression larger (or leave it the same)—never smaller.

One thing that makes the expression difficult to work with is the sum under the square root. We would be much better off without it. To add the two expressions together, it would be nice to have a common denominator. This can be arranged, since

$$\frac{1}{2n+1} < \frac{1}{n}$$

Because the square function is increasing for positive numbers,

$$\frac{1}{(2n+1)^2} = \left(\frac{1}{2n+1}\right)^2 < \left(\frac{1}{n}\right)^2 = \frac{1}{n^2}$$

Thus

$$\frac{36}{(2n+1)^2} + \frac{1}{n^2} < \frac{37}{n^2}$$

---

[2]. If $N = 100$ works, then $N = 1000$ and $N = 1,000,000$ will work, too. There are many possible choices for N. Our job is to give only one of them.

This and the fact that the square root function is increasing imply that

$$\sqrt{\frac{36}{(2n+1)^2} + \frac{1}{n^2}} < \sqrt{\frac{37}{n^2}} = \frac{\sqrt{37}}{n}$$

Now *this* is a much more tractable expression! We can easily see that the inequality

$$\frac{\sqrt{37}}{n} < \epsilon$$

will hold for all values of $n > N$ where $N > \frac{\sqrt{37}}{\epsilon}$.

Now we can write down the proof. The meat of the proof is, of course, in the scratch work we have already done. In the actual proof, however, the choice of N must come at the beginning as an immediate response to the choice of ϵ rather than as a natural conclusion arising from a long calculation. This often makes it seem as though the way that we chose to manipulate the expressions came *as a result* of the choice of N rather than the other way around! No wonder epsilonics can seem like voodoo at first!

Problems D.2

1. Decide whether or not the following sequences of real numbers converge. Prove your conjecture.

    (a) $-3, -3, -3, -3, -3, \ldots$.
    (b) $1, -1, 1, -1, 1, -1, 1, -1, \ldots$
    (c) $1, \frac{1}{2}, \frac{1}{3}, \frac{1}{4}, \frac{1}{5}, \frac{1}{6}, \ldots$
    (d) $\frac{1}{2}, -\frac{1}{3}, \frac{1}{4}, -\frac{1}{5}, \frac{1}{6}, \ldots$
    (e) $1, 10, 100, 1000, 10000, 100000, 1000000, \ldots$
    (f) The sequence (a_n) where for each $n \in \mathbb{N}$ $a_n = \frac{10-n}{10n}$

2. Show, using Definition 3.3.1, that each of the following sequences in $\mathbb{R}^2$ converge. Start your work by drawing an appropriate picture along the lines of that shown in Figure D.1.

    (a) $\mathbf{a}_n = \left(1, \frac{1}{n}\right)$

(b) $\mathbf{a}_n = \left(\dfrac{1}{n}, \dfrac{1}{n^2}\right)$

(c) $\mathbf{a}_n = (a_{1n}, a_{2n})$ where

$$a_{1n} = \begin{cases} \dfrac{(-1)^{\frac{1}{2}(n-1)}}{n} & \text{if } n \text{ is odd} \\ 0 & \text{if } n \text{ is even} \end{cases}$$

$$a_{2n} = \begin{cases} 0 & \text{if } n \text{ is odd} \\ \dfrac{(-1)^{\frac{1}{2}n}}{n} & \text{if } n \text{ is even} \end{cases}$$

3. Decide whether the following sequences converge. In either case, prove your conjecture. Compute the estimates explicitly. Do not appeal to Theorem D.1.5.

   (a) In $\mathbb{R}^2$: $\mathbf{a}_n = \left((-1)^n, \dfrac{n}{n+1}\right)$

   (b) In $\mathbb{R}^3$: $\mathbf{a}_n = \left(\dfrac{1}{n}, \dfrac{1}{n^2}, \dfrac{1}{n^3}\right)$

4. Prove Theorem D.1.5. (*Hint*: Make each coordinate closer to "its" limit than $\dfrac{\epsilon}{\sqrt{n}}$.)

D.3 Infinite Limits

Exercise D.3.1 Give examples of sequences (a_n) and (b_n) of real numbers for which it would be reasonable to say that

$$\lim_{n \to \infty} a_n = \infty \text{ and } \lim_{n \to \infty} b_n = -\infty$$

■

Exercise D.3.2 Complete the following definitions:

- We say that $\lim\limits_{n \to \infty} a_n = \infty$ if for every ...

- We say that $\lim\limits_{n \to \infty} b_n = -\infty$ if for every ...

■

Exercise D.3.3 Would it make sense to say that a sequence in $\mathbb{R}^2$ (or even $\mathbb{R}^n$) "goes to infinity"? If so, what might one mean by this? ■

D.4 Some Important Special Sequences

A few specific limits show up again and again in important calculations. Hints are provided here and there, but the proofs are left to you. Some of the proofs rely on numerical facts set out in Excursion B.

We start with an easy, but useful lemma that allows us to deduce "unknown" limits from related "known" limits.

Lemma D.4.1 Let (a_n) and (b_n) be sequences of real numbers with $0 \leq a_n \leq b_n$ for all $n \in \mathbb{N}$. If $b_n \to 0$, then $a_n \to 0$ as well.

Theorem D.4.2 Let r be a real number. Then

$$\lim_{n \to \infty} r^n = \begin{cases} 0 & \text{if } -1 < r < 1 \\ 1 & \text{if } r = 1 \\ \infty & \text{if } r > 1 \end{cases}.$$

The sequence fails to converge if $r \leq -1$. But if $r < -1$, then $|r^n| \to \infty$. (*Hint*: If you're stumped, look at Problem 7 in Excursion B.)

You may have heard it said that "factorials grow faster than exponentials." Theorem D.4.3 makes this statement precise. The limit crops up frequently in the study of power series. (The proof of the theorem is sketched for you. Your job is to fill in the details.)

Theorem D.4.3 Let x be a real number. Then

$$\lim_{n \to \infty} \frac{|x|^{n+1}}{(n+1)!} = 0.$$

Proof Sketch:
The key is to think of the expression in the limit as a product of fractions:

$$\frac{|x|^n}{(n+1)!} = \frac{|x|}{1} \cdot \frac{|x|}{2} \cdot \frac{|x|}{3} \cdot \ldots \cdot \frac{|x|}{n} \cdot \frac{|x|}{n+1}.$$

As $n \to \infty$, we have more and more terms in the product. For k larger than some fixed $K \in \mathbb{N}$,

$$\frac{|x|}{k} < \frac{1}{2}.$$

It follows that for $k > K$,

$$\frac{|x|^k}{(k+1)!} \leq \frac{|x|^K}{K!} \cdot \left(\frac{1}{2}\right)^{k+1-K}.$$

Apply Theorem D.4.2 and Lemma D.4.1. □

For the next several proofs, you will need Bernoulli's inequality, which is Problem 9 in Excursion B. It is restated here for your reference.

Lemma D.4.4 [Bernoulli's Inequality] Let x be any real number with $x > -1$. Let $n \in \mathbb{N}$. Then

$$(1 + x)^n \geq 1 + nx.$$

Feel free to appeal to the standard properties of exponents, as they are needed to establish the remaining limits. Hints on how to prove them can be found in Excursion C.

In the excursion on power series, we will need the following limit.

Theorem D.4.5 $\lim_{n \to \infty} n^{\frac{1}{n}} = 1$

The following is a proof that uses a specialized, but useful, trick. Fill in the details in the proof.

Proof Sketch:

Step 1. Note that if $n \in \mathbb{N}$ and $n > 1$, then $n^{\frac{1}{n}} > 1$. For each $n \in \mathbb{N}$, choose $k_n = n^{\frac{1}{n}} - 1$.

Step 2. Use the binomial theorem (Theorem 3.1.8 in Excursion 3) to expand

$$n = (1 + k_n)^n.$$

Step 3. Deduce the fact that

$$n - 1 \geq \frac{1}{2} n(n-1) k_n^2.$$

Step 4. Use this result to show that $k_n \to 0$ as $n \to \infty$.

□

Here's another useful limit.

Theorem D.4.6 Let c be a real number with $c > 1$. Then $\lim_{n \to \infty} c^{\frac{1}{n}} = 1$.

I leave the proof to you, but here is a *hint*: The proof uses a simpler version of the trick used to prove Theorem D.4.5. Note that $c^{\frac{1}{n}} > 1$ for all n. Let (k_n) be the sequence of positive real numbers such that $c^{\frac{1}{n}} = 1 + k_n$ for all n. Apply Bernoulli's inequality to $(1 + k_n)^n$.

Exercise D.4.7 Modify the proof sketched in Theorem D.4.5 to prove that
$$\lim_{n \to \infty} (n+1)^{\frac{1}{n}} = 1.$$

Excursion E
Limits of Functions from $\mathbb{R}$ to $\mathbb{R}$

> *Required Background*
> The definition of the limit of a function at a point given in Section 4.2.

E.1 Example Proofs

Example E.1.1 In this example, we will show explicitly, using Definition 4.2.1 that

1. $\lim_{x \to -2}(5x + 3) = -7$.

2. $\lim_{x \to 2}(x^2 + 3x) = 10$.

3. $\lim_{x \to 1}(x^4 - 3x + 2) = 0$.

4. $\lim_{x \to 3}\dfrac{x^2}{1 + x^2} = \dfrac{9}{10}$.

The amount of thought that goes into the choice of δ increases with each of these examples.

Proof

1. Let $\epsilon > 0$. Choose $\delta = \frac{\epsilon}{5}$. Suppose that $0 < |x+2| < \delta$. Then
$$\begin{aligned} d(f(x), L) &= |(5x+3) + 7| \\ &= |5x + 10| \\ &= 5|x+2| \\ &< 5\left(\frac{\epsilon}{5}\right) = \epsilon. \end{aligned}$$

2. Let $\epsilon > 0$. Choose $\delta = \min\left(1, \frac{\epsilon}{8}\right)$. Suppose now that $0 < |x-2| < \delta$. Notice in particular that, because $\delta \leq 1$, $x \in (1, 3)$.
$$\begin{aligned} d(f(x), L) &= |x^2 + 3x - 10| \\ &= |x+5||x-2| \\ &< 8\left(\frac{\epsilon}{8}\right) = \epsilon. \end{aligned}$$

3. Let $\epsilon > 0$. Choose $\delta = \min\left(1, \frac{\epsilon}{30}\right)$. Now suppose that $0 < |x-1| < \delta$. Then
$$\begin{aligned} d(f(x), L) &= |x^4 - 3x + 2| = |x^4 - 1 - 3x + 3| \\ &\leq |x^4 - 1| + |-3x + 3| \\ &= |x^2 + 1||x-1||x+1| + 3|x-1| \\ &< (5)\left(\frac{\epsilon}{30}\right)(3) + (3)(\frac{\epsilon}{30}) \\ &< \frac{\epsilon}{2} + \frac{\epsilon}{2} = \epsilon. \end{aligned}$$

4. Let $\epsilon > 0$. Choose $\delta = \min(1, \epsilon)$. Suppose that $0 < |x-3| < \delta$. Then
$$\begin{aligned} d(f(x), L) &= \left|\frac{x^2}{1+x^2} - \frac{9}{10}\right| \\ &= \left|\frac{x^2 - 9}{10(1+x^2)}\right| \\ &= \frac{1}{10(1+x^2)}|x-3||x+3| \\ &< \frac{1}{50}(\epsilon)(7) \\ &< \epsilon. \end{aligned}$$

□

> We tend to think of ϵ as a "very small" number because it is this range of values that usually presents the greatest challenge. Thus, if we take $\delta = \min(1, \epsilon)$, it is reasonable to think that in an actual game of epsilonics δ will very likely end up being ϵ rather than 1. In reality, there is no *mathematical* requirement that ϵ be a small number. Thus we have to build into our strategy a plan for what to do when our worst enemy gets especially diabolical and gives us a large value for ϵ.

E.2 Epsilonics: Some General Principles

The proofs given in Example E.1.1 illustrate some general principles about epsilonics arguments. It should be useful to examine these points more explicitly.

In the proof that $\lim_{x \to 1}(x^4 - 3x + 2) = 0$, our goal is to show that $|(x^4 - 3x + 2) - 0|$ is "small" provided that $|x - 1|$ is "small." Therefore, our initial task is to rewrite $|x^4 - 3x + 2|$ somehow in terms that involve $|x - 1|$ and then to use this expression as a lever to help us get what we want.

To get our hands on $|x - 1|$, we use the triangle inequality to split the expression into $(x^4 - 1) - 3x + 3$. Then treat the two parts separately.

$$\begin{aligned} d(f(x), L) = |x^4 - 3x + 2| &= |x^4 - 1 - 3x + 3| \\ &\leq |x^4 - 1| + |-3x + 3| \\ &= |x^2 + 1||x + 1||x - 1| + 3|x - 1| \end{aligned}$$

We chose $x^4 - 1$ because we could see that $x - 1$ was a factor. The expression $-3x + 3$ just followed from this choice. The good news is that, if you break the expression down so that $x - 1$ is a factor of the first part, then it will be a factor of the second part, too. (You might ponder why this is.)

Here you see the first big principle involved in showing that something is small:

> **The sum of small things is small.** You can break up an estimate into a sum with as many terms as necessary *provided* that you can make each summand as small as you like.

If we require that $|x - 1| < \dfrac{\epsilon}{6}$, the second term in the expression

$$|x^2 + 1||x + 1||x - 1| + 3|x - 1|$$

will be smaller than $\dfrac{\epsilon}{2}$. Thus we will ultimately want to choose $\delta < \frac{\epsilon}{6}$.

Our real problem seems to lie with $|x^2 + 1|$ and $|x + 1|$, which are most definitely *not* small! However, our ability to choose δ in any way we please allows us to place a *maximum bound* on these expressions.

For instance, if we choose $\delta < 1$, then any x such that $|x-1| < \delta$, must be in the interval $(0, 2)$. For x's in this interval, $|x^2+1| < 5$ and $|x+1| < 3$.[1] Then if we pick $\delta < \min\left(1, \frac{\epsilon}{30}, \frac{\epsilon}{6}\right)$,[2] we deduce that $|x^2+1||x+1||x-1| < \frac{\epsilon}{2}$ and that $3|x-1| < \frac{\epsilon}{2}$. Thus the original expression will be less than ϵ, as needed.

Our strategy for dealing with the quantity $|x^2+1||x+1||x-1|$ is illustrated by our second big principle:

The product of something small and something bounded is small. We can handle products *provided* that we can make at least one of the factors as small as we like and that we can find upper bounds for each of the others.

The Multitask δ

As you have seen in a number of instances, the δ that we ultimately choose must perform several tasks. The order in which these tasks are handled is crucial to the ultimate success of our argument. For instance, in working with the product
$$|x^2+1||x+1||x-1|$$
we needed to find bounds on $|x^2+1|$ and $|x+1|$ *before* we could decide how small the other factor should be. The decision to make $\delta < \frac{\epsilon}{30}$ depended on our knowing that
$$|x^2+1||x+1| < 15$$
Thus it was necessary to decide first that δ would be less than 1.

When making a product small, we first need to establish an upper bound for all "large" quantities. Only then can we determine how small we will have to force the "small" quantities to be.

Problems E.2

1. Let $f : \mathbb{R} \to \mathbb{R}$ be a function. Rewrite Definition 4.2.1 in language that takes into account the fact that both the domain and the range of the

---

[1]. Of course, there is nothing magic about using 1. If we require that $\delta < 10$, then we could conclude that
$$|x^2+1| < 101 \text{ and } |x+1| < 11$$
which would work just as well.

[2]. Notice that in the proof we just specified $\delta < \min\left(1, \frac{\epsilon}{30}\right)$, because $\frac{\epsilon}{30} < \frac{\epsilon}{6}$.

function are ℝ under the usual metric. In each of the following cases, use your definition to show that $\lim_{x \to a} f(x)$ exists.

(a) $f(x) = 2x + 1$; $a = 3$
(b) $f(x) = x^2$; $a = -1$
(c) $f(x) = x^3 - x$; $a = 0$
(d) $f(x) = x^4 + x + 2$; $a = 2$
(e) $f(x) = \dfrac{1}{x-4}$; $a = 2$

2. **Infinite Limits and Limits at Infinity** In your study of functions, you have probably heard it said that "a certain function tends to infinity at a certain point" or that "as x tends to infinity, a function f tends to some number a." This language is valuable for describing the geometric shape of a graph or the "long-term" behavior of a function.[3]

Consider the graphs shown in Figure E.1. It might seem reasonable to say that as x tends toward zero, $f(x) = (x^4 + x^2 + 1)/x^2$ tends toward infinity.

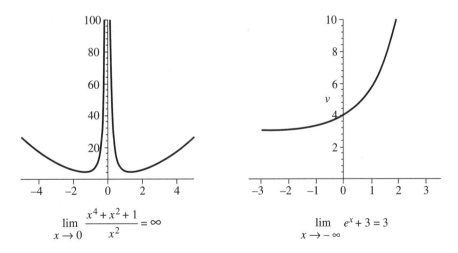

Figure E.1 Graphs for Problem 2.

---

3. For example, if a function describes the number of trees in a forest as a function of time, we might want to know what will happen to the forest over the long term. Will the trees all die out? Will they reach some sort of "steady state" in which there are trees and the number of trees stays constant?

Likewise, you might say that as x goes to $-\infty$, $f(x) = e^x + 3$ goes to 3.

Let $f : \mathbb{R} \to \mathbb{R}$ be a function. Let $a \in \mathbb{R}$.

(a) Give a careful mathematical definition of what it means to say that $\lim_{x \to a} f(x) = \infty$.

(b) Modify the definition you gave in part (a) to define $\lim_{x \to a} f(x) = -\infty$.

(c) Define what it means to say that $\lim_{x \to \infty} f(x) = a$.

(d) Modify the definition given in part (c) to define $\lim_{x \to -\infty} f(x) = a$.

(e) Suppose $f : X \to Y$ is a function from the metric space X to the metric space Y. Under what circumstances can the definitions you gave in parts (a)–(d) be generalized so that they apply to the function f? If you don't think they can, explain your reasoning.

Excursion F

Doubly Indexed Sequences

> *Required Background*
> This excursion could, in principle, be studied immediately after considering the convergence of sequences (Sections 3.3 and 3.4). But the interplay between the limit of a doubly indexed sequence and its iterated limit uses a uniform convergence idea, which will be best explored after having considered another uniformity condition such as uniform continuity or uniform convergence of sequences of functions. It will be challenging but quite doable after experience with uniform continuity (as it will foreshadow ideas studied in Chapter 12). It will serve as a nice reinforcement of the connection between uniform convergence and the interchange of limit operations if explored after a study of Chapter 12.

We are frequently faced, for one reason or another, with the need to take the limit of a sequence that is indexed by two parameters. This excursion gives the needed definitions and explores the relevant facts. The excursion is written as a long set of interconnected problems and exercises. The proofs of the theorems are left to you.

F.1 Double Sequences and Convergence

Definition F.1.1 Let X be a set. An array

$$(a_{ij}) = \begin{bmatrix} x_{11} & x_{12} & x_{13} & x_{14} & x_{15} & \cdots \\ x_{21} & x_{22} & x_{23} & x_{24} & x_{25} & \cdots \\ x_{31} & x_{32} & x_{33} & x_{34} & x_{35} & \cdots \\ x_{41} & x_{42} & x_{43} & x_{44} & x_{45} & \cdots \\ x_{51} & x_{52} & x_{53} & x_{54} & x_{55} & \cdots \\ \vdots & \vdots & \vdots & \vdots & \vdots & \ddots \end{bmatrix}$$

of elements of X indexed by the two variables $i = 1, 2, 3, \ldots$ and $j = 1, 2, 3, \ldots$ is called a **doubly indexed sequence in X** or just a **double sequence**.

In keeping with our usual notion for convergence of sequences, we would like to say that a double sequence converges in a metric space X if there is a point $x \in X$ such that if n and m are "sufficiently large," the distance between x_{nm} and x will be "small."

Definition F.1.2 Let X be a metric space. Let (a_{nm}) be a doubly indexed sequence in X, and let a be a real number. We say that (a_{nm}) converges to a provided that for all $\epsilon > 0$, there exists $N \in \mathbb{N}$ such that if $n > N$ and $m > N$, $d(a_{nm}, a) < \epsilon$. We denote this by

$$\lim_{n,m \to \infty} a_{nm}.$$

Theorem F.1.3 [Uniqueness of Limits] Let X be a metric space. Let (a_{nm}) be a doubly indexed sequence in X. Suppose that (a_{nm}) converges to a and also to b. Then $a = b$.

Theorem F.1.4 [Cauchy Criterion] Let X be a complete metric space. Let (a_{nm}) be a doubly indexed sequence in X. Then (a_{nm}) converges if and only if for every $\epsilon > 0$, there exists $N \in \mathbb{N}$ such that if n, m, t, and s are all larger than N, then $d(a_{nm}, a_{ts}) < \epsilon$.

We have two indices that will have to "get large." We can conceivably achieve this goal in one of three ways: We can make both indices get large simultaneously, and see whether all terms approach the same value, as we did in the definition, or we can let one index increase at a time. Provided that all of the necessary limits exist, we can think of taking the limits of the row sequences, and then the limit of those limits. Alternatively, we can take the limits of the column sequences, and then take the limits of those limits. These are called iterated limits. (See Figure F.1.)

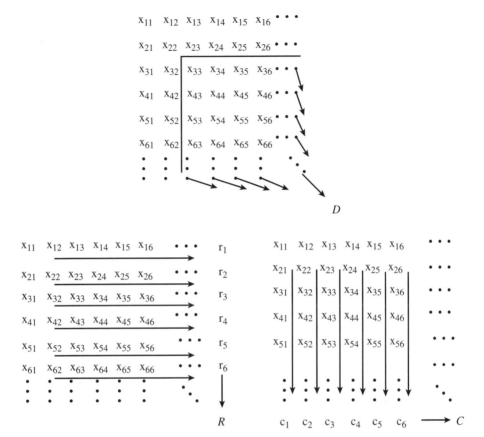

Figure F.1 Provided that all the limits exist, we could make both indices get large at once, or we can think of taking the limits of the row or column sequences and then the limit of those limits. Are D, R, and C all equal?

Dealing with one index at a time is sometimes easier than dealing with both indices simultaneously. Thus it is useful to ask about the relationship between the double limits and iterated limits. Some natural questions immediately come to mind:

- If the double limit exists, do both iterated limits exist?
- If both of the iterated limits exist, must they be equal?
- If the iterated limits exist and are equal to each other, does this imply the existence of the double limit and, if so, is the double limit equal to the iterated limits?

- If the answer to any of the preceding questions is "no," what additional hypotheses are needed to make things "work nicely"?

In each of the following problems, let X be a fixed metric space, and let (a_{nm}) be a doubly indexed sequence in X.

1. Surprisingly, the iterated limits may not exist, even if the double limit does exist. Find an example of a doubly indexed sequence of real numbers for which the double limit exists and yet the iterated limits do not.

2. Suppose that the double limit of (a_{nm}) exists and is equal to a. Prove that if each row sequence converges, then the limit of those limits exists and is equal to a. That is, $\lim_{n\to\infty} a_{nm}$ exists, implies that $\lim_{m\to\infty} \lim_{n\to\infty} a_{nm} = a$. A parallel fact holds for the other iterated limit.

This establishes the following theorem.

Theorem F.1.5 Let X be a metric space, let (a_{nm}) be a doubly indexed sequence in X, and let $a \in X$. Suppose that $\lim_{n,m\to\infty} a_{nm}$ exists and is equal to a. Suppose also that $\lim_{n\to\infty} a_{nm}$ and $\lim_{m\to\infty} a_{nm}$ both exist.

Then both iterated limits exist and

$$\lim_{m\to\infty} \lim_{n\to\infty} a_{nm} = \lim_{n\to\infty} \lim_{m\to\infty} a_{nm} = a.$$

3. Our next question was whether the existence of the two iterated limits implied their equality. You decide. Give a proof or a counterexample. (*Hint*: If you have trouble figuring out how to go about this, looking at the statements of the subsequent theorems might help.)

Definition F.1.6 Let X be a metric space, and let (a_{nm}) be a doubly indexed sequence in X. Suppose that for each $n \in \mathbb{N}$, $a_{nm} \to a_n$. Then we say that the row sequences **converge uniformly** provided that for every $\epsilon > 0$, there exists $N \in \mathbb{N}$ such that for all $m > N$ and all $N \in \mathbb{N}$,

$$d(a_{nm}, a_n) < \epsilon.$$

4. Using Definition F.1.5 as a model, formulate the definition of uniform convergence for the collection of column sequences of a double sequence.

5. Suppose that $\lim_{n,m\to\infty} a_{nm}$ and $\lim_{m\to\infty} a_{nm}$ all exist. Prove that the collection $\{(a_{nm}) : n \in \mathbb{N}\}$ is uniformly convergent.

This brings us to the big, final result. It seems a good idea to set it off by itself as a theorem.

Theorem F.1.7 Let X be a metric space, and let (a_{nm}) be a doubly indexed sequence in X. Suppose that for all $i, j \in \mathbb{N}$, $\lim_{n \to \infty} a_{ni}$ and $\lim_{m \to \infty} a_{jm}$ exist, and that the convergence of either the row sequences or the column sequences is uniform. Then both iterated limits exist, the double limit exists, and all the limits are equal.

Many additional questions can be considered in regard to limits of doubly indexed sequences. Here are some things you might want to think about:

- Are convergent doubly indexed sequences bounded?

- If (a_{nm}) is a doubly indexed sequence of real numbers that converges to a, is it true that $|a_{nm}| \to |a|$?

- What about convergence and order? (generalizing results from Section 3.4)

- Suppose we have two convergent double sequences (a_{nm}) and (b_{nm}) in $\mathbb{R}$. What can we say about the convergence of the double sequences $(a_{nm} + b_{nm})$ and $(a_{nm} b_{nm})$? What about other arithmetic combinations of this type?

Remark: Is this the right way to think about adding and/or multiplying these sequences? Might there be other reasonable ways to do it?

Excursion G

Subsequences and Convergence

> *Required Background*
> All of the necessary ideas are covered by the end of Chapter 3.

This excursion consists of a series of numbered exercises that are meant to lead you through a set of ideas. Work the exercises as you go along.

G.1 Subsequential Limits

Definition G.1.1 Let (x_n) be a sequence in a metric space X. If (x_n) has a subsequence that converges to x, we say that x is a **limit point** of the sequence (x_n).[1]

1. Prove that every *non*-convergent bounded sequence of real numbers has at least two limit points.

2. Give an example of a sequence in $\mathbb{R}$ that has exactly two limit points.

3. Prove that the set of limit points of a sequence in metric space X is a closed subset of the metric space.

4. Give an example of a sequence in $\mathbb{R}$ whose set of limit points is the closed interval $[0, 1]$.

---

[1]. Problems 6 and 8 at the end of Section 3.3 and Theorem 3.4.10 are relevant to this idea and are worth reviewing now.

G.2 Limits Superior and Inferior

In sequences of real numbers, two subsequential limits play especially important roles. They are called the limit superior and the limit inferior of a sequence.

Definition G.2.1 Let (a_n) be a bounded sequence of real numbers. We define two sequences based on the sequence (a_n). The **upper sequence** of (a_n) is defined as follows:

$$\text{For each } n \in \mathbb{N}, \qquad \overline{a_n} = \sup_{k \geq n} a_k.$$

The **lower sequence** of (a_n) is defined similarly:

$$\text{For each } n \in \mathbb{N}, \qquad \underline{a_n} = \inf_{k \geq n} a_k.$$

1. Find the upper and lower sequences for the following sequences:

   $$2, -1, 0, 1, 1, 2, -1, 0, 1, 1, 2, -1, 0, 1, 1, 2, -1, 0, 1, 1, \ldots,$$

   $$\frac{1}{2}, \frac{2}{3}, \frac{3}{4}, \frac{4}{5}, \frac{5}{6}, \ldots,$$

   $$\frac{1}{2}, -\frac{2}{3}, \frac{3}{4}, -\frac{4}{5}, \frac{5}{6}, \ldots$$

2. Show that the upper and lower sequences of a bounded real sequence (a_n) are monotonic and bounded and, therefore, convergent.

Definition G.2.2 Let (a_n) be a bounded, sequence of real numbers. Then

$$\lim_{n \to \infty} \overline{a_n}$$

is called the **limit superior** (or lim sup) of the sequence (a_n). It is denoted by $\limsup a_n$. Likewise,

$$\lim_{n \to \infty} \underline{a_n}$$

is called the **limit inferior** (or lim inf) of the sequence (a_n). It is denoted by $\liminf a_n$.

Notice that every bounded sequence of real numbers has both a lim sup and a lim inf. This is an excellent thing. There are many times when it would be convenient for the limit of a sequence to exist, but it doesn't. The good news is that it is often just as good to consider the lim sup and/or the lim inf.

3. Find the lim inf and the lim sup of the sequences given in Problem 1.

4. Let (a_n) be a bounded sequence of real numbers.

   (a) Prove that $\limsup a_n = L$ if and only if the following two conditions hold:

      i. For every $\epsilon > 0$, there exists $N \in \mathbb{N}$ such that for all $n > N$, $a_n < L + \epsilon$.

      ii. For every $\epsilon > 0$ and for every $N \in \mathbb{N}$, there exists $n > N$ such that $a_n > L - \epsilon$.

   (b) Similarly, prove that $\liminf a_n = L$ if and only if the following two conditions hold:

      i. For every $\epsilon > 0$, there exists $N \in \mathbb{N}$ such that for all $n > N$, $a_n > L - \epsilon$.

      ii. For every $\epsilon > 0$ and for every $N \in \mathbb{N}$, there exists $n > N$ such that $a_n < L + \epsilon$.

5. Let (a_n) be a bounded sequence of real numbers.

   (a) Prove that (a_n) has a subsequence converging to $\liminf a_n$ and a subsequence converging to $\limsup a_n$.

   (b) Show that $\limsup a_n$ and $\liminf a_n$ are the largest and smallest subsequential limits of (a_n), respectively. (*Hint*: Prove that if (a_n) has a subsequence converging to $k \in \mathbb{R}$, then $\liminf a_n \leq k \leq \limsup a_n$.)

   (c) Prove that (a_n) converges if and only if $\limsup a_n = \liminf a_n$.

6. Let (a_n) be a sequence of real numbers. Prove the following closely related results:

   (a) Let $K \in \mathbb{R}$. Suppose that $\limsup a_n \leq k$ for all $k > K$. Prove that $\limsup a_n \leq K$.

   (b) Let S be a subset of $\mathbb{R}$ with greatest lower bound K. Suppose that $\limsup a_n \leq k$ for all $k \in S$. Prove that $\limsup a_n \leq K$.

   (c) Consider the results that were proved in parts (a) and (b). Formulate and prove parallel results for $\liminf a_n$.

7. Suppose that (a_n) and (b_n) are positive, bounded sequences of real numbers, and let k be a positive real number.

   (a) Prove that $\limsup k a_n = k \limsup a_n$.

   (b) Prove that $\limsup a_n b_n \leq \limsup a_n \limsup b_n$.

   (c) Give an example to show that the inequality in part (b) may be strict.

(d) Part (a) says that if one of the sequences is constant, we get equality in part (b). Is there a weaker condition that could be imposed upon one of the sequences that would ensure that $\limsup a_n b_n = \limsup a_n \limsup b_n$? Prove your conjecture.

(e) Consider the results in parts (a)–(d). Formulate and prove parallel results for the lim inf.

(f) Consider the results in parts (a)–(d). The sequences were assumed to be positive and bounded. The real number k was assumed to be positive. Can any of these hypotheses be removed or relaxed?

8. Suppose that (a_n) and (b_n) are bounded sequences of real numbers with $a_n \leq b_n$ for all $n \in \mathbb{N}$. Prove that $\limsup a_n \leq \limsup b_n$ and $\liminf a_n \leq \liminf b_n$.

9. Our definitions for lim sup and lim inf so far apply only to bounded sequences. If we allow $\pm\infty$ as answers, we can extend the definitions to arbitrary sequences. You should try to do this. Here are some things to think about:

- Come up with examples of sequences of real numbers that (your intuition tells you) have
  - Infinite lim sup and finite lim inf.
  - Finite lim sup and infinite lim inf.
  - Both lim sup and lim inf infinite. (Try to find a sequence where both are $+\infty$, a sequence where both are $-\infty$, and a sequence where one is $+\infty$ and the other is $-\infty$.)
- Consider sequences that are bounded above and not below (or below and not above). What can be said about the lim sup and the lim inf of such sequences? What about sequences that are bounded neither above nor below?
- Which of the theorems proved earlier for finite lim sup and lim inf still apply if one or the other of these is infinite?

Excursion H

Series of Real Numbers

> *Required Background*
> This excursion relies on ideas from Chapter 3 and Excursions F and G. To be more precise, most of the excursion can be understood using only ideas from Chapter 3; however, the discussion of the root and the ratio tests assumes familiarity with the limit supremum and the limit infimum, as discussed in Excursion G. Excursion F, on doubly indexed sequences, is needed only for Section H.5, in which two series are multiplied together.

H.1 Definition and Basic Properties

Definition H.1.1 Let (a_n) be a sequence of real numbers. We define a new sequence (s_n) based on (a_n) as follows:

For each $n \in \mathbb{N}$,
$$s_n = \sum_{i=0}^{n} a_i.$$

- This sequence is denoted by $\sum_{n=0}^{\infty} a_n$ and is called a **series**.[1]

- The terms of the sequence (a_n) are called the **terms** of the series.

---

1. We read $\sum_{n=0}^{\infty} a_n$ as "the sum from zero to infinity of a_n" or as "the sum of a_n from zero to infinity."

- The terms of the sequence (s_n) are called the **partial sums** of the series.

- If (s_n) converges, we say that the **series** $\sum_{n=0}^{\infty} a_n$ **converges**.

- If (s_n) converges to L, we call L the **sum of the series**, and we denote L by $\sum_{n=0}^{\infty} a_n$.

- If (s_n) fails to converge, we say that the series **diverges**.

Remark: Note that Definition H.1.1 contains an ambiguity. We use the notation

$$\sum_{n=0}^{\infty} a_n$$

to refer both to the sequence (s_n) and to its limit. This is rarely a problem, because the context usually makes it clear which meaning is intended. Nevertheless, it is a good idea to keep notational ambiguities in mind to avoid confusion.

Series with *positive* terms are very well behaved. The facts given in Exercise H.1.2 are quite useful and follow easily from previously proved facts about sequences of real numbers.

Exercise H.1.2 Let (a_n) be a sequence of positive real numbers. Prove the following facts:

1. The sequence of partial sums of the series $\sum_{n=0}^{\infty} a_n$ is strictly increasing.

2. The series $\sum_{n=0}^{\infty} a_n$ converges if and only if its sequence of partial sums is bounded above.

3. If the series $\sum_{n=0}^{\infty} a_n$ converges to S, then $\sum_{n=0}^{N} a_n < S$ for all $n \in \mathbb{N}$.

It follows that if $\sum_{n=0}^{\infty} a_n$ diverges, it does so because its sequence of partial sums goes to infinity. ∎

Geometric Series

One of the most important examples of a series is the geometric series. It crops up naturally in all sorts of places.

Definition H.1.3 Let a be a nonzero real number and let r be any real number. The series

$$\sum_{i=0}^{\infty} ar^i$$

is called a **geometric series**. The number r is called the **common ratio** of the geometric series. The number a is called the **leading term** of the series.

Theorem H.1.4 A geometric series converges if and only if $|r| < 1$. When $|r| < 1$,

$$\sum_{i=0}^{\infty} ar^i = \frac{a}{1-r}.$$

Proof: The proof is Problem 2 at the end of this section. $\square$

Cauchy Criterion for Series Convergence

With some notable exceptions like the geometric series, it is usually difficult to compute the sum of a series exactly. It is much easier to determine whether or not the series converges. Thus our theoretical emphasis (at least initially) will be to determine the circumstances under which series converge and to do so without having to compute the limit in the process. Our first task is to derive a convergence criterion that does not refer directly to a specific limit.

Theorem H.1.5 [Cauchy Criterion for Series Convergence] Let (a_n) be a sequence of real numbers. Prove that $\sum_{n=1}^{\infty} a_n$ converges if and only if for all positive real numbers, ϵ there exists $N \in \mathbb{N}$ such that for all integers m and n that satisfy $n > m > N$,

$$\left| \sum_{i=m}^{n} a_i \right| < \epsilon.$$

Proof: The proof is Problem 3 at the end of this section. $\square$

The importance of the Cauchy criterion is theoretical rather than practical. Indeed, we rarely use it to test the convergence of a specific series. Instead, it is the mathematical tool that we use to develop more practical tests for the convergence of series.

N^{th} Term Test for Divergence

Consider the divergent series $\sum_{i=1}^{\infty} i$. This series fails to converge because we keep adding on larger and larger numbers. The partial sums will clearly "blow up" and cannot possibly converge. If a series is to converge, the partial sums beyond some point cannot change very much. That is, we must add on smaller and smaller numbers or there is no hope of convergence. This bit of intuition is made precise in Theorem H.1.6, which is the first big consequence of the Cauchy criterion.

Theorem H.1.6 Let (a_n) be a sequence of real numbers. If $\sum_{n=0}^{\infty} a_n$ converges, then $a_n \to 0$.

Proof: The proof is Problem 4 at the end of this section. $\square$

Exercise H.1.7 Theorem H.1.6 is sometimes called "the Nth term test for divergence." Explain the use of the word divergence. ∎

In some sense, Theorem H.1.6 is fairly obvious. The more delicate question is whether the converse of this theorem is true. That is, if the terms of the series go to zero, does the series converge? The answer to this question is "no."

Example H.1.8 The quintessential example of a divergent series whose terms go to zero is the **harmonic series**:

$$1 + \frac{1}{2} + \frac{1}{3} + \frac{1}{4} + \cdots = \sum_{n=1}^{\infty} \frac{1}{n}.$$

The easiest way to see that this series diverges is to think about "grouping" the terms of the series in blocks of length a power of 2:

$$1 + \frac{1}{2} + \underbrace{\frac{1}{3} + \frac{1}{4}} + \underbrace{\frac{1}{5} + \frac{1}{6} + \frac{1}{7} + \frac{1}{8}}$$
$$+ \underbrace{\frac{1}{9} + \frac{1}{10} + \frac{1}{11} + \frac{1}{12} + \frac{1}{13} + \frac{1}{14} + \frac{1}{15} + \frac{1}{16}} + \cdots.$$

Notice that each group gives a result that is greater than or equal to $\frac{1}{2}$ and that a similar pattern can be continued indefinitely. Thus the partial sums of the harmonic series are unbounded. It follows that the series cannot converge.

This idea is captured by Lemma H.1.9.

Lemma H.1.9 [Grouping Lemma] Let (a_n) be a decreasing sequence of positive real numbers. Then $\sum_{n=1}^{\infty} a_n$ converges if and only if $\sum_{n=0}^{\infty} 2^{n-1} a_{2^n}$ converges.

Proof: The proof is Problem 5a at the end of this section. □

The grouping lemma not only proves that the harmonic series diverges, but also establishes the convergence criterion for a larger class of series, of which the harmonic series is one. These series are frequently called p-series.

Corollary H.1.10 [p-Series] Let p be a positive real number. The series

$$\sum_{n=1}^{\infty} \frac{1}{n^p}$$

converges whenever $p > 1$. It diverges when $p \leq 1$.

Proof: The proof is Problem 5b at the end of this section. □

Experiments

Computing a bunch of partial sums is not very helpful for deciding whether a series converges. It takes more than 12,000 terms of the harmonic series to get a partial sum that is greater than 10. Ten times that many terms get us to only about 12.3. If we were to guess, based on an experiment in which we computed partial sums, whether the harmonic series converges or diverges, we would likely guess the former. For series (like the harmonic series) that diverge but very slowly, numerical experiments are more likely to mislead than to help. Thus we really need the theory.

Absolute versus Conditional Convergence

One might assume that a series containing both positive and negative terms would be more likely to converge than a series containing only one or the other. The idea is that the negative terms provide a "counterbalance" to the positive terms, keeping the partial sums from getting too large. This assumption proves to be correct and the result follows easily from the Cauchy criterion.

Theorem H.1.11 Let (a_n) be a sequence of real numbers. If $\sum_{n=0}^{\infty} |a_n|$ converges, so does $\sum_{n=0}^{\infty} a_n$.

Proof: The proof is Problem 7 at the end of this section. □

As we shall see, the converse of Theorem H.1.11 is false. There are series for which $\sum_{n=0}^{\infty} a_n$ converges, but $\sum_{n=0}^{\infty} |a_n|$ diverges. This motivates Definition H.1.12.

Definition H.1.12 Let (a_n) be a sequence of real numbers. Suppose that $\sum_{n=0}^{\infty} a_n$ is a convergent series.

- If $\sum_{n=0}^{\infty} |a_n|$ converges, then $\sum_{n=0}^{\infty} a_n$ is said to **converge absolutely**.

- If $\sum_{n=0}^{\infty} |a_n|$ diverges, then $\sum_{n=0}^{\infty} a_n$ is said to **converge conditionally**.

As noted earlier, series with some positive terms and some negative terms are more likely to converge than those with only positive terms or only negative terms. Thus it would seem that (from the point of view of convergence) the most propitious circumstance would be to have the terms alternate in sign.

Definition H.1.13 A series of real numbers in which all the even terms are positive and all the odd terms are negative, or vice versa, is called an **alternating series**.

Theorem H.1.14 [Alternating Series Test] Let $\sum_{n=0}^{\infty} a_n$ be an alternating series of real numbers. If $(|a_n|)$ is monotonically decreasing and $a_n \to 0$, then the series converges.

Proof: The proof is Problem 12 at the end of this section. Stop right now to draw a picture, however. This will give you insight into what is going on in the theorem. □

Exercise H.1.15 Show by giving an example that not all alternating series converge. ■

Example H.1.16 [Alternating Harmonic Series] Notice that the alternating series test gives us an easy way to show that some alternating series converge: Just show that the absolute values of their terms decrease to zero. Applying the alternating series test to

$$1 - \frac{1}{2} + \frac{1}{3} - \frac{1}{4} + \frac{1}{5} + \cdots$$

tells us that the **alternating harmonic series** converges. This gives us the simplest example of a series that converges conditionally.

Problems H.1

1. Let (a_n) and (b_n) be sequences of real numbers, and let k be a real number. Suppose that $\sum_{i=0}^{\infty} a_n$ and $\sum_{i=0}^{\infty} b_n$ both converge.

    (a) Prove that $\sum_{i=0}^{\infty} k\, a_n$ converges and
    $$\sum_{i=0}^{\infty} k\, a_n = k \sum_{i=0}^{\infty} a_n$$

    (b) Prove that $\sum_{i=0}^{\infty}(a_n + b_n)$ converges and
    $$\sum_{i=0}^{\infty} (a_n + b_n) = \sum_{i=0}^{\infty} a_n + \sum_{i=0}^{\infty} b_n$$

2. Prove the theorem that characterizes the convergence of geometric series (Theorem H.1.4). (*Hint*: Use Problem 8c on page 297 to compute the partial sums of the series. Then apply Theorem 3.4.11 and Theorem D.4.2.)

3. Establish the Cauchy criterion for the convergence of series by proving Theorem H.1.5.

4. Establish the Nth term test for divergence by proving Theorem H.1.6.

5. **Convergence of p-Series.**

    (a) Prove the grouping lemma (Lemma H.1.9). (*Hint*: Since these are series with positive terms, convergence is equivalent to the boundedness of the partial sums.)

    (b) Use the grouping lemma to establish the criterion, given in Corollary H.1.10, for the convergence of p-series.

6. Here is a possible alternative definition for Cauchy sequence:

    Let (a_n) be a sequence in a metric space X. We say that (a_n) is Cauchy provided that for all $\epsilon > 0$, there exists $N \in \mathbb{N}$ such that for all $n > N$, $d(a_n, a_{n+1}) < \epsilon$.

This is certainly less complicated and seems very reasonable at first glance. After all, it requires that things arbitrarily far out are getting arbitrarily close to each other.

Use the harmonic series to explain why this is not sufficiently strong a definition.

7. Prove that absolute convergence implies convergence (Theorem H.1.11).

8. Prove that if a geometric series converges, it converges absolutely.

9. Suppose that (a_n) is a sequence of real numbers. What is the relationship between the convergence of the two series:

$$\sum_{n=0}^{\infty} a_n \quad \text{and} \quad \sum_{n=0}^{\infty} (-1)^n a_n?$$

That is, could both converge absolutely? Could both diverge? Could one converge absolutely and one converge conditionally? (Consider all such possible pairings. In each case, give an example to show that a pairing can occur or a proof that shows it cannot.)

10. Let (a_n) and (b_n) be sequences of real numbers. Find a proof or a counterexample for each of the following statements:

    (a) If $\sum_{n=0}^{\infty} a_n$ and $\sum_{n=0}^{\infty} b_n$ both converge absolutely, then $\sum_{n=0}^{\infty} (a_n + b_n)$ must converge absolutely.

    (b) If $\sum_{n=0}^{\infty} a_n$ and $\sum_{n=0}^{\infty} b_n$ both converge conditionally, then $\sum_{n=0}^{\infty} (a_n + b_n)$ must converge conditionally.

    (c) If $\sum_{n=0}^{\infty} a_n$ converges absolutely and $\sum_{n=0}^{\infty} b_n$ converges conditionally, then $\sum_{n=0}^{\infty} (a_n + b_n)$ must converge conditionally.

    (d) If $\sum_{n=0}^{\infty} a_n$ and $\sum_{n=0}^{\infty} b_n$ both diverge, then $\sum_{n=0}^{\infty} (a_n + b_n)$ must diverge.

11. Let (a_n) be a monotonically decreasing sequence of positive real numbers. Suppose the alternating series $\sum_{n=0}^{\infty} (-1)^n a_n$ converges. Prove that the sum of the alternating series lies between $a_0 - a_1$ and a_0.

12. Establish the alternating series test by proving Theorem H.1.14. (*Hint:* By all means, draw a picture! If you are having difficulty, try doing Problem 11 first.)

13. Give an example of a divergent alternating series whose terms go to zero. (Why doesn't the existence of such a series contradict the alternating series test?)

14. Can the requirement that $(|a_n|)$ be a decreasing sequence be eliminated from Theorem H.1.14? If not, can it be relaxed somehow? Explain.

H.2 Comparing Series

The following test allows us to use previous knowledge about convergence or divergence of specific series to deduce the convergence or divergence of a wider class of series.

Theorem H.2.1 [Comparison Test] Let (a_n) and (b_n) be sequences of real numbers. Suppose that $0 \leq a_n \leq b_n$ for all $n \in \mathbb{N}$. Then

$$\text{if } \sum_{n=1}^{\infty} b_n \text{ converges, then } \sum_{n=1}^{\infty} a_n \text{ converges.}$$

Equivalently,

$$\text{if } \sum_{n=1}^{\infty} a_n \text{ diverges, then } \sum_{n=1}^{\infty} b_n \text{ diverges, also.}$$

Proof: The proof is Problem 1 at the end of this section. □

Exercise H.2.2 Note that the comparison test, as stated in Theorem H.2.1, requires that (a_n) and (b_n) be sequences of positive real numbers. Is this requirement necessary, or could it be removed? Explain. ■

There is a useful variant of the comparison test that is often easier to apply.

Theorem H.2.3 [Limit Comparison Test] Let (a_n) and (b_n) be sequences of positive real numbers. Suppose that there exists a positive real number K such that

$$\lim_{n \to \infty} \frac{a_n}{b_n} = K.$$

Then

$$\sum_{n=0}^{\infty} a_n \text{ converges if and only if } \sum_{n=0}^{\infty} b_n \text{ converges.}$$

Proof: Suppose that $\sum_{n=0}^{\infty} a_n$ converges. I will prove that $\sum_{n=0}^{\infty} b_n$ converges. (The converse is left to you in problem 2 at the end of the section.) Choose $N \in \mathbb{N}$ such that for all $n > N$,

$$\left| \frac{a_n}{b_n} - K \right| < \frac{K}{2}.$$

Thus, in particular, $0 < K - \frac{K}{2} < \frac{a_n}{b_n}$, which implies that $0 < \frac{K}{2} b_n < a_n$. It follows that

$$\sum_{n=0}^{\infty} \frac{K}{2} b_n$$

converges by the comparison test. Finally,

$$\sum_{n=0}^{\infty} b_n = \sum_{n=0}^{\infty} \frac{2}{K} \frac{K}{2} b_n$$

converges by Problem 1 on page 340. $\square$

The comparison test and the limit comparison test are *extrinsic* tests. That is, they allow us to determine the convergence of a series only if we have prior knowledge about the convergence of other series. The comparison test, however, can also be used to derive some very useful tests that require us to use only information that is *intrinsic* to the series in which we are interested.

Problems H.2

1. Establish the comparison test by proving Theorem H.2.1.

2. Let (a_n) and (b_n) be sequences of positive real numbers. Suppose that there exists a positive real number K such that

$$\lim_{n \to \infty} \frac{a_n}{b_n} = K.$$

   Prove that if $\sum_{n=0}^{\infty} b_n$ converges, then so does $\sum_{n=0}^{\infty} a_n$. (This is one half of Theorem H.2.3. The converse was proved in the section.)

3. **Extending the limit comparison test.** Let (a_n) and (b_n) be sequences of positive real numbers.

(a) Suppose that
$$\lim_{n \to \infty} \frac{a_n}{b_n} = 0.$$
What (if anything) can be said about the convergence of $\sum_{n=0}^{\infty} a_n$ given the convergence of $\sum_{n=0}^{\infty} b_n$? What if you know that $\sum_{n=0}^{\infty} a_n$ converges? Does this tell you that $\sum_{n=0}^{\infty} b_n$ converges?

(b) Suppose that
$$\lim_{n \to \infty} \frac{a_n}{b_n} = +\infty.$$
What (if anything) can be said about the convergence of $\sum_{n=0}^{\infty} a_n$ given the convergence of $\sum_{n=0}^{\infty} b_n$? What if you know that $\sum_{n=0}^{\infty} a_n$ converges? Does this tell you that $\sum_{n=0}^{\infty} b_n$ converges?

(c) By reconsidering the proofs given for the case when the limit exists, determine what you can say if you know only that
$$\limsup \frac{a_n}{b_n} = K?$$
Consider the cases when $0 < K < \infty$, $K = 0$, and $K = \infty$.

H.3 Relatives of the Geometric Series

Recall that for a real number r, the geometric series $\sum_{n=0}^{\infty} r^n$ converges if $|r| < 1$ and diverges otherwise. Let us think of the terms of the geometric series in two different ways. Consider the ratios of adjacent terms in the series and the nth root of the nth term:

$$\frac{r^{n+1}}{r^n} = r \quad \text{and} \quad \sqrt[n]{r^n} = r.$$

In either case, we get a constant. This allows us to give two slightly different (albeit equivalent) descriptions of the behavior of geometric series:

- **The ratio of adjacent terms is constant.** When $-1 < r < 1$, the key to the convergence of this series is that each term is a factor of r closer to zero than the previous term. This makes the terms of the series go to zero sufficiently fast so that the series converges.

- **The nth root of the nth term is constant.** When $|r| < 1$, the powers of r go to zero exponentially fast, causing the series to converge.

The fact that this behavior occurs for every term of the series is not significant. Any series that behaved in one of these two equivalent ways *in the long run* should converge or diverge, as does the corresponding geometric series. This insight gives us two tests for the absolute convergence of a series. The ratio test looks at the long-term behavior of the ratios of adjacent terms of a series. The root test looks at the long term behavior of the nth root of the nth term.

Theorem H.3.1 [Ratio Test] Let (a_n) be a sequence of non-zero real numbers.

1. The series $\sum_{n=0}^{\infty} a_n$ converges absolutely if $\limsup \left| \dfrac{a_{n+1}}{a_n} \right| < 1$.

2. The series $\sum_{n=0}^{\infty} a_n$ diverges if $\liminf \left| \dfrac{a_{n+1}}{a_n} \right| > 1$.

Proof: The proof is Problem 3a at the end of this section. □

Note: If $\lim_{n \to \infty} \left| \dfrac{a_{n+1}}{a_n} \right|$ exists, the lim sup and the lim inf in the theorem become ordinary limits. It is likely you have seen the ratio test stated in this slightly less general form. Sometimes, however, it is convenient to be able to apply the test when the limit does not exist. (See, for instance, Problem 2 at the end of this section.)

The "Gap" in the Ratio Test
Consider the series

$$\sum_{n=1}^{\infty} \frac{1}{n^2}, \quad \sum_{n=1}^{\infty} \frac{1}{n}, \quad \text{and} \quad \sum_{n=1}^{\infty} \frac{(-1)^n}{n}.$$

What do you know about the convergence or divergence of these series? In each case, what can you say about the long-term behavior of $\left| \dfrac{a_{n+1}}{a_n} \right|$? Sequences for which

$$\limsup \left| \frac{a_{n+1}}{a_n} \right| \geq 1 \quad \text{and} \quad \liminf \left| \frac{a_{n+1}}{a_n} \right| \leq 1$$

sometimes converge and sometimes diverge. The ratio test doesn't tell us anything at all about such series; it simply does not apply. We must look elsewhere for information.

Question to Ponder: Statements made at the beginning of the section ought to give you some insight into the behavior of series for which the ratio test gives a conclusive answer. They behave, at least in the long run, sort of like geometric series—either convergent or divergent geometric series. What, if anything, can you conclude about the convergence behavior of series for which the ratio test gives no answer? For instance, think about what it means to say that

$$\lim_{n \to \infty} \left| \frac{a_{n+1}}{a_n} \right| = 1.$$

Notice also that the ratio test fails for all conditionally convergent series. Does this give you any insight into conditional convergence?

Theorem H.3.2 [Root Test] Let (a_n) be a sequence of real numbers.

1. If $\limsup \sqrt[n]{|a_n|} < 1$, then $\sum_{n=0}^{\infty} a_n$ converges absolutely.

2. If $\limsup \sqrt[n]{|a_n|} > 1$, then $\sum_{n=0}^{\infty} a_n$ diverges.

Proof: The proof is Problem 4 at the end of this section. □

Comparing the Root Test and the Ratio Tests

Ordinarily, the ratio test is easier to apply than the root test, because it is easier to compute ratios than it is to compute roots. Nevertheless, the root test has a distinct advantage over the ratio test: It is a stronger test. As you know, the ratio test and the root test have "gaps," in the sense that there are series for which they cannot give us an answer. But the gaps in the root test are "less wide," in the sense that there are series on which the root test is effective but the ratio test isn't. On the other hand, every time that the ratio test is conclusive, the root test is as well.

Theorem H.3.3 Let (a_n) be a sequence of non-zero real numbers. If the ratio test applies, then so does the root test. That is,

1. If $\limsup \left| \frac{a_{n+1}}{a_n} \right| < 1$, then $\limsup \sqrt[n]{|a_n|} < 1$.

2. If $\liminf \left| \frac{a_{n+1}}{a_n} \right| > 1$, then $\limsup \sqrt[n]{|a_n|} > 1$.

Proof: The proof is Problem 5 at the end of this section. □

Problems H.3

1. Use the ratio test to determine whether the following series converge or diverge.

    (a) $\sum_{n=0}^{\infty} \dfrac{(-8)^n}{n!}$ (b) $\sum_{n=0}^{\infty} \dfrac{2(n+1)}{3^n}$ (c) $\sum_{n=0}^{\infty} \dfrac{5^{n-12}}{(12n)3^{n+12}}$

2. Although in practice we often find that

$$\lim_{n \to \infty} \left| \dfrac{a_{n+1}}{a_n} \right| \text{ exists,}$$

   there are some perfectly ordinary series for which it doesn't. Consider, for instance, the following sequence:

$$1 + \dfrac{2}{3} + \dfrac{1}{2} + \dfrac{1}{3} + \dfrac{1}{4} + \dfrac{1}{6} + \dfrac{1}{8} + \dfrac{1}{12} + \dfrac{1}{16} + \dfrac{1}{24} + \cdots + \dfrac{1}{2^n} + \dfrac{1}{3(2^{n-1})} \cdots$$

   (a) Show that $\lim_{n \to \infty} \left| \dfrac{a_{n+1}}{a_n} \right|$ does not exist.

   (b) What does the ratio test tell us about the convergence or divergence of the series?

3. **The ratio test**

   (a) Prove Theorem H.3.1 (the ratio test).

   (b) Generalizing the ratio test: The divergence provision of the ratio test can be generalized slightly. Replace the statement:

   2. The series $\sum_{n=0}^{\infty} a_n$ diverges if $\liminf \left| \dfrac{a_{n+1}}{a_n} \right| > 1$.

   by

   2'. The series $\sum_{n=0}^{\infty} a_n$ diverges if there exists $N \in \mathbb{N}$ such that for all $n > N$, $\left| \dfrac{a_{n+1}}{a_n} \right| \geq 1$.

First prove the theorem with the more general hypothesis (it is likely that your basic argument will still work), and then explain why we cannot just replace the original hypothesis with the following statement:

2″. The series $\sum_{n=0}^{\infty} a_n$ diverges if $\liminf \left| \dfrac{a_{n+1}}{a_n} \right| \geq 1$.

4. Prove Theorem H.3.2 (the root test).

5. Prove that if the ratio test applies, so does the root test (Theorem H.3.3).

6. Give an example of a series on which the ratio test is inconclusive but the root test isn't. Can you give examples of both convergent and divergent series?

H.4 Rearranging the Terms of a Series

Suppose that $\sum_{n=0}^{\infty} a_n$ is a convergent series of real numbers. If $\sum_{n=0}^{\infty} |a_n|$ converges as well, it may seem obvious why we say that the series converges *absolutely*. If $\sum_{n=0}^{\infty} |a_n|$ diverges, our choice of the words "conditional convergence" may not be quite so natural. The use of the word "conditional" comes from the fact that both the sum of the series and the convergence of the series depend very strongly on the order of the terms in the series. To be more explicit, we know that if we are adding finitely many real numbers, the order in which we add them is immaterial. Add them in any order and the answer is always the same. This ability to rearrange the order of the summands does not extend to infinite series, unless the series is absolutely convergent.

Rearranging the Terms of a Sequence (a_n)

What we want is a mechanism for obtaining a sequence that has the same terms as (a_n) but in a different order. Say, for the sake of argument, that the tenth term of (a_n) will be the zeroth term of the new sequence. The third term of (a_n) will be the first term of the new sequence, the term a_{12} will come next, followed by a_{37} .... One way of thinking about this is to think about positions within the sequence as "slots" into which the terms are placed. In our case the tenth term will get the zeroth slot, the third term will get the first slot, and so forth.

> In effect, there is a function from the slots to the terms and, by composition, from the "new" slots to the "old" slots.
>
> $$0 \to 10$$
> $$1 \to 3$$
> $$2 \to 12$$
> $$3 \to 37$$
> $$\vdots \; \vdots \; \vdots$$
>
> This is a function $\pi : \mathbb{N} \cup \{0\} \to \mathbb{N} \cup \{0\}$. We want to have each term of the old sequence appear exactly once in the new sequence. That is, we want π to be one-to-one and onto.
>
> Our new sequence will be
>
> $$a_{\pi(0)}, a_{\pi(1)}, a_{\pi(2)}, a_{\pi(3)}, \ldots = a_{10}, a_3, a_{12}, a_{37}, \ldots$$

Definition H.4.1 Let $\pi : \mathbb{N} \cup \{0\} \to \mathbb{N} \cup \{0\}$ be a one-to-one and onto function. If (a_n) is a sequence of real numbers, the sequence $(a_{\pi(n)})$ is said to be a **rearrangement** of (a_n). Likewise, the series

$$\sum_{n=0}^{\infty} a_{\pi(n)} \text{ is said to be a rearrangement of } \sum_{n=0}^{\infty} a_n.$$

Theorem H.4.2 Let $\sum_{n=0}^{\infty} a_n$ be an absolutely convergent series of non-zero real numbers, and let $\sum_{n=0}^{\infty} a_{\pi(n)}$ be a rearrangement of $\sum_{n=0}^{\infty} a_n$. Then $\sum_{n=0}^{\infty} a_{\pi(n)}$ also converges absolutely and the two series have the same sum.

Proof: Suppose that $\sum_{n=0}^{\infty} a_n$ converges to the real number S. We want to show that $\sum_{n=0}^{\infty} a_{\pi(n)}$ also converges to S.

- Choose $N_1 \in \mathbb{N}$ such that if $m > N_1$, $\left| \sum_{n=0}^{m} a_n - S \right| < \dfrac{\epsilon}{2}$.

- Choose $N_2 \in \mathbb{N}$ such that $N_2 > N_1$ and if $i \geq j > N_2$, $\sum_{n=i}^{j} |a_n| < \dfrac{\epsilon}{2}$.

- Choose $N_3 \in \mathbb{N}$ such that $\{0,1,2,3,\ldots,N_2\} \subseteq \{\pi(0), \pi(1), \ldots \pi(N_3)\}$.

Note that $N_1 < N_2 \leq N_3$. Fix $m > N_3$. Set $T = \max_{n \leq m} \pi(n)$. Observe that $T > N_2$.

$$\left|\sum_{n=0}^{m} a_{\pi(n)} - S\right| = \left|\sum_{n=0}^{N_2} a_n - S + \sum_{\substack{n \leq m \\ \pi(n) > N_2}} a_{\pi(n)}\right|$$

$$\leq \left|\sum_{n=0}^{N_2} a_n - S\right| + \sum_{\substack{n \leq m \\ \pi(n) > N_2}} |a_{\pi(n)}|$$

$$< \frac{\epsilon}{2} + \sum_{n=N_2+1}^{T} |a_n|$$

$$< \epsilon.$$

Thus the rearranged series also converges to S. $\square$

A rearrangement of a conditionally convergent series $\sum_{n=0}^{\infty} a_n$ need not converge to the same limit as the original series. Indeed we have an amazing fact: Given any real number, a rearrangement of $\sum_{n=0}^{\infty} a_n$ can be found that will converge to that number.[2]

Definition H.4.3 Let (a_n) be a sequence of real numbers. Define two sequences of real numbers as follows:

$$a_n^+ = \frac{1}{2}(|a_n| + a_n) \quad \text{and} \quad a_n^- = \frac{1}{2}(|a_n| - a_n).$$

Exercise H.4.4 This exercise will help you to get a feel for the sequences (a_n^+) and (a_n^-).

1. For the sequence (a_n)

$$1, -1, -\frac{1}{2}, \frac{1}{2}, \frac{1}{3}, -\frac{1}{3}, -\frac{1}{4}, \frac{1}{4}, \ldots$$

Compute the sequences (a_n^+) and (a_n^-).

2. Let (a_n) be a sequence of real numbers. Show that all terms of the sequences (a_n^+) and (a_n^-) are non-negative.

---

[2] Moreover, some rearrangements don't converge at all. See Problem 3 at the end of this section.

3. Let (a_n) be a sequence of real numbers. Guess an algebraic relationship between a_n, a_n^+, and a_n^-. Modify your guess to get a relationship between $|a_n|$, a_n^+, and a_n^-. Verify your conjectures by using the formulas given for a_n^+ and a_n^-.

4. Suppose that $\sum_{n=0}^{\infty} a_n$ converges. Show that (a_n^+) and (a_n^-) both go to 0 as n goes to ∞.

∎

Lemma H.4.5 Let (a_n) be a sequence of real numbers. Suppose that $\sum_{n=0}^{\infty} a_n$ converges conditionally, then both $\sum_{n=0}^{\infty} a_n^+$ and $\sum_{n=0}^{\infty} a_n^-$ diverge. In particular, each of these series has infinitely many non-zero terms.

Proof: The proof is Problem 2 at the end of this section. □

Theorem H.4.6 Let (a_n) be a sequence of real numbers, and let $K \in \mathbb{R}$. Suppose that $\sum_{n=0}^{\infty} a_n$ converges conditionally. Then there exists a rearrangement

$$\sum_{n=0}^{\infty} a_{\pi(n)} \text{ of } \sum_{n=0}^{\infty} a_n$$

that converges to K.

Proof Sketch: Let (p_n) be the subsequence of (a_n^+) that contains all of the non-zero terms but omits those that are zero. Likewise, let (m_n) be the subsequence of $(-a_n^-)$ that contains all of the non-zero terms and omits those that are zero. [These sequences constitute the positive and negative terms of (a_n), respectively.] Note that both of these sequences go to 0 as n goes to ∞.

Let's consider the case where $K \geq 0$. The case where $K < 0$ is parallel. We will choose "blocks" of positive terms and negative terms, alternately, according to the following scheme.

Choose $n_1 \in \mathbb{N}$ such that

$$\sum_{n=0}^{n_1-1} p_n \leq K < \sum_{n=0}^{n_1} p_n.$$

(In other words, choose the smallest number of positive terms so that the sum exceeds K.) Define the sum over the first block to be

$$s_1 = \sum_{n=0}^{n_1} p_n.$$

Choose $n_2 \in \mathbb{N}$ such that

$$s_1 + \sum_{n=0}^{n_2-1} m_n \geq K > s_1 + \sum_{n=0}^{n_2} m_n.$$

(In other words, choose n_2 so that when we add the negative terms $m_0, m_1, \ldots, m_{n_2}$ to s_1, the total will slip below K but no smaller number of negative terms will give an overall total that is less than K.) Define the sum over the second block to be

$$s_2 = \sum_{n=0}^{n_2} m_n.$$

Now we choose positive terms once again. Choose $n_3 > n_1$ so that

$$s_1 + s_2 + \sum_{n=n_1+1}^{n_3-1} p_n \leq K < s_1 + s_2 + \sum_{n=n_1+1}^{n_3} p_n.$$

Set

$$s_3 = \sum_{n=n_1+1}^{n_3} p_n.$$

Now "go backwards" once again by picking some negative terms. Choose $n_4 > n_2$ so that

$$s_1 + s_2 + s_3 + \sum_{n=n_2+1}^{n_4-1} m_n \geq K > s_1 + s_2 + s_3 + \sum_{n=n_2+1}^{n_4} m_n.$$

Set

$$s_3 = \sum_{n=n_2+1}^{n_4} m_n.$$

Continue this inductive process. Because at least one positive term is chosen at each odd stage and at least one negative term at each even stage, it is clear that every term of the original series is eventually chosen. Thus the new series constitutes a rearrangment of the original series.

The rearranged series converges to K. Here is a sketch of the argument that establishes this result. It is clear that $s_1 + s_2 + s_3 + \ldots + s_i$ is a partial sum of the rearranged series. Suppose that i is odd. Since

$$s_1 + s_2 + \cdots + (s_i - p_{n_i}) < K < s_1 + s_2 + \cdots + s_i,$$

the error with which $s_1 + s_2 + \cdots + s_i$ estimates K is no more than p_{n_i}. Because the sequence (p_k) goes to zero, this error can be made as small as we like provided that we choose a sufficiently large value for i. Likewise, if i is even, the error will be no more than $|m_{n_i}|$, which is small if i is large enough.

We can conclude that $s_1 + s_2 + s_3 + \cdots + s_i$ converges to K as i goes to ∞. Thus a subsequence of the partial sums of $\sum_{n=0}^{\infty} a_{\pi(n)}$ converges to K.

How well do arbitrary partial sums approximate K? An arbitrary partial sum will consist of a finite number of completed "blocks" of positive or negative terms, along with a partial block. That is, the partial sums are of the following form:

$$s_1 + s_2 + \cdots + s_i + \text{ a few more terms, all positive or all negative.}$$

Once again we have two cases. If i is odd and $t < n_i$,

$$s_1 + s_2 + \cdots + s_i > s_1 + s_2 + \cdots + s_i + \sum_{n=n_{i-1}+1}^{t} m_n > K.$$

It follows that the error with which the arbitrary partial sum approximates K is smaller than the error with which $s_1 + s_2 + \cdots + s_i$ approximates K. We have already established that this can be made as small as desired. Similarly, if i is even and $t < n_{i+1}$,

$$s_1 + s_2 + \cdots + s_i < s_1 + s_2 + \cdots + s_i + \sum_{n=n_{i-1}+1}^{t} p_n < K.$$

As before, the error with which the arbitrary partial sum approximates K is smaller than the error with which $s_1 + s_2 + \cdots + s_i$ approximates K. We have already established that this error can be made as small as desired.

Thus the rearranged series converges to K. □

Problems H.4

1. Let (a_i) be a sequence in a metric space X. Suppose that (a_i) has a subsequence converging to $x \in X$. Show that if $(b_i) = (a_{\pi(i)})$ is a rearrangement of the sequence (a_i), then (b_i) also has a subsequence converging to x. In other words, rearranging a sequence does not change its set of limit points. (*Hint*: Show that every subsequence of (b_i) has a subsequence in common with (a_i). Then use Problem 6 at the end of Section 3.3.)

2. Prove that if $\sum_{n=0}^{\infty} a_n$ converges conditionally, then $\sum_{n=0}^{\infty} a_n^+$ and $\sum_{n=0}^{\infty} a_n^-$ both converge (Lemma H.4.5).

3. Let (a_n) be a sequence of real numbers. Suppose that $\sum_{n=0}^{\infty} a_n$ converges conditionally. Theorem H.4.6 says that we can rearrange the series to make it converge to any arbitrary real number.

   (a) The case when $K > 0$ was proved in the text. That proof asserted that a parallel argument would produce a rearrangement of the sequence that converges to $K < 0$. Modify the argument to obtain this result.

   In fact, much more is true. By rearranging the terms of a conditionally converging series, we can obtain almost any convergence result.

   (b) Modify the proof of Theorem H.4.6 to obtain a rearrangement of the series that goes to $+\infty$. How would you modify your proof to get a rearrangement that goes to $-\infty$?

   (c) Fix two real numbers $S < R$. Modify the proof of Theorem H.4.6 to obtain a rearrangement of the series whose limit supremum is R and whose limit infimum is S.

   (d) Modify the proof of Theorem H.4.6 to obtain a rearrangement of the series whose limit supremum is $+\infty$ and whose limit infimum is $-\infty$.

H.5 Multiplying Series

We have talked about adding series term by term. The problem of multiplying series is a bit more complicated. The complication starts with the following question:

Just what is meant by

$$\left(\sum_{n=0}^{\infty} a_n\right) \left(\sum_{k=0}^{\infty} b_k\right)?$$

If the series converge to $A = \sum_{n=0}^{\infty} a_n$ and $B = \sum_{k=0}^{\infty} b_k$, we clearly want

$$\left(\sum_{n=0}^{\infty} a_n\right) \left(\sum_{k=0}^{\infty} b_k\right) = AB.$$

But this begs the question of how we actually multiply the *series* together, as opposed to multiplying their *sums*. What we want is to find a series whose limit will be AB and whose terms can reasonably be thought of as coming from a product of the two factor series.

Our scheme must clearly work if all but finitely many terms in each series is 0. Thus we take our cue from the multiplication of finite sums:

$$\left(\sum_{n=0}^{N} a_n\right)\left(\sum_{k=0}^{K} b_k\right) = a_0(b_0 + b_1 + b_2 + \cdots + b_K) + a_1(b_0 + b_1 + b_2 + \cdots + b_K)$$
$$+ \cdots + a_N(b_0 + b_1 + b_2 + \cdots + b_K).$$

In this vein, we try to make sense of a formal sum (which at the moment really has no meaning):

$$\left(\sum_{n=0}^{\infty} a_n\right)\left(\sum_{k=0}^{\infty} b_n\right) = a_0(b_0 + b_1 + b_2 + \cdots) + a_1(b_0 + b_1 + b_2 + \cdots)$$
$$+ a_3(b_0 + b_1 + b_2 + \cdots) + \cdots.$$

This is a start, but the profusion of ellipses make it impossible to interpret this as the limit of a sequence of partial sums. So, by way of motivation, we do some judicious rearranging and grouping of the terms in these formal sums:

$$a_0(b_0 + b_1 + b_2 + \cdots) + a_1(b_0 + b_1 + b_2 + \cdots) + a_3(b_0 + b_1 + b_2 + \cdots) + \cdots$$
$$= a_0 b_0 + (a_0 b_1 + a_1 b_0) + (a_0 b_2 + a_1 b_1 + a_2 b_0) + (a_0 b_3 + a_1 b_2 + a_2 b_1 + a_3 b_0) + \cdots.$$

This motivates Definition H.5.1.

Definition H.5.1 Let (a_n) and (b_n) be sequences of real numbers. For each $k \in \mathbb{N}$, set

$$c_k = a_0 b_k + a_1 b_{k-1} + a_2 b_{k-3} + \cdots + a_k b_0.$$

Then the series $\sum_{k=0}^{\infty} c_k$ is called the **Cauchy product** or just the **product** of the two series

$$\left(\sum_{n=0}^{\infty} a_n\right) \text{ and } \left(\sum_{k=0}^{\infty} b_k\right).$$

In the motivating discussion preceding Definition H.5.1, we did some rearranging and grouping of the formal product that led to our notion of the Cauchy product. In view of Section H.4, in which we saw that rearranging the order of terms in a series is not stable except in the case where the series converges absolutely, it should not surprise you to see that absolute convergence plays a part in the desired result

$$\sum_{k=0}^{\infty} c_k = \left(\sum_{n=0}^{\infty} a_n\right)\left(\sum_{k=0}^{\infty} b_k\right).$$

Indeed, the Cauchy product of two convergent series of real numbers converges, provided that at least one of the factor series converges absolutely. Furthermore, the sum of the Cauchy product is the product of the sums of its factor series. Theorem H.5.2 sets this out carefully.

Theorem H.5.2 Let (a_n) and (b_n) be sequences of real numbers. For each $n \in \mathbb{N}$, set

$$c_n = a_0 b_n + a_1 b_{n-1} + a_2 b_{n-2} + \cdots + a_n b_0.$$

Suppose that $\sum_{n=0}^{\infty} a_n$ converges to A, that $\sum_{n=0}^{\infty} b_n$ converges to B, and that at least one of these series converges absolutely. Then the Cauchy product $\sum_{n=0}^{\infty} c_n$ of these two series converges to AB.

The proof is outlined next. It is up to you to justify each statement made in the outline and to fill in any missing details in the proof.

Proof Outline: Assume $\sum a_n$ converges absolutely. The other case is proved similarly.

- Set $A_m = \sum_{n=0}^{m} a_n$, $B_m = \sum_{n=0}^{m} b_n$, and $D_m = B_m - B$. Note that $A_m \to A$, $B_m \to B$, and $D_m \to 0$. We want to show that $\sum_{n=0}^{m} c_n \to AB$.

- To see this, first show that

$$\sum_{n=0}^{m} c_n = A_m B + a_0 D_m + a_1 D_{m-1} + \cdots + a_m D_0.$$

- Note that it is enough to show that

$$a_0 D_m + a_1 D_{m-1} + \cdots + a_m D_0 \to 0 \text{ as } m \to \infty. \text{ Why?}$$

- Observe that the partial sums of the series $\sum |a_n|$ are bounded by some positive number K, and that the terms of the sequence $(|D_i|)$ are bounded by some positive real number M. Why?

- Let $\epsilon > 0$. Now choose $N_1 \in \mathbb{N}$ such that

$$\text{for all } j > i > N_1, \quad \sum_{n=i}^{j} |a_n| < \frac{\epsilon}{2M}.$$

- Next, choose $N_2 \in \mathbb{N}$ such that
$$\text{for all } n \geq N_2, \quad |D_n| < \frac{\epsilon}{2K}.$$

- Let $N = N_1 + N_2$. Fix $m > N$. Note that $m > N_2$ and that $m - N_2 > N_1$. Use these facts to show that
$$|a_0 D_m + a_1 D_{m-1} + \cdots + a_m D_0| < \epsilon.$$

□

Excursion I

Probing the Definition of the Riemann Integral

> *Required Background*
> The co-requisite for this excursion is Section 11.2.

The definition of the Riemann integral contains some interesting nuances. This excursion examines the definition with a critical eye.

I.1 Regular Riemann Sums

Definition I.1.1 Let a and b be real numbers with $a < b$. Let

$$P = \{x_0, x_1, x_2, \ldots, x_n\}$$

be a partition of $[a, b]$. The set P is called a **regular** partition if each of its subintervals has the same length.

Exercise I.1.2 Let a and b be real numbers with $a < b$. Let $n \in \mathbb{N}$. Suppose that P is a regular partition of $[a, b]$ with n subintervals.

1. Explain why there is only one such partition.

2. Explain further why

$$\|P\| = \frac{b-a}{n} = x_i - x_{i-1}$$

for all $i = 1, 2, 3, \ldots, n$.

■

Excursion I ▪ Probing the Definition of the Riemann Integral

Definition I.1.3 Let $f : \mathbb{R} \to \mathbb{R}$ be a function, and let $P = \{x_0, x_1, x_2, \ldots, x_n\}$ be a *regular* partition of $[a, b]$. Then the sum

$$\sum_{i=1}^{n} f(x_i)(x_i - x_{i-1})$$

is called the **right endpoint Riemann sum** for f corresponding to P.

Similarly,

$$\sum_{i=0}^{n-1} f(x_i)(x_i - x_{i-1})$$

is called the **left endpoint Riemann sum** for f corresponding to P.

Exercise I.1.4 Let $f : \mathbb{R} \to \mathbb{R}$ be a function, and let P_n be the regular partition of $[a, b]$ with n subintervals. Explain why the right endpoint Riemann sum for f corresponding to P_n is

$$R_n(f) = \sum_{i=1}^{n} f\left(a + i\left(\frac{b-a}{n}\right)\right)\left(\frac{b-a}{n}\right)$$

Find a similar expression for the left endpoint Riemann sum $L_n(f)$. ∎

The partitions and the sets of sampling points referred to in the definition of the Riemann integral are quite general. The partitions are not assumed to be regular, nor are the sampling points chosen in any systematic way. In fact, the only thing that is assumed about x_i^* is that it lies in the interval $[x_{i-1}, x_i]$. However, when we actually compute Riemann sums (say, as approximations to an actual integral), we often use regular partitions and corresponding left or right (or possibly midpoint) Riemann sums. Such systematic choices make the computation of the Riemann sums much easier. Theorem I.1.5 proves the legitimacy of using left or right Riemann sums to approximate the integral of a function that is known to be Riemann integrable.

Theorem I.1.5 Let a and b be real numbers with $a < b$. Let $f : [a, b] \to \mathbb{R}$ be a function. Suppose that either $(R_n(f))$ or $(L_n(f))$ is a convergent sequence. Then the other sequence converges as well, and the limits of two sequences are equal.

If f is Riemann integrable on $[a, b]$, then $(R_n(f))$ and $(L_n(f))$ both converge, and

$$\lim_{n \to \infty} R_n(f) = \lim_{n \to \infty} L_n = \int_a^b f$$

Proof: The proof is Problem 2 at the end of this excursion. □

I.2 Why the Generality?

Theorem I.1.5 suggests that the definition of the Riemann integral might be greatly simplified by defining $\int_a^b f$ as the limit of one or the other of the sequences $(R_n f)$ or $(L_n(f))$. So why not simplify the theory in this way? To explore this idea further, we start with a definition.

Definition I.2.1 Fix an interval $[a, b]$ where $a < b$. Let us say that f is *right integrable* if the sequence $(R_n(f))$ converges. In this case, we denote the *right integral of f* by

$$R_a^b(f) = \lim_{n \to \infty} R_n(f).$$

Exercise I.2.2 Prove that every function that is Riemann integrable is also right integrable and that, for any Riemann integrable function f,

$$\int_a^b f = R_a^b(f)$$

∎

The converse of this result is false. As you saw in Example 11.2.6, the characteristic function of the rational numbers is not Riemann integrable on $[0, 1]$. It *is*, however, right integrable on $[0, 1]$.

Theorem I.2.3 Let $f : \mathbb{R} \to \mathbb{R}$ be the characteristic function of the rational numbers. That is, f is given by

$$f(x) = \begin{cases} 0 & \text{if } x \text{ is irrational} \\ 1 & \text{if } x \text{ is rational} \end{cases}$$

Then f is right integrable on $[0, 1]$.

Proof: The proof is Problem 4a at the end of this section. □

We have shown that whenever the Riemann integral gives an answer, the right integral gives the same answer. Moreover, Theorem I.2.3 tells us that the right integral may give us an answer even if the Riemann integral does not.[1] Given that the right integral can calculate *more* integrals than the Riemann integral and given that they agree when both give an answer, one might think that the right integral is, in fact, superior to the Riemann integral. This is not the case, however. In fact, the right integral is badly behaved in some ways. Theorem I.2.4 illustrates this bad behavior.

---

1. It does *not*, of course, tell us that the right integral will always give us an answer. We still don't know whether the right integral will assign a value to every function.

Excursion I ■ Probing the Definition of the Riemann Integral

Theorem I.2.4 There exist a non-negative function $f : \mathbb{R} \to \mathbb{R}$ and real numbers $a < c < b$ such that
$$R_a^b(f) \neq R_a^c(f) + R_c^b(f).$$

Proof: The proof is Problem 4c at the end of this section. □

Exercise I.2.5 Explain why the function that is referred to in Theorem I.2.4 cannot possibly be Riemann integrable. ■

Exercise I.2.6 Write a short paragraph that summarizes why it is unwise to substitute the (simpler) definition of the right integral for the standard Riemann integral. Comment on why the idea of the "right integral" is still a useful one. ■

Problems I.2

1. Regular partitions. Let a and b be real numbers with $a < b$. Let P_n be the regular partition of $[a, b]$ with n subintervals.[2]

   (a) Prove that if a and b are both rational, then every point in the partition is rational.

   (b) Suppose that either a or b is irrational. Prove that there is at most one rational point in the partition P_n.

   (c) Give an example of a partition in which the endpoints are both irrational but there is one rational partition point in the interior of the interval.

2. Prove that if f is integrable, then the right and left Reimann sums for f will converge to $\int_a^b f$ (Theorem I.1.5).

3. Use Theorem I.1.5 to prove that the right Riemann integral gives the "correct" answer for the area of the triangle determined by the points $(-1, 1)$, $(3, 1)$, and $(0, 7)$. (Be sure to indicate how you know the function(s) in question are Riemann integrable.)

---

2. Facts about the arithmetic of rational and irrational numbers are laid out in Excursion B. (See page 296.) Those facts will be useful in this problem.

4. Right integrals of the characteristic function of the rational numbers.

   (a) Prove that the characteristic function of the rationals is right integrable on $[0, 1]$ (Theorem I.2.3).

   (b) If f is the characteristic function of the rational numbers, what can you say about the right integrability of f on an arbitrary interval $[a, b]$? (*Hint:* Separate the case in which both a and b are rational from the case when at least one of them is irrational.)

   (c) Prove Theorem I.2.4. (*Hint:* Let c be an irrational number in the interval $[0, 1]$.)

Excursion J

Power Series

> *Required Background*
> Chapters 9 and 12 and Excursion H.

Polynomials, functions of the form

$$P(x) = a_n(x-a)^n + a_{n-1}(x-a)^{n-1} + \cdots + a_1(x-a) + a_0$$

are well-behaved and well-understood functions. Thus, when we have a function that is hard to work with or whose behavior we do not fully understand, it is often desirable to look for a polynomial approximation. In Chapter 12, we talked about approximating a function by a sequence (or series) of functions that converged to it. Putting these two ideas together, we can think of looking for a sequence of polynomials that approximate a function increasingly well, in the sense that the sequence of polynomials converges to the function in question. Power series are sequences of polynomial functions with special properties not shared by most sequences of functions. They play an important role in the study of many common elementary functions.

In this excursion we define power series, examine the conditions under which they converge, and discuss some of the properties of the resulting limit functions. At the end, we will return to the question of approximating a function using power series.

J.1 Definitions and Convergence of Power Series

Definition J.1.1 Given a sequence (a_n) of real numbers and any real number a, the series of functions $\sum_{n=0}^{\infty} a_n(x-a)^n$ is called a **power series**.[1] The terms of the sequence (a_n) are called the **coefficients** of the power series. The series is said to be **centered** or **based** at a.

The convergence of power series is very regular and easy to understand.

Theorem J.1.2 Let (a_n) be a sequence of real numbers, and let a be any real number. Then exactly one of three things is true:

1. If $\limsup \sqrt[n]{|a_n|} = 0$, then $\sum_{n=0}^{\infty} a_n(x-a)^n$ converges absolutely for all $x \in \mathbb{R}$.

2. If $0 < \limsup \sqrt[n]{|a_n|} < \infty$, then $\sum_{n=0}^{\infty} a_n(x-a)^n$ converges absolutely for all x in the open interval $(a-R, a+R)$, where $R = 1/(\limsup \sqrt[n]{|a_n|})$. $\sum_{n=0}^{\infty} a_n(x-a)^n$ diverges if $|x-a| > R$.

3. If $\limsup \sqrt[n]{|a_n|} = \infty$, then $\sum_{n=0}^{\infty} a_n(x-a)^n$ converges absolutely to a_0 when $x = a$ and diverges otherwise.

Proof: The proof is Problem 1 at the end of this section. □

It is always true that $0 \leq \limsup \sqrt[n]{|a_n|} \leq \infty$, so Theorem J.1.2 can be restated more simply as follows.

Corollary J.1.3 Let (a_n) be a sequence of real numbers, and let a be any real number. Then exactly one of three things is true:

1. $\sum_{n=0}^{\infty} a_n(x-a)^n$ converges absolutely for all $x \in \mathbb{R}$.

2. There exists a positive real number R such that $\sum_{n=0}^{\infty} a_n(x-a)^n$ converges absolutely for all x in the open interval $(a-R, a+R)$ and diverges if $|x-a| > R$.

3. $\sum_{n=0}^{\infty} a_n(x-a)^n$ converges absolutely to a_0 when $x = a$ and diverges otherwise.

---

1. We want to interpret $\sum_{n=0}^{\infty} a_n(x-a)^n$ as the limit of the sequence of functions
$$a_0, a_0 + a_1(x-a), a_0 + a_1(x-a) + a_2(x-a)^2, \ldots$$
Thus, for ease of notation, we set $(x-a)^0 = 1$, even in the case when $x = a$, which gives (the ordinarily ill-defined) 0^0. We will hold to this convention whenever we discuss power series.

Exercise J.1.4 Let (a_n) be a sequence of real numbers. In light of Theorem J.1.2, classify each of the following situations as "possible" (meaning the theorem doesn't exclude the possibility) or "impossible" (meaning that the theorem shows it cannot happen). Briefly justify your answers.

1. $\sum_{n=0}^{\infty} a_n(x-3)^n$ converges absolutely at $x=3$ but diverges at $x=4$.
2. $\sum_{n=0}^{\infty} a_n(x-3)^n$ converges conditionally at $x=6$ but diverges at $x=-2$.
3. $\sum_{n=0}^{\infty} a_n(x-3)^n$ converges absolutely at $x=6$ and conditionally at $x=0$.
4. $\sum_{n=0}^{\infty} a_n(x-3)^n$ diverges at $x=5$ but converges conditionally at $x=1$.
5. $\sum_{n=0}^{\infty} a_n(x-3)^n$ converges conditionally at $x=5$ but converges absolutely at $x=-3$.

∎

Definition J.1.5 [Radius and Interval of Convergence] Let (a_n) be a sequence of real numbers, and let a be any real number.

- If $\sum_{n=0}^{\infty} a_n(x-a)^n$ converges for all $x \in \mathbb{R}$, we say that its radius of convergence is infinite.

- If $\sum_{n=0}^{\infty} a_n(x-a)^n$ converges only when $x=a$, we say that its radius of convergence is zero.

- Suppose $R > 0$. If $\sum_{n=0}^{\infty} a_n(x-a)^n$ converges on the interval $(a-R, a+R)$ and diverges if $|x-a| > R$, then we say that its radius of convergence is R.

The set of points at which a power series converges is called its **interval of convergence**.

For convenience in speaking, we will join the three cases given in Definition J.1.5 and allow R to take on all values from 0 to $+\infty$, inclusive. In other words, if we say that the radius of convergence of a power series is R, it is understood that R may be 0 or $+\infty$, unless we specify otherwise. Note that Theorem J.1.2 tells us that for a positive but finite radius of convergence R,

$$R = \frac{1}{\limsup \sqrt[n]{|a_n|}}.$$

Exercise J.1.6 Use either the root test or the ratio test to compute the radius of convergence of the following series. Then find the intervals of convergence. (The root and ratio tests will tell you about convergence only on the interior of the intervals. If the radius of convergence is finite but not zero, check separately to see whether the series converges absolutely, conditionally, or not at all at the endpoints.)

1. $\sum_{n=0}^{\infty} \frac{1}{n!}(x-8)^n$

2. $\sum_{n=0}^{\infty} \frac{3}{n^2}(x-5)^n$

3. $\sum_{n=0}^{\infty} \left(\frac{1}{3}\right)^n (x-2)^n$

4. $\sum_{n=0}^{\infty} \frac{1}{n}(x-5)^n$

∎

As you can see, power series whose radii of convergence are finite and non-zero exhibit a variety of behaviors at the endpoints of their intervals of convergence. In Problem 2 at the end of this section, you will be asked to determine which combinations are actually possible.

Theorem J.1.7 Let (a_n) be a sequence of real numbers, and let a be any real number. Suppose $\sum_{n=0}^{\infty} a_n(x-a)^n$ has a non-zero radius of convergence R. Then the power series converges uniformly on any closed, bounded subinterval of $(a-R, a+R)$.

Proof: Let I be a closed, bounded subinterval of $(a-R, a+R)$. We can find a positive real number $t < r$ such that $I \subseteq [a-t, a+t] \subseteq (a-R, a+R)$. Thus it is sufficient to prove that $\sum_{n=0}^{\infty} a_n(x-a)^n$ converges uniformly on closed, bounded subintervals that are centered at a.

Let t be any positive real number that is less than R. Let s be any real number in (t, R). Because $a + s$ lies inside $(a-R, a+R)$,

$$\sum_{n=0}^{\infty} a_n(a+s-a)^n = \sum_{n=0}^{\infty} a_n s^n$$

is a convergent series of real numbers. Thus we can find $M \in \mathbb{R}$ such that $|a_n|s^n < M$ for all $n \in \mathbb{N}$.

Let $x \in [a-t, a+t]$. Then

$$|a_n(x-a)^n| = |a_n||x-a|^n \leq |a_n|t^n$$
$$= |a_n|s^n\left(\frac{t}{s}\right)^n$$
$$< M\left(\frac{t}{s}\right)^n$$

Since $t/s < 1$,

$$\sum_{n=0}^{\infty} M\left(\frac{t}{s}\right)^n$$

is a convergent geometric series. The uniform convergence of the power series $\sum_{n=0}^{\infty} a_n(x-a)^n$ on the interval $[a-t, a+t]$ follows from the Weierstrass M-test (Theorem 12.3.4).

$\square$

Problems J.1

1. Prove the theorem that establishes the radius of convergence of a power series (Theorem J.1.2).

2. Suppose that $\sum_{n=0}^{\infty} a_n(x-a)^n$ has a radius of convergence R with $0 < R < \infty$. The purpose of this problem is to further explore the convergence of the series at the endpoints of its interval of convergence. Show that the only possibilities are as follows:

    - The series converges absolutely at both endpoints.
    - The series diverges at both endpoints.
    - The series diverges at one endpoint and converges conditionally at the other endpoint.
    - The series converges conditionally at both endpoints.

    This is related to the question posed in Problem 9 at the end of Section H.1.

3. Let (a_n) be a sequence of real numbers, and let a be any real number. Suppose that $\sum_{n=0}^{\infty} a_n(x-a)^n$ has a radius of convergence $R > 0$. If $f : (a - R, a + R) \to \mathbb{R}$ is given by

$$f(x) = \sum_{n=0}^{\infty} a_n(x-a)^n.$$

prove that f is continuous on $(a - R, a + R)$. (*Careful*: The convergence is not uniform on $(a - R, a + R)$!)

4. Let $\epsilon > 0$, and let (a_n) and (b_n) be sequences of real numbers. Suppose that

$$\sum_{n=0}^{\infty} a_n(x-s)^n \quad \text{and} \quad \sum_{n=0}^{\infty} b_n(x-s)^n$$

converge and are equal on $(s - \epsilon, s + \epsilon)$. Prove that $a_n = b_n$ for all $n \in \mathbb{N}$. What does this result tell you?

J.2 Integration and Differentiation of Power Series

Lemma J.2.1 Let (a_n) be a sequence of real numbers, and let a be any real number. Suppose that the power series $\sum_{n=0}^{\infty} a_n(x-a)^n$ has non-zero radius of convergence R. Then each of the power series

$$\sum_{n=1}^{\infty} na_n(x-a)^{n-1} \quad \text{and} \quad \sum_{n=0}^{\infty} \frac{a_n}{n+1}(x-a)^{n+1}$$

has a radius of convergence of at least R.

Proof: The proof is Problem 1 at the end of this section. □

Theorem J.2.2 Let (a_n) be a sequence of real numbers, and let a be any real number. Suppose that the power series $\sum_{n=0}^{\infty} a_n(x-a)^n$ has a radius of convergence $R > 0$. Then the series can be differentiated term by term on $(a - R, a + R)$. In other words, if

$$f(x) = \sum_{n=0}^{\infty} a_n(x-a)^n \text{ for all } x \in (a - R, a + R)$$

then f is differentiable on $(a - R, a + R)$ and

$$f'(x) = \sum_{n=1}^{\infty} na_n(x-a)^{n-1} \text{ for all } x \in (a - R, a + R).$$

Proof: The proof is Problem 2 at the end of this section. □

Exercise J.2.3 [Notational Aside] Notice that the power series for f in Theorem J.2.2 starts with the index 0, whereas the series that gives f' starts with the index 1. *Why?* What would happen if the "counting" started at 0? ■

Corollary J.2.4 Let (a_n) be a sequence of real numbers, and let a be any real number. Suppose that the power series $\sum_{n=0}^{\infty} a_n (x-a)^n$ has a radius of convergence $R > 0$. Then the function

$$f(x) = \sum_{n=0}^{\infty} a_n (x-a)^n$$

has derivatives of all orders on the interval $(a - R, a + R)$. That is, f', f'', f''',... all exist on $(a - R, a + R)$.

Furthermore, for each $n \in \mathbb{N}$,

$$a_n = \frac{f^{(n)}(a)}{n!}$$

Proof: The proof is Problem 3 at the end of this section. □

Corollary J.2.5 Let (a_n) be a sequence of real numbers, and let a be any real number. Suppose that the power series $\sum_{n=0}^{\infty} a_n (x-a)^n$ has a radius of convergence $R > 0$. Then

$$\sum_{n=0}^{\infty} \frac{a_n}{n+1} (x-a)^{n+1}$$

is an antiderivative for $\sum_{n=0}^{\infty} a_n (x-a)^n$ on $(a - R, a + R)$.

Proof: The proof is Problem 4 at the end of this section. □

Problems J.2

1. Prove Lemma J.2.1.

   You may make use of the following facts: $\lim_{n \to \infty} n^{\frac{1}{n}} = 1$ and $\lim_{n \to \infty} (n+1)^{-\frac{1}{n}} = 1$. These limits (with substantial hints) are established in other parts of the book.[2]

---

2. See Theorem D.4.5 on page 317, and Problem 11 on page 93.

2. Prove that a power series can be differentiated term by term on the interior of its interval of convergence (Theorem J.2.2). (*Hint*: You need to make judicious use of Theorem J.1.7, Lemma J.2.1, and Corollary 12.4.5.)

3. Prove that a function that is given by a power series has derivatives of all orders (Corollary J.2.4).

4. Prove that a power series can be antidifferentiated term by term (Corollary J.2.5).

5. Suppose that the radius of convergence of $\sum_{n=0}^{\infty} a_n(x-a)^n$ is R. Prove that the term-by-term derivative and the term-by-term antiderivative of this series

$$\sum_{n=1}^{\infty} n a_n (x-a)^{n-1}$$

$$\sum_{n=0}^{\infty} \frac{a_n}{n+1} (x-a)^{n+1}$$

also have radius of convergence equal to R.

J.3 Taylor Series

We began the excursion by talking about approximating an arbitrary function f by a sequence of polynomial functions that converges to f. It might seem that the rest of the excursion has digressed from this topic by speaking only about the convergence of power series and the properties of functions to which they converge. In fact, this discussion has not been a digression at all, because it has given us a great deal of information about functions for which such an approximation scheme is feasible. Before we elaborate on this observation, we introduce some useful language.

Definition J.3.1 Let $K \subseteq \mathbb{R}$, and let $f : K \to \mathbb{R}$ be a function. Fix $a \in K$. If there exist $\epsilon > 0$ and a sequence (a_n) of real numbers such that

$$f(x) = \sum_{n=0}^{\infty} a_n (x-a)^n$$

for all $x \in (a - \epsilon, a + \epsilon)$, we say that f is **given by the power series** $\sum_{n=0}^{\infty} a_n(x-a)^n$ **on** $(a - \epsilon, a + \epsilon)$ or, alternatively, that f is **analytic at** a.

Suppose that f is some function. In raising the question of approximating f around a in the context of power series, what we have really been asking is whether f is analytic at a. The theorems we have proved in this excursion shed some immediate light on this question. Corollary J.2.4 tells us that if f is given by a power series, it must have derivatives of all orders. Furthermore, there is only one power series that could possibly work:

$$f(a) + f'(a)(x-a) + \frac{f''(a)}{2!}(x-a)^2 + \frac{f^{(3)}(a)}{3!}(x-a)^3 \cdots$$

Note that this is the limiting series obtained from the sequence of Taylor polynomial approximations to the function f. That is, the polynomial approximation scheme we discussed in Chapter 9 is the *only possible choice*!

Definition J.3.2 Let f be a real-valued function of a real variable that is defined and has derivatives of all orders at $x = a$. Then the series

$$f(a) + f'(a)(x-a) + \frac{f''(a)}{2!}(x-a)^2 + \frac{f^{(3)}(a)}{3!}(x-a)^3 + \cdots$$

is called the **Taylor series for f centered (or based) at a**.

If f is defined and has derivatives of all orders at $x = 0$, we can define the **Maclaurin series for f**:

$$f(0) + f'(0)\,x + \frac{f''(0)}{2!}x^2 + \frac{f^{(3)}(0)}{3!}x^3 + \cdots$$

Note that f is analytic at a if and only if f has derivatives of all orders at a and it agrees with its Taylor series centered at a on some open interval centered at a.

Remark: Suppose we have a function f that has derivatives of all orders at some point a. Corollary J.2.4 tells us that if f is given by a power series on some interval centered at a, then that power series must be its Taylor series based at a. Just as important to note is what is *not* guaranteed by the corollary:

- It *does not* guarantee that the Taylor series for f will converge except at a.

- Even if the radius of convergence of the Taylor series is $R > 0$, there is no guarantee that $(a - R, a + R)$ is the entire domain of f.

- If the radius of convergence of the Taylor series is $R > 0$, there is no guarantee that it converges to f on $(a - R, a + R)$. Indeed, the sum of the series need not equal f except at $x = a$. (See Problem 3 at the end of this section.)

Corollary J.3.3 is a recasting of Corollary 9.7.4 that incorporates our current language.

Corollary J.3.3 Let a and b be real numbers with $a < b$, and let K be a positive real number. Let $f : [a,b] \to \mathbb{R}$ be a function with derivatives of all orders on $[a,b]$. Suppose that
$$|f^{(n+1)}(x)| \leq K \text{ for all } x \in [a,b].$$
Assume further that t and s are any two points in $[a,b]$. Then
$$\left| f(t) - \sum_{k=0}^{n} \frac{f^k(s)}{k!}(t-s)^k \right| \leq \frac{K}{(n+1)!} |t-s|^{n+1}.$$

In other words, if we approximate f by its nth Taylor polynomial approximation based at s, the error at a point $t \in [a,b]$ is no more than
$$\frac{K}{(n+1)!} |t-s|^{n+1}.$$
It follows that the error at any arbitrary point x in $[a,b]$ is no more than
$$\frac{K}{(n+1)!} |b-a|^{n+1}.$$

□

Problems J.3

1. Assuming standard facts about the algebra and calculus of sines and cosines, show that both the sine and the cosine are given by their respective Maclaurin series on all of $\mathbb{R}$. (*Hint:* Use Corollary J.3.3 and Theorem D.4.3.)

2. Assuming standard facts about the algebra and calculus of $f(x) = e^x$, show that the exponential function is given by its Maclaurin series on all of $\mathbb{R}$. (*Hint:* Use Corollary J.3.3 and Theorem D.4.3.)

> Problem 1 says something really amazing. The value of the sine function at 0 and the values of all its successive derivatives at zero determine *completely* the values of the sine everywhere on the real line. A similar fact holds for the cosine function and for the exponential function. If a function f is analytic at a point a, then the value of f at a, along with all the derivatives of f at a, determine the values of f on an interval around a, sometimes a very large interval around a!

3. Consider the function

$$f(x) = \begin{cases} \exp\left(-\frac{1}{x^2}\right) & \text{if } x \neq 0 \\ 0 & \text{if } x = 0 \end{cases}.$$

In this problem you will show that f has derivatives of all orders and that $f^n(0) = 0$ for all $n \in \mathbb{N}$. The differentiation rules tell us that f has derivatives of all orders for values of $x \neq 0$. Thus the key is to show that f has derivatives of all orders at $x = 0$.

(a) For non-zero x values, prove that for every $n \in \mathbb{N}$, there exists a function g_n for which $f^{(n)}(x) = g_n(x) f(x)$, and which is the sum of terms of the form k/x^m with k a real constant and m an integer greater than or equal to 3.[3]

(b) Prove that for all $n \in \mathbb{N} \cup \{0\}$,

$$\lim_{x \to 0} \frac{\exp\left(-\frac{1}{x^2}\right)}{x^n} = 0.$$

(c) Use parts (a) and (b) to prove that $f^n(0) = 0$ for all $n \in \mathbb{N}$.

(d) Explain the significance of this result. In particular, what does it tell you about the Taylor series expansion of f based at $x = 0$?

---

3. For instance, a quick computation shows that

$$f^{(3)}(x) = \left(\frac{24}{x^5} - \frac{36}{x^7} + \frac{8}{x^9}\right) f(x)$$

so

$$g_3(x) = \left(\frac{24}{x^5} - \frac{36}{x^7} + \frac{8}{x^9}\right)$$

Excursion K

Everywhere Continuous, Nowhere Differentiable

> *Required Background*
> Chapters 9 and 12.

K.1 Introduction

You already know that differentiable functions are continuous (Theorem 9.2.5). You also know that the converse is not true. Functions can be continuous without being differentiable. A continuous function can fail to have a derivative at a point either because its graph has a "corner" at the point or because the tangent to the graph of the function is vertical at the point.

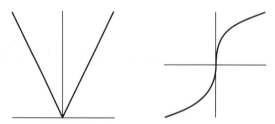

It is easy to visualize continuous functions with infinitely many points of non-differentiability—for instance, a "zigzag" function that has a corner at each integer.

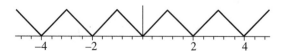

Excursion K ■ Everywhere Continuous, Nowhere Differentiable

In 1861, Karl Weierstrass made a discovery that stunned the mathematical community: He proved the existence of functions that are continuous at every point of the real line and differentiable nowhere![1] This excursion leads you through the construction of such a function and helps you show that it has the desired properties.

We need a continuous function that has either a corner or an infinite slope at every point of the real line. We achieve this goal by successive approximation using a series of functions. Our first term will be the zigzag curve. The general scheme is to take the zigzag and make each of its straight pieces into a zigzag. To each of those smaller zigs, we will add tiny zags of their own, and so on. Each successive term will add more corners to the partial sum and make the remaining parts steeper.

Each of the terms of the series will be continuous. To guarantee that the limit function be continuous, we will need to be sure our series converges uniformly. Uniform convergence will follow from the Weierstrass M-test (Theorem 12.3.4), if we make sure that the terms of the series hug the x-axis more and more closely.

K.2 Constructing the Function

We can take any generic curve like the parabola $f(x) = x^2$

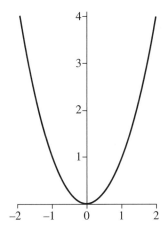

---

1. Such curves don't seem quite as pathological to us these days. Fractal curves, which play a central role in the mathematical theory of chaos, have this property.

and make it wiggle up and down by adding on a periodic function with a large frequency and a small amplitude. Here is $f(x) = x^2 + \sin(10\pi x)/5$:

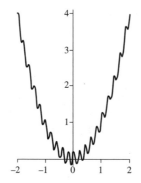

We are going to use this idea over and over again in the construction of our everywhere continuous, nowhere differentiable function. We start with the zigzag function that we have already seen, and we construct zigzag functions that move up and down more frequently but don't go so high. Then we add them together. (The first four terms of our series are shown in Figure K.1, and the first four partial sums of the series are shown in Figure K.2.)

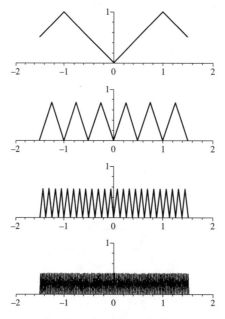

Figure K.1 These are the first four terms of a series that converges to an everywhere continuous, nowhere differentiable function.

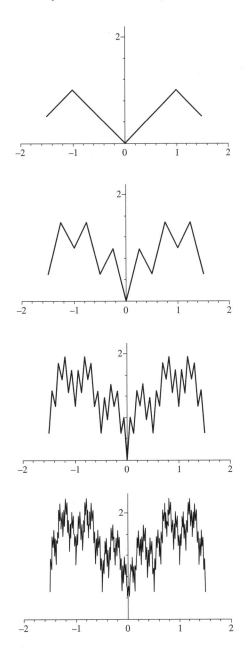

Figure K.2 These are the first four partial sums of a series that converges to an everywhere continuous, nowhere differentiable function.

Excursion K ■ Everywhere Continuous, Nowhere Differentiable

Theorem K.2.1 Let $f_0 : \mathbb{R} \to \mathbb{R}$ be the function obtained by taking the absolute value function on $[-1, 1]$ and extending it periodically to all of $\mathbb{R}$. Then the series

$$\sum_{n=0}^{\infty} \left(\frac{3}{4}\right)^n f_0(4^n x)$$

converges uniformly on all of $\mathbb{R}$ and is, therefore, continuous. In addition, the limit of the series fails to be differentiable at every point.

Your task is to prove this theorem using the outline and hints that follow.

1. Use the Weierstrass M-test (Theorem 12.3.4) to prove that the series converges uniformly on $\mathbb{R}$. Call the limit function f and conclude that it is continuous everywhere.

   Showing that f fails to be differentiable everywhere is more complicated because it requires that you deal explicitly with the shape of the functions in the series. To begin, fix $x \in \mathbb{R}$. The remaining exercises outline a procedure for showing that f is not differentiable at x.

2. Set up the difference quotient

   $$\frac{f(x + \Delta x) - f(x)}{\Delta x}.$$

   for f at x.

   Explain why it is sufficient to show that there is sequence (Δx_m) converging to 0, such that

   $$\left| \sum_{n=0}^{\infty} \left(\frac{3}{4}\right)^n \frac{f_0(4^n(x + \Delta x_m)) - f_0(4^n x)}{\Delta x_m} \right| \to \infty \text{ as } m \to \infty.$$

3. Fix $n \in \mathbb{N}$. (*Hint*: Drawing detailed graphs is absolutely essential in this problem.)

   (a) Show that for each $m \in \mathbb{N}$ with $m < n$, if $\Delta x = \pm \frac{1}{2} 4^{-m}$, then

   $$f_0(4^n(x + \Delta x)) - f_0(4^n x) = 0$$

   (b) Show that one of the following identities holds:

   $$\left| \frac{f_0(4^n(x + \frac{1}{2} 4^{-n})) - f_0(4^n x)}{\frac{1}{2} 4^{-n}} \right| = 4^n.$$

   or

   $$\left| \frac{f_0(4^n(x - \frac{1}{2} 4^{-n})) - f_0(4^n x)}{-\frac{1}{2} 4^{-n}} \right| = 4^n.$$

Set $t = +1$ if the first holds, and $t = -1$ if it doesn't.

(c) Show that for each $m \in \mathbb{N}$ with $m > n$, if $\Delta x = t\frac{1}{2}4^{-m}$, then
$$\left| \frac{f_0(4^n(x + \Delta x)) - f_0(4^n x)}{\Delta x} \right| = 4^n.$$

4. Find a sequence (Δx_m) that goes to zero so that the following condition is satisfied for all $m \in \mathbb{N}$:
$$\left| \frac{f_0(4^n(x - \Delta x_m)) - f_0(4^n x)}{\Delta x_m} \right| = \begin{cases} 0 & \text{if } n > m \\ 4^n & \text{if } n \leq m \end{cases}.$$

5. Show that for each $m \in \mathbb{N}$,
$$\left| \sum_{n=0}^{\infty} \left(\frac{3}{4}\right)^n \frac{f_0(4^n(x + \Delta x_m)) - f_0(4^n x)}{\Delta x_m} \right| \geq \frac{1}{2}(3^m - 1).$$

Now that you have worked out all the details, write up a "clean" proof of Theorem K.2.1. The ideas in the proof are, of course, in the work that you have already done. Your final job is to work on connecting the ideas with clear transitions.

Excursion L

Newton's Method

> Required Background
> Chapters 9 and 10.

A synecdoche is a rhetorical device in which a part of something is used to represent the whole. In some ways, that is what this excursion tries to do. Indeed, the purpose of the excursion is twofold. Newton's method is a familiar numerical technique for approximating solutions to real equations. Under reasonable conditions it converges very fast, and its convergence behavior is well understood. One purpose of the excursion is to prove some useful theorems about Newton's method. But it also has a more general objective.

Real analysis is the key to understanding numerical approximation techniques. I hope to give you a sense for the kinds of questions that arise when we solve problems numerically and for the way in which real analysis can be used to answer those questions. Newton's method is my synecdoche, a context in which to tell you a bit about the role of real analysis in numerical methods.

L.1   Setting the Stage

Many times, in the course of solving a mathematical problem, we face the need to solve an algebraic equation that cannot be solved using analytical methods. For instance,

$$x^2 e^{-x^2} = 1/10$$

has four solutions, as a graph will easily tell you (see Figure L.1). Unfortunately, analytical methods are useless for actually finding them.

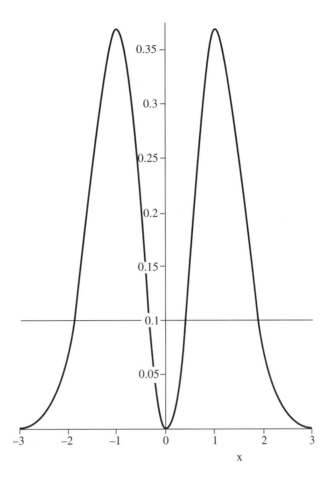

Figure L.1 $x^2 e^{-x^2} = 1/10$ has exactly four solutions.

Our only option is to obtain approximate solutions using numerical methods. One of the best known is the iterative method called the Newton–Raphson method or simply Newton's method.

First, Newton's method takes advantage of the fact that solving equations is the same thing as finding roots of functions. Finding the values where $x^2 e^{-x^2} = 1/10$ is the same as finding the places where the graph of $f(x) = x^2 e^{-x^2} - 1/10$ crosses the x-axis. (See Figure L.2.)

Newton's method finds roots of functions using a very simple idea. Suppose we are looking for r, a root of the differentiable function f. Suppose further that we have an estimate p for r. How can we refine our estimate to get a better one?

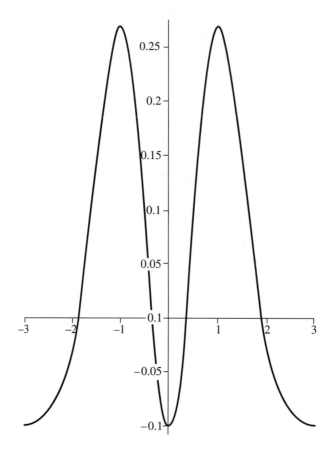

Figure L.2 $x^2 e^{-x^2} = 1/10$ where this function crosses the x-axis.

Newton noted that if we replace the graph of f by its tangent line at the point $(p, f(p))$ and follow the tangent line to *its* root at q, q will frequently be closer to r than p was. (See Figure L.3.) If we want an even better approximation, we can repeat the procedure to improve the estimate q.

Exercise L.1.1 Show that

$$q = p - \frac{f(p)}{f'(p)}.$$

■

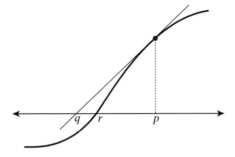

Figure L.3 Newton's method improves the estimate p of r by finding q.

Regarding Numerical Methods

Any iterative numerical method raises three very important questions: Can we be sure it converges? If so, does it converge to the right thing? And if the answer to both of these questions is "yes," how good is our approximation after n iterations?

The importance of the first two questions is obvious. We are trying to accomplish some task, such as solving a differential equation or finding the root of a function. If we adopt a strategy for solving the problem, we want to be sure that our approach will yield a result and that it answers the question we wanted to ask. The third question is also crucial in view of the fact that numerical methods are implemented on computers. We need to be able to tell the computer when to stop, so we need to know how many iterations will ensure that the error in our approximation is acceptably small.

In the case of Newton's method, we can answer all of these questions satisfactorily, requiring only mild restrictions on the function f.

L.2 Iterating the Newton Function

Let $f : \mathbb{R} \to \mathbb{R}$ be a differentiable function.[1] Newton's method consists of making an initial guess x_0 for a root r of the function f and iterating the **Newton function**

$$N(x) = x - \frac{f(x)}{f'(x)}$$

---

[1]. It is possible to discuss Newton's method for functions that are not defined or not differentiable on all of $\mathbb{R}$, but this introduces technical difficulties that I choose to sidestep in this excursion.

beginning at x_0. Assuming we never reach a point where $f'(x_n) = 0$, the iteration generates a sequence of points

$$x_0, x_1, x_2, x_3, \ldots$$

which we hope will converge to r. This sequence is called the **Newton sequence based at x_0**.

> Note that if $f'(x_n) = 0$ for some $n \in \mathbb{N}$, the iteration fails because the Newton function is undefined at x_n. (Explain graphically why the procedure fails if $f'(x_n) = 0$.)

In the case of real numbers x_0 for which $f'(x_n) \neq 0$ for $n \in \mathbb{N}$, we will say that the Newton iteration is "well defined" at x_0.

Given that we are iterating a real function, we can use the results of the chapter on iteration (Chapter 10) to get a great deal of (almost) free information about Newton's method.

Exercise L.2.1 Let $f : \mathbb{R} \to \mathbb{R}$ be a differentiable function at $t \in \mathbb{R}$ with $f'(t) \neq 0$. Prove that t is a fixed point of the Newton function for f if and only if t is a root of f. ■

Exercise L.2.2 Suppose that $f : \mathbb{R} \to \mathbb{R}$ has a continuous derivative. Prove that if the Newton sequence based at x_0 is well defined and convergent, it must converge to a root of the function f. ■

Theorem L.2.3 Suppose that $f : \mathbb{R} \to \mathbb{R}$ has a continuous second derivative. Let r be a root of f such that $f'(r) \neq 0$. Then there exists an open interval I centered at r such that if $x_0 \in I$, then the Newton sequence based at x_0 is well defined and will converge to r.

Proof: The proof is Problem 2 at the end of this section. □

Problems L.2

1. Suppose that $f : \mathbb{R} \to \mathbb{R}$ is twice differentiable, and let r be a root of f with $f'(r) \neq 0$. Prove that r is a super-attracting fixed point for the Newton function N. (See Problem 11 at the end of Section 10.1 for a definition of a super-attracting fixed point.)

2. Prove that if f has a continuous second derivative and a root r such that $f'(r) \neq 0$, then choosing an initial guess sufficiently close to r guarantees that Newton's method will converge to r (Theorem L.2.3).

L.3 Experimenting with Newton's Method

We are lucky to live in an age when computers are cheap, powerful, and easily available. If we want to understand a numerical procedure, the computer makes it easy to conduct experiments that allow us to test the strengths and weaknesses of the method. The insight that we gain is invaluable when we look for theorems that give us conclusive answers about the behavior of the numerical approximation method. For this section, you will need access to a computer program that implements Newton's method. Some of the experiments will be considerably clearer if you also have a computer program that implements a graphical display of the Newton iteration—that is, a program that will produce a picture similar to that in Figure L.4. (It would be even better if you could manage a program that *animates* the iteration procedure.)

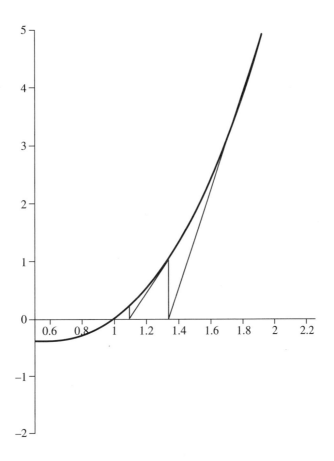

Figure L.4 Several iterations of Newton's method.

1. Just to get used to the Newton program, use it to find all solutions to the equation $x^2 e^{-x^2} = 1/10$. Try to get 10 decimal places correct to your satisfaction. (What method could you use to be reasonably sure you have 10 decimal places?)

We know that Newton's method will fail if the derivative of any iterate is zero. The next exercises show other cases in which difficulties arise. For each situation you are asked not only to describe what happens, but to give your best guess about *why* it happens. The idea is to use a combination of numerical results and geometric intuition to explain why the Newton scheme fails to work.

2. Try applying Newton's method to the function $f(x) = \dfrac{x}{1+x^2}$ with $x_0 = 6$. What happens? Why does it happen?

3. Try applying Newton's method to the function $f(x) = x^3 - x$ with an initial guess of $\dfrac{1}{\sqrt{5}}$. (For this problem only, do the calculations analytically, using "exact" methods—don't use decimals or approximate answers. Feel free to use a computer algebra system to do the algebra, if you wish.) What happens? Why does it happen?

4. Consider again the function $f(x) = x^3 - x$. This time try an initial guess of $x_0 = 0.55$. What happens? Why does it happen?

5. Try applying Newton's method to the function $f(x) = -x^3 + 8x^2 - 80$ with $x_0 = 4.7$. What happens? Why does it happen?

L.4 On Choosing x_0

Theorem L.2.3 tells us that if $f : \mathbb{R} \to \mathbb{R}$ is reasonably well behaved and our initial guess x_0 is sufficiently close to a root r of f, then the Newton sequence based at x_0 will converge to r. Indeed, some experimentation tells you that when the initial estimate is chosen well, Newton's method converges *very* fast. When it works, it works beautifully! Experimentation also tells you that not just any choice of x_0 will do. The question of how to choose x_0 is, therefore, a crucial one.

How close to r does x_0 have to be to ensure that Newton's method will work? Obviously, we cannot have a theorem that reads anything like this: "If $|r - x_0| < 0.72$, then the Newton sequence based at x_0 will converge to r." The behavior of the Newton function depends on the function f and the root r. Any theorem that gives us precise information about how to make an initial guess must deal explicitly with the geometry of the function f.

We know that Newton's method fails if the derivative of f at any iterate is zero. The results may not be as catastrophic, but may nevertheless be unsatisfactory, if the derivative of f at a guess is close to zero. For example, the function $f(x) = x^3 - x$ has zeros at $x = 0$, $x = -1$, and $x = 1$. We have seen that an initial guess of $x = 0.55$, whose closest root is $x = 1$, nevertheless generates a Newton sequence that converges to $x = -1$. This is because $f'(0.55)$ is very close to zero; the tangent to the graph of f at $(0.55, f(0.55))$ intersects the x-axis a long way from $x = 0.55$. In general, if $f'(x_0) \approx 0$, x_0 and $N(x_0)$ will be very far apart. Even if our first guess was fairly close to a root, subsequent guesses may not be. To ensure reasonable behavior for Newton's method, we need some control over the size of the derivative. In particular, we need to have the derivative bounded away from zero.

When we replace the graph of f by its tangent line, we make the tacit assumption that f is close to a straight line. When a function curves sharply, Newton's method can be "fooled" into thinking there is a root where one does not actually exist.

You observed this phenomenon when you looked at $f(x) = -x^3 + 8x^2 - 80$ beginning at $x_0 = 4.7$. Newton's method began by hopping back and forth around a couple of "virtual roots" that it imagined should occur near the maximum at $x = 5\frac{1}{3}$. To obtain greater control over the convergence of our numerical procedure, we will need a bound on the size of $|f''|$.

We are looking for a theorem that takes into account both the size of $|f'|$ and the size of $|f''|$.

L.5 Convergence Rate

Suppose $f : \mathbb{R} \to \mathbb{R}$ is differentiable at a. For real numbers close to a, the values given by the tangent line to f at $(a, f(a))$ approximate the values of the function f itself. That is, if $b \in \mathbb{R}$ is close to a,

$$f(b) \approx f'(a)(b-a) + f(a).$$

How good this approximation is depends on how much the graph of f curves between a and b. That is, the error depends on the values of f'' between a and b.

This error is made precise by using Taylor's remainder theorem for linear approximations (Theorem 9.7.1).

It says that when we have a twice-differentiable function $f : \mathbb{R} \to \mathbb{R}$ and numbers $a < b$, there exists a real number c lying between a and b such that

$$f(b) = f(a) + f'(a)(b-a) + \frac{f''(c)}{2!}(b-a)^2.$$

Taylor's theorem lets us see precisely how the values of $|f'|$ and $|f''|$ affect the error in the Newton sequence. Suppose we have an estimate x for the root r. The basic question is this: How does the distance from $N(x)$ to r compare to the distance from x to r? (We hope it will be smaller; preferably a *lot* smaller!)

Lemma L.5.1 Suppose that $f : \mathbb{R} \to \mathbb{R}$ is a twice-differentiable function. Let r be a root of the function f, and let I be an open interval containing r. Let M and R be fixed positive real numbers.

Suppose that f satisfies the following properties:

- $|f'(t)| \geq R$ for all $t \in I$ and
- $|f''(t)| \leq M$ for all $t \in I$.

Then, if $x \in I$,
$$|r - N(x)| \leq \frac{M}{2R}|r - x|^2.$$

Proof: The proof is Problem 1 at the end of this section. □

Lemma L.5.1 allows us to derive an explicit error estimate for Newton's method. That is, we have a theorem that tells us how to pick x_0 so as to guarantee convergence and that gives us an upper bound for the error when x_n approximates r.

Theorem L.5.2 Suppose that $f : \mathbb{R} \to \mathbb{R}$ is a twice-differentiable function. Let r be a root of the function f, and let P, M, and R be fixed positive real numbers.

Suppose that f satisfies the following properties:

- $|f'(x)| \geq R$ for all $x \in (r - P, r + P)$ and
- $|f''(x)| \leq M$ for all $x \in (r - P, r + P)$.

Fix
$$\delta < \min\left(P, \frac{2R}{M}\right).$$

Then, if $x_0 \in (r - \delta, r + \delta)$, the Newton sequence based at x_0 will converge to r. In fact, if
$$K = \frac{M}{2R}|r - x_0|$$
and x_n is the nth term in the Newton sequence based at x_0, then
$$|r - x_n| \leq \frac{2R}{M} K^{2^n}.$$

Proof: The proof is Problem 2 at the end of this section. □

The upshot of this error estimate is that when we start with a "reasonable" first guess, Newton's method converges *very* fast.

Problems L.5

1. Prove Lemma L.5.1. (*Hint*: Taylor's theorem tells you that there exists c between x and r such that
$$0 = f(r) = f(x) + f'(x)(r-x) + f''(c)\frac{(r-x)^2}{2}.$$
Use this information to find an upper estimate for $|r - N(x)|$.)

2. Prove Theorem L.5.2.

Excursion M
The Implicit Function Theorem

> *Required Background*
> Chapters 12 and 13. (A general knowledge of Newton's method for finding roots, such as might be gotten from any calculus text, is also needed to understand the discussion relating quasi-Newton's methods to the proof of the implicit function theorem.)

M.1 Solving Systems of Equations

Suppose we want to solve a system of several equations in several unknowns. The rule of thumb is that if we have more equations than unknowns, the system is unlikely to have a solution. If we have the same number of equations as unknowns, the system is likely to have a unique solution. And if we have fewer equations than unknowns, the system is likely to have infinitely many solutions.[1] In the last case, we generally expect to have as many "free variables" as the difference between the number of unknowns and the number of equations.

Let us begin by considering several examples.

---

[1] We use the word "likely" because these outcomes may or may not hold for a particular system, but the rule of thumb is a useful one.

Example M.1.1 [A System of Linear Equations] Consider the system of equations given by

$$\begin{aligned} 2y_1 + y_2 + 4x_1 + 7x_2 &= 0 \\ y_1 + y_2 + x_1 + 5x_2 &= 0 \\ -y_1 + y_3 - 3x_1 &= 0 \end{aligned}.$$

We have three equations in five unknowns, so we expect two free variables. Because these equations are linear, linear algebra tells us exactly what to do. We can use standard matrix techniques to solve for three of the unknowns, each in terms of the other two. This process yields

$$\begin{aligned} y_1 &= -3x_1 - 2x_2 \\ y_2 &= 2x_1 - 3x_2 \\ y_3 &= -2x_2 \end{aligned}.$$

Note that our solution induces a function $\mathbf{f} : \mathbb{R}^2 \to \mathbb{R}^3$:

$$\mathbf{f}(x_1, x_2) = (-3x_1 - 2x_2,\ 2x_1 - 3x_2,\ 2x_2).$$

Example M.1.2 Consider the following system of equations:

$$\begin{aligned} y_1 - 3y_2 + \sin(x) &= 1 \\ 4y_1 + \cos(x) + 3x &= 4 \end{aligned}.$$

As you can see for yourself, it is easy to solve for y_1 and y_2 in terms of x. Once again, we get a function. The function $\mathbf{g} = (g_1, g_2) : \mathbb{R} \to \mathbb{R}^2$ is given by

$$\begin{aligned} y_1 &= g_1(x) = -\tfrac{3}{4}x - \tfrac{1}{4}\cos(x) + 1 \\ y_2 &= g_2(x) = -\tfrac{1}{4}x + \tfrac{1}{3}\sin(x) - \tfrac{1}{12}\cos(x) \end{aligned}.$$

Examples M.1.1 and M.1.2 suggest that an algebraic solution to a "typical" system of n equations in $n + m$ unknowns induces a function that takes values in $\mathbb{R}^m$ and gives values in $\mathbb{R}^n$. In this case, we say that the function is given **implicitly** by the system of equations. In Example M.1.2, the system of equations gives the function $\mathbf{g}$ *im*plicitly, whereas the formula

$$\mathbf{g}(x) = \left(-\tfrac{3}{4}x - \tfrac{1}{4}\cos(x) + 1,\ -\tfrac{1}{4}x + \tfrac{1}{3}\sin(x) - \tfrac{1}{12}\cos(x) \right)$$

gives it *explicitly*.

Thus using algebra to solve a system with fewer equations than unknowns amounts to finding an explicit formula for an implicitly defined function. Unfortunately, not all such systems are solvable using algebraic methods; as you know, even a single equation in two unknowns can be tricky or impossible to solve.

Example M.1.3 Consider the following equation:

$$\cos(x - y) - \sin(xy) = 0.$$

Suppose we define $h : \mathbb{R}^2 \to \mathbb{R}$ by $f(x,y) = \cos(x-y) - \sin(xy)$. The solutions to our equation are the points at which $h(x, y) = 0$; that is, the solutions are the points where the graph of h intersects the xy plane. Figure M.1 clearly shows that this set of points is not the graph of a function. However, if we restrict our attention to a carefully chosen portion of the solution set, we do get the graph of a function.

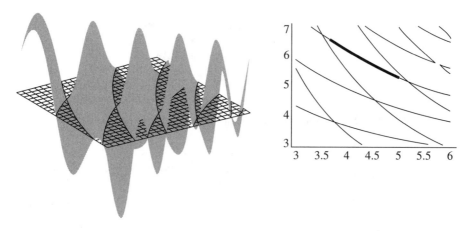

Figure M.1 These figures show the intersection of the xy plane with the surface $h(x, y) = \cos(x - y) - \sin(xy)$. The left figure shows the surface and the plane in $\mathbb{R}^3$. The right figure shows the intersection of the two projected into $\mathbb{R}^2$. Although the solution set is not the graph of a function, "pieces" of it can be interpreted as such. One possibility is highlighted on the right.

Example M.1.3 suggests a new perspective. With any system that has n equations in $n + m$ unknowns, we can associate a function $\mathbf{F}$ that takes values in $\mathbb{R}^{n+m}$ and returns values in $\mathbb{R}^n$. We choose $\mathbf{F}$ so that the solution set for the system of equations coincides with the set of all vectors $(y_1, \ldots, y_n, x_1, \ldots, x_m)$ for which

$$\mathbf{F}(y_1, \ldots, y_n, x_1, \ldots, x_m) = \mathbf{0}.$$

We will call this set the **0**-level set[2] of **F**. Our rule of thumb says that we ought to be able to solve for the "y-variables" in terms of the "x-variables" and get a function $\mathbf{f} = (f_1, f_2, \ldots, f_n)$ given by

$$\begin{aligned} y_1 &= f_1(x_1, \ldots, x_m) \\ y_2 &= f_2(x_1, \ldots, x_m) \\ &\vdots \\ y_n &= f_n(x_1, \ldots, x_m) \end{aligned}.$$

Example M.1.3 tells us that we can't necessarily find this solution on the whole of $\mathbb{R}^m$, but that we can hope to do it in some neighborhood of a known solution.

In the concrete setting of Example M.1.2, a vector (y_1, y_2, x) solves the system if and only if $\mathbf{G}(y_1, y_2, x) = \mathbf{0}$ where $\mathbf{G} : \mathbb{R}^3 \to \mathbb{R}^2$ is given by

$$\mathbf{G}(y_1, y_2, x) = (\ y_1 - 3y_2 + \sin(x) - 1,\ 4y_1 + \cos(x) + 3x - 4\ ).$$

That is, (y_1, y_2, x) solves the system if and only if it lies on the **0**-level set of **G**. We found a function **g** from $\mathbb{R}$ to $\mathbb{R}^2$ that gave (y_1, y_2) in terms of x so that $\mathbf{G}(\mathbf{g}(x), x) = \mathbf{0}$.

In the general case (unlike in Example M.1.2), it will not always be possible to find an explicit algebraic form for the function **f** describing the **0**-level set of **F**. Furthermore, we may not be able to give a single **f** that describes the set everywhere. Nevertheless, we may hope to show the existence of a function describing the **0**-level set in a small neighborhood of a particular solution. This is the task addressed by the implicit function theorem.

M.2 The Implicit Function Theorem

Before we go any further, we need to explore a useful bit of notation. If we have a vector **y** with n components and a vector **x** with m components, we can form a new vector with $n+m$ components by concatenating **y** and **x**. We will denote this new vector by $(\mathbf{y}, \mathbf{x})$. For instance, if $\mathbf{y} = (-1, 0)$ and $\mathbf{x} = (-3, 1, -2)$, the concatenated vector $(\mathbf{y}, \mathbf{x})$ is $(-1, 0, -3, 1, -2)$. The vector field associated with a system of n equations in $n + m$ unknowns is associated with a vector field in $n + m$ variables: $(y_1, \ldots, y_n, x_1, \ldots, x_m)$. Our convention will be to write this vector in the more compact form $(\mathbf{y}, \mathbf{x})$, where $\mathbf{y} = (y_1, \ldots, y_n)$ and $\mathbf{x} = (x_1, \ldots, x_m)$.[3]

In this vein, suppose we have a vector field $\mathbf{F} : \mathbb{R}^{n+m} \to \mathbb{R}^m$. Suppose also that we have a function **f** from some subset of $\mathbb{R}^n$ to $\mathbb{R}^m$ that solves the

---

2. The name is motivated by functions from $\mathbb{R}^2$ to $\mathbb{R}$. The word "level" is used to indicate that all the function values are at the same "height."

3. Not incidentally, the symbols **y** and **x** are chosen to remind our well-acculturated brains that we ultimately want a function with the first n variables as dependent variables and the last m variables as independent variables.

associated system of equations. If $\mathbf{x}$ is any element of the domain of $\mathbf{f}$, we know that the concatenated vector $(\mathbf{f}(\mathbf{x}), \mathbf{x})$ lies on the $\mathbf{0}$-level set of $\mathbf{F}$. That is, $\mathbf{F}(\mathbf{f}(\mathbf{x}), \mathbf{x}) = \mathbf{0}$.

The implicit function theorem asserts that if $\mathbf{F} : \mathbb{R}^{m+n} \to \mathbb{R}^m$ is a vector field and $(\mathbf{b}, \mathbf{a}) \in \mathbb{R}^{n+m}$ is such that $\mathbf{F}(\mathbf{b}, \mathbf{a}) = \mathbf{0}$, some fairly mild continuity and differentiability conditions on $\mathbf{F}$ will guarantee that the equation $\mathbf{F}(\mathbf{y}, \mathbf{x}) = \mathbf{0}$ (which is really a system of equations!) will give rise to a unique solution function $\mathbf{y} = \mathbf{f}(\mathbf{x})$. We can choose the domain of $\mathbf{f}$ to be an open ball about the point $\mathbf{a}$ in $\mathbb{R}^m$. In addition, we can expect a "well-behaved" solution function in the sense that $\mathbf{f}$ will share many of the nice properties of the vector field $\mathbf{F}$.

Theorem M.2.1 [Implicit Function Theorem] Let $m, n \in \mathbb{N}$, $\mathbf{a} \in \mathbb{R}^m$, and $\mathbf{b} \in \mathbb{R}^n$. Suppose that W is an open subset of $\mathbb{R}^{n+m}$ containing the point $(\mathbf{b}, \mathbf{a})$, and that
$$\mathbf{F} = (f_1, f_2, f_3, \ldots, f_n) : \mathbb{R}^{n+m} \to \mathbb{R}^n$$
is a vector field. Further, assume that $\mathbf{F}$ satisfies the following properties:

1. $\mathbf{F}(\mathbf{b}, \mathbf{a}) = \mathbf{0}$.

2. $\mathbf{F}$ is continuous on W.

3. The partial derivatives of $\mathbf{F}$ with respect to the first n variables
$$\frac{\partial f_1}{\partial y_1} \quad \frac{\partial f_1}{\partial y_2} \quad \cdots \quad \frac{\partial f_1}{\partial y_n}$$
$$\frac{\partial f_2}{\partial y_1} \quad \frac{\partial f_2}{\partial y_2} \quad \cdots \quad \frac{\partial f_2}{\partial y_n}$$
$$\vdots \quad \vdots \quad \cdots \quad \vdots$$
$$\frac{\partial f_n}{\partial y_1} \quad \frac{\partial f_n}{\partial y_2} \quad \cdots \quad \frac{\partial f_n}{\partial y_n}$$
exist and are continuous in W.

4. The matrix
$$D = \begin{bmatrix} \frac{\partial f_1}{\partial y_1}(\mathbf{b}, \mathbf{a}) & \frac{\partial f_1}{\partial y_2}(\mathbf{b}, \mathbf{a}) & \cdots & \frac{\partial f_1}{\partial y_n}(\mathbf{b}, \mathbf{a}) \\ \frac{\partial f_2}{\partial y_1}(\mathbf{b}, \mathbf{a}) & \frac{\partial f_2}{\partial y_2}(\mathbf{b}, \mathbf{a}) & \cdots & \frac{\partial f_2}{\partial y_n}(\mathbf{b}, \mathbf{a}) \\ \vdots & \vdots & \cdots & \vdots \\ \frac{\partial f_n}{\partial y_1}(\mathbf{b}, \mathbf{a}) & \frac{\partial f_n}{\partial y_2}(\mathbf{b}, \mathbf{a}) & \cdots & \frac{\partial f_n}{\partial y_n}(\mathbf{b}, \mathbf{a}) \end{bmatrix}$$
and (therefore) the linear transformation it represents are invertible.

Then there exist $t > 0$, $r > 0$, and a unique function $\mathbf{f} : B_t(\mathbf{a}) \to B_r(\mathbf{b})$ such that $\mathbf{F}(\mathbf{f}(\mathbf{x}), \mathbf{x}) = \mathbf{0}$ for all $x \in B_t(\mathbf{a})$.

Proof: There are several different proofs of the implicit function theorem. But this is a very deep result, so all of the arguments are ... involved. One of the most beautiful approaches to the problem involves fixed point theory. This approach is outlined for you in Problem 3 at the end of this section. □

What on Earth?!?

It is daunting just to read the statement of the implicit function theorem. There are so many hypotheses! Given the discussion preceding the theorem, some of these hypotheses probably seem natural to you. Others, by contrast, may seem so completely mysterious that you find yourself asking the question in the heading above. For instance, you might wonder why the differentiability condition is needed—but only for the first n variables.[4] Here's the intuition.

Consider the equation $x^2 + y^2 = 1$. (See Figure M.2.) We don't need the implicit function theorem to tell us that we can solve the equation for y in

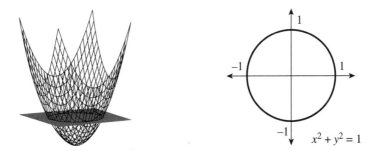

Figure M.2 The left diagram shows the **0**-level set of $F(x,y) = x^2 + y^2 - 1$. The right diagram projects this set into $\mathbb{R}^2$.

---

4. I won't say much about the hypothesis that the partial derivatives with respect to the first n variables must be continuous except to point out that, by this time, the requirement shouldn't surprise you. The issue was stated very succinctly by Walter Rudin in his lovely book *Principles of Mathematical Analysis* [RUD]:

> In cases where the existence of a derivative is sufficient when dealing with functions of one variable, continuity or at least boundedness of the partial derivatives is needed for functions of several variables.

(Remember that existence of the partial derivatives doesn't even guarantee the continuity of the function!)

terms of x in a neighborhood of, say, $(0, 1)$. The formula for this function is $y = +\sqrt{1 - x^2}$, and its graph is shown in Figure M.3.[5] The second part of Figure M.3, however, shows that we cannot solve for y in terms of x in a neighborhood of $(1, 0)$.

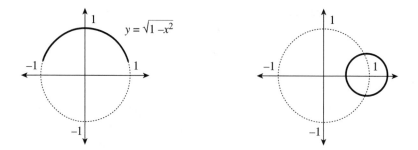

Figure M.3 The darkened portion on the left diagram shows that we can solve for y in terms of x in a neighborhood of the point $(0, 1)$. In the diagram on the right, we see that no matter how small a circle we draw about $(1, 0)$, we will never get the graph of a function.

How are the points $(1, 0)$ and $(0, 1)$ on the unit circle different? The tangent line to the circle at $(0, 1)$ is *vertical*. In terms of the three-dimensional picture shown in Figure M.2,

$$\frac{\partial F}{\partial y}(1, 0) = 0.$$

(Be sure you see the connection between the vertical tangent in two-dimensions and the zero partial derivative in three-dimensions.) Our example, therefore, gives us some insight into why the hypotheses of the implicit function theorem impose conditions on only *some* of the partial derivatives of $\mathbf{F}$. For the example $x^2 + y^2 = 1$, the partial derivative with respect to x is irrelevant. Only the partial derivative with respect to y plays a role in determining whether we can solve the equation in a neighborhood of a given point. (In this simplest case, the partial derivative with respect to y must not be zero.)

Now we expand our thinking into many dimensions. Suppose that we have a differentiable vector field $\mathbf{F} : \mathbb{R}^{n+m} \to \mathbb{R}^n$. We know that, in a small neighborhood about the point $(\mathbf{b}, \mathbf{a})$, the geometry of the graph of $\mathbf{F}$ is very much like the geometry of a linear transformation from $\mathbb{R}^{n+m}$ to $\mathbb{R}^n$ in a neighborhood

---

[5] We need to locate our neighborhood around a point *on the graph*, rather than a value of the independent variable alone. If we had only specified a neighborhood around $x = 0$, we would not have determined whether we wanted the top or the bottom of the circle.

about **0**. To be specific, it mirrors the geometry of the linear transformation represented by the Jacobian matrix

$$J_{\mathbf{F}}(\mathbf{b},\mathbf{a}) = \begin{bmatrix} \frac{\partial f_1}{\partial y_1}(\mathbf{b},\mathbf{a}) & \cdots & \frac{\partial f_1}{\partial y_n}(\mathbf{b},\mathbf{a}) & \frac{\partial f_1}{\partial x_1}(\mathbf{b},\mathbf{a}) & \cdots & \frac{\partial f_1}{\partial x_m}(\mathbf{b},\mathbf{a}) \\ \vdots & \ddots & \vdots & \vdots & \ddots & \vdots \\ \frac{\partial f_n}{\partial y_1}(\mathbf{b},\mathbf{a}) & \cdots & \frac{\partial f_n}{\partial y_n}(\mathbf{b},\mathbf{a}) & \frac{\partial f_n}{\partial x_1}(\mathbf{b},\mathbf{a}) & \cdots & \frac{\partial f_n}{\partial x_m}(\mathbf{b},\mathbf{a}) \end{bmatrix}.$$

It stands to reason, then, that the geometric conditions necessary to make $\mathbf{F}(\mathbf{y},\mathbf{x}) = \mathbf{0}$ solvable near $(\mathbf{b},\mathbf{a})$ would be the same as the conditions necessary to make $[J_{\mathbf{F}}(\mathbf{b},\mathbf{a})](\mathbf{y},\mathbf{x}) = \mathbf{0}$ solvable. What makes a linear system solvable? Let us look once again at the linear system of equations we considered earlier in this excursion:

$$\begin{aligned} 2y_1 + y_2 + 4x_1 + 7x_2 &= 0 \\ y_1 + y_2 + x_1 + 5x_2 &= 0 \\ -y_1 + y_3 - 3x_1 &= 0 \end{aligned}$$

or, in matrix form,

$$\begin{bmatrix} 2 & 1 & 0 & 4 & 7 \\ 1 & 1 & 0 & 1 & 5 \\ -1 & 0 & 1 & -3 & 0 \end{bmatrix} \begin{bmatrix} y_1 \\ y_2 \\ y_3 \\ x_1 \\ x_2 \end{bmatrix} = \begin{bmatrix} 0 \\ 0 \\ 0 \\ 0 \\ 0 \end{bmatrix}$$

Linear algebra tells us that we can solve for the variables y_1, y_2, and y_3 in terms of x_1 and x_2 if and only if the leftmost 3×3 submatrix

$$\begin{bmatrix} \boxed{\begin{matrix} 2 & 1 & 0 \\ 1 & 1 & 0 \\ -1 & 0 & 1 \end{matrix}} & \begin{matrix} 4 & 7 \\ 1 & 5 \\ -3 & 0 \end{matrix} \end{bmatrix} \Rightarrow \begin{bmatrix} 2 & 1 & 0 \\ 1 & 1 & 0 \\ -1 & 0 & 1 \end{bmatrix}$$

is invertible. This submatrix contains the coefficients of the first three variables y_1, y_2, and y_3.

In terms of the system $\mathbf{J_F}(\mathbf{b},\mathbf{a})(\mathbf{y},\mathbf{x}) = \mathbf{0}$, we will require the invertibility of the leftmost $n \times n$ submatrix corresponding to the partial derivatives with respect to the first n variables:

$$\begin{bmatrix} \frac{\partial f_1}{\partial y_1}(\mathbf{b},\mathbf{a}) & \frac{\partial f_1}{\partial y_2}(\mathbf{b},\mathbf{a}) & \cdots & \frac{\partial f_1}{\partial y_n}(\mathbf{b},\mathbf{a}) \\ \frac{\partial f_2}{\partial y_1}(\mathbf{b},\mathbf{a}) & \frac{\partial f_2}{\partial y_2}(\mathbf{b},\mathbf{a}) & \cdots & \frac{\partial f_2}{\partial y_n}(\mathbf{b},\mathbf{a}) \\ \vdots & \vdots & \cdots & \vdots \\ \frac{\partial f_n}{\partial y_1}(\mathbf{b},\mathbf{a}) & \frac{\partial f_n}{\partial y_2}(\mathbf{b},\mathbf{a}) & \cdots & \frac{\partial f_n}{\partial y_n}(\mathbf{b},\mathbf{a}) \end{bmatrix}.$$

But this is precisely the matrix D in the hypothesis of the implicit function theorem!

How does this relate to the two-variable example $x^2 + y^2 = 1$? That example required $\frac{\partial F}{\partial y}$ to be non-zero at (a,b). If we interpret the number

$$\frac{\partial F}{\partial y}(a,b)$$

as a 1×1 matrix, we see that it has to be *invertible* (that is, non-zero) to avoid the vertical tangent line that prevents the solution set of $x^2 + y^2 = 1$ from being the graph of a function defined on a neighborhood of $(1,0)$.

Properties of the Solution Function

The introductory material to this section noted that the solution functions guaranteed by the implicit function theorem "share many of the nice properties of the vector field $\mathbf{F}$." No additional hypotheses are needed to guarantee that the solution function is continuous.

Theorem M.2.2 Let m, n, $\mathbf{a}$, $\mathbf{b}$, $\mathbf{F}$, and W all be exactly as in the statement of the implicit function theorem. Then the function $\mathbf{f}$ given in the conclusion of the theorem is continuous.

Proof: The proof is left to you as Problem 4 at the end of this section (but I give you a pretty hefty hint!). □

If we strengthen the hypotheses of the implicit function theorem so that the partial derivatives of $\mathbf{F}$ all exist and are continuous in W, then the solution function is, as well.

Excursion M ■ The Implicit Function Theorem 399

Theorem M.2.3 Let m, n, **a**, **b**, **F**, and W all be exactly as in the statement of the implicit function theorem except that all the partial derivatives of **F** exist and are continuous on W. Then all the partial derivatives of **f** exist and are continuous on $B_t(\mathbf{a})$.

Proof: The proof is left to you as Problem 4 at the end of this section. (I could give you a hint for this problem, but I will leave it as a challenge problem for you to "chew on.") □

Problems M.2

1. Let $\mathbf{z} \in \mathbb{R}^n$, $\mathbf{w} \in \mathbb{R}^m$, and let $R > 0$. Set $r = \dfrac{R}{\sqrt{2}}$. Prove that if $\mathbf{x} \in B_r(\mathbf{z})$ and $\mathbf{y} \in B_r(\mathbf{w})$, then $(\mathbf{x}, \mathbf{y}) \in B_R((\mathbf{z}, \mathbf{w}))$.

2. Consider the function $F : \mathbb{R}^2 \to \mathbb{R}$ given by $F(x, y) = (x^2 + y^2)^2 - x^2 + y^2$. Figure M.4 shows the solution to the system $F(x, y) = 0$. Note that there is no open set around $(0, 0)$ for which the 0-level set is a function of x in terms of y. Without doing any calculations, explain why $\frac{\partial F}{\partial y}(0, 0)$ must be 0. (*Hint:* The function F is locally planar at $(0, 0)$. What plane will approximate it there?) Explicitly connect this fact to the implicit function theorem.

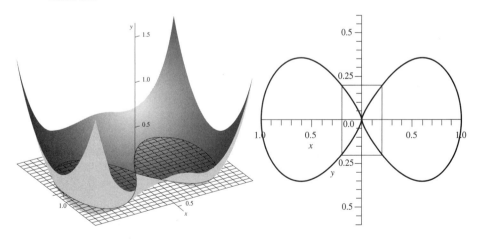

Figure M.4 These figures show the intersection of the xy plane with the surface $F(x, y) = (x^2 + y^2)^2 - x^2 + y^2$.

3. Prove the implicit function theorem (Theorem M.2.1). Here's a detailed proof outline to help you along:

Proof Outline

Strategy: For each $\mathbf{x}$ in some neighborhood of $\mathbf{a}$, we need to know how to define $\mathbf{y} = \mathbf{f}(\mathbf{x})$. (Our target will be a point in $\mathbb{R}^n$ somewhere close to $\mathbf{b}$.) We will identify $\mathbf{y}$ by iterating a map that is cleverly constructed so that $\mathbf{y}$ satisfies $\mathbf{F}(\mathbf{y}, \mathbf{x}) = \mathbf{0}$ if and only if it is a fixed point of the map. If we can show the function has a unique fixed point, we will know how to define $\mathbf{f}(\mathbf{x})$. (Figure M.5 is a schematic that may help you keep track of the pieces in the proof, and Section M.3 discusses the intuition behind the proof.) Eventually, we will concentrate on points that are close to $\mathbf{a}$, but for now fix $\mathbf{x} \in \mathbb{R}^m$, arbitrary.

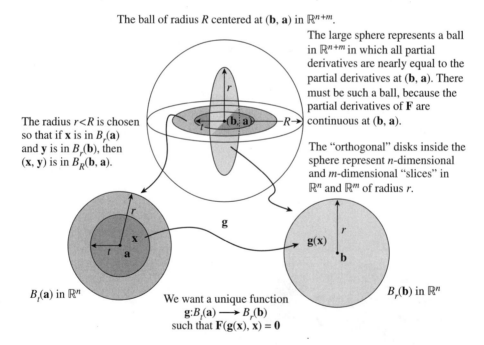

Figure M.5 We get the value $g(x)$, by finding the fixed point of a cleverly constructed map. But the radius of r around a will not be small enough to guarantee that this map has a unique fixed point in $B_r(b)$. To guarantee the existence of a fixed point, we will need to restrict our attention to a ball of smaller radius t.

Step 1. Define the map that will be iterated to identify f(x):

$$\phi_{\mathbf{x}} : \mathbb{R}^n \to \mathbb{R}^n \text{ by } \phi_{\mathbf{x}}(\mathbf{y}) = \mathbf{y} - D^{-1}(\mathbf{F}(\mathbf{y}, \mathbf{x}))$$

Step 2. Show ϕ_x contracts distances on some ball centered at b.

Notation: for each $(\mathbf{v}, \mathbf{w})$ in W, let

$$d\mathbf{F}(\mathbf{v}, \mathbf{w}) = \begin{bmatrix} \frac{\partial f_1}{\partial y_1}(\mathbf{v}, \mathbf{w}) & \frac{\partial f_1}{\partial y_2}(\mathbf{v}, \mathbf{w}) & \cdots & \frac{\partial f_1}{\partial y_n}(\mathbf{v}, \mathbf{w}) \\ \frac{\partial f_2}{\partial y_1}(\mathbf{v}, \mathbf{w}) & \frac{\partial f_2}{\partial y_2}(\mathbf{v}, \mathbf{w}) & \cdots & \frac{\partial f_2}{\partial y_n}(\mathbf{v}, \mathbf{w}) \\ \vdots & \vdots & \cdots & \vdots \\ \frac{\partial f_n}{\partial y_1}(\mathbf{v}, \mathbf{w}) & \frac{\partial f_n}{\partial y_2}(\mathbf{v}, \mathbf{w}) & \cdots & \frac{\partial f_n}{\partial y_n}(\mathbf{v}, \mathbf{w}) \end{bmatrix}$$

[Observe that $D = d\mathbf{F}(\mathbf{b}, \mathbf{a})$.]

We know that D is invertible, so $\|D^{-1}\| \neq 0$. Now, appealing to the continuity of the partial derivatives with respect to the first n variables (condition 3), we choose a real number $R > 0$ small enough so that $(\mathbf{y}, \mathbf{x}) \in B_R(\mathbf{b}, \mathbf{a})$ implies that

$$\|D - d\mathbf{F}(\mathbf{y}, \mathbf{x})\| < \frac{1}{2\|D^{-1}\|} \quad \text{(Theorem 13.3.26)}$$

Use this to show that $\|\phi'_{\mathbf{x}}(\mathbf{y})\| < \frac{1}{2}$ whenever $(\mathbf{y}, \mathbf{x}) \in B_R(\mathbf{b}, \mathbf{a})$.

Technical detail: Set $r = \frac{R}{\sqrt{2}}$. Problem 1 at the end of this section tells us that if $\mathbf{y} \in B_r(\mathbf{b})$ and $\mathbf{x} \in B_r(\mathbf{a})$, then $(\mathbf{y}, \mathbf{x}) \in B_R(\mathbf{b}, \mathbf{a})$.

Now use Corollary 13.5.5 to show that if $\mathbf{x} \in B_r(\mathbf{a})$, then $\phi_{\mathbf{x}}$ contracts distances with a contraction constant of $\frac{1}{2}$ on $B_r(\mathbf{b})$.

Step 3. Restrict the radius of points around a: If x is close enough to a, $\phi_{\mathbf{x}}$ has a unique fixed point in $B_r(\mathbf{b})$:

The key is Corollary 10.3.4. It says, in a nutshell, that if $\phi_{\mathbf{x}}(\mathbf{b})$ is "not too far" from $\mathbf{b}$, then $\phi_{\mathbf{x}}$ will have a unique fixed point in $B_r(\mathbf{b})$. To show that $\phi_{\mathbf{x}}$ satisfies the conditions of Corollary 10.3.4, we will appeal to the fact that

$$\phi_{\mathbf{x}}(\mathbf{b}) - \mathbf{b} = D^{-1}(\mathbf{F}(\mathbf{b}, \mathbf{x})) = D^{-1}(\mathbf{F}(\mathbf{b}, \mathbf{x}) - \mathbf{F}(\mathbf{b}, \mathbf{a}))$$

You will also need the fact that $\mathbf{F}$ is continuous at $(\mathbf{b}, \mathbf{a})$. To make use of this hypothesis, you will need to make sure that $\mathbf{x}$ is "sufficiently close" to $\mathbf{a}$.

Step 4. Define the function $\mathbf{f} : \mathrm{B}_t(\mathbf{a}) \to \mathrm{B}_r(\mathbf{b})$. Be sure to show that it satisfies all the necessary conditions.

$\square$

4. Prove that the function given in the conclusion of the implicit function theorem (Theorem M.2.2) is continuous.

    Hint: To prove that $\mathbf{f}$ is continuous on $\mathrm{B}_t(\mathbf{a})$, we fix $\mathbf{x}_1 \in \mathrm{B}_t(\mathbf{a})$. We need to guarantee that if $\mathbf{x}_2 \in \mathrm{B}_t(\mathbf{a})$ is "close" to $\mathbf{x}_1$, then the fixed points of the maps $\phi_{\mathbf{x}_1}$ and $\phi_{\mathbf{x}_2}$ will be "close" in $\mathrm{B}_r(\mathbf{b})$. To do so, we use a property of our contractive mapping:

    > A point in $\mathrm{B}_r(\mathbf{b})$ is no more than twice as far from the fixed point as it is from its image under the contraction.

    - First, fix $\mathbf{y} \in \mathrm{B}_r(\mathbf{b})$ and $\mathbf{x} \in \mathrm{B}_t(\mathbf{a})$. Prove that
      $\|\mathbf{y} - \mathbf{f}(\mathbf{x})\| \leq 2\,\|\mathbf{y} - \phi_{\mathbf{x}}(\mathbf{y})\|$.
    - Don't forget that $\mathbf{f}(\mathbf{x})$ is a fixed point of the map $\phi_{\mathbf{x}}$. Also, use the fact that $\phi_{\mathbf{x}}$ contracts distances on $\mathrm{B}_r(\mathbf{b})$.
    - Finally, appeal to the continuity of the function $\mathbf{T}(\mathbf{x}) = \mathbf{F}(\mathbf{f}(\mathbf{x}_1), \mathbf{x})$ at $\mathbf{x}_1$ to show that $\mathbf{f}$ is continuous at $\mathbf{x}_1$.

5. Prove Theorem M.2.3.

M.3 Connections: Quasi-Newton's Methods

We proved the implicit function theorem by reformulating the problem as a fixed point problem. We defined a family of maps whose fixed points had to be solutions to the system of equations we were trying to solve (and conversely). By proving that each map had a unique fixed point, we were able to define our solution function.

The proof outline that was provided in Problem 3 simply gave you the maps. No attempt was made to explain where those maps came from. The purpose of this discussion is to explain that provenance. In effect, you will see that we can think of the approach as a "quasi-Newton's method." To make things as clear as possible, I will need to come at this in a roundabout way. Bear with me.

When we discussed iteration, we considered a (very naive!) root-finding method. (See page 201.) We started by supposing that we wished to solve the equation
$$0.005x^4 + 0.0281x^3 - 0.051x^2 - 0.3x + 0.3 = 0.$$
We obtained an equivalent equation by adding x to both sides, and then we found a solution to the equation by iterating the function
$$g(x) = (0.005x^4 + 0.0281x^3 - 0.051x^2 - 0.3x + 0.3) + x$$
given by the left side of the new equation. Starting at $x = 0$, the iteration yielded a convergent sequence. The limit $r \approx 0.94059$ had to be a root of the original equation, as it was a fixed point of the continuous function g.

When I first mentioned this idea, I rhetorically asked whether such an approach would always work, gave the short answer "no," and then went on to other matters. Now I want to take up the issue again and examine it in some detail.

It is easy to come up with equations for which the suggested trick fails because the resulting iterated map doesn't converge. Suppose that we have a continuously differentiable function $f : \mathbb{R} \to \mathbb{R}$ and that we want to solve the equation $f(x) = 0$. We try iterating $g(x) = x + f(x)$ to find a fixed point r. The problem is that r may be a repelling fixed point if g is too "steep" near r. We must have $|g'(r)| < 1$ to guarantee that r is an attracting fixed point.

Using this insight, we can improve our approach to the problem. Let K be any nonzero real number. Note that r is a root of $f(x) = 0$ if and only if it is also a root of $Kf(x) = 0$. So we can apply our iteration trick to any equation of the form
$$g(x) = x + Kf(x) = x.$$
The idea is to choose K cleverly so as to "flatten" the graph of g at r. If we can do this sufficiently well, r will be an attracting fixed point and the iterated map will then converge to it.

How should we choose K? The rate at which the iterated maps of g converge to r is governed by the size of $|g'(r)|$. The closer this value is to 0, the faster the convergence. So we might as well choose K to get the best possible convergence rate. In other words, we choose K such that $g'(r) = 0$. We can do this provided that $f'(r) \neq 0$:
$$g'(r) = 1 + Kf'(r) = 0 \quad \Longrightarrow \quad K = -\frac{1}{f'(r)}.$$
That is, if we want to solve the equation $f(x) = 0$ by this iteration scheme, our absolute best bet is to iterate the function
$$g(x) = x - \frac{f(x)}{f'(r)}$$

where r is a solution to the equation. Let's call this function and the associated iteration scheme the "ideal root finder."

You may already have noticed something interesting. The ideal root finder is very reminiscent of the Newton function for f:

$$N(x) = x - \frac{f(x)}{f'(x)}.$$

When f' is continuous, $N(x)$ and $g(x)$ should be nearly equal, provided x is close to r. This gives us an insight into why Newton's method works so well if we start with a reasonable guess.

Newton's method has an advantage over the ideal iteration scheme in that implementing it doesn't require any knowledge of the root. The "ideal root finder," after all, requires us to know the value of $f'(r)$ to find r itself! Of course, we compromise on the convergence rate. Also, we have to recompute the derivative at every step, so this increases the number of computations we have to do to implement it. (This might be a significant issue if f or f' were extremely complicated functions.)

Perhaps there is a middle ground. Suppose we find a positive real number D that is close to the value of $f'(r)$. Defining the function

$$p(x) = x - \frac{f(x)}{D}$$

we see that $|p'(x)|$, although perhaps not zero, is nevertheless close to zero on some interval I containing r (say, $|p'(x)| < \frac{1}{2}$ on I). Then we know that r is an attracting fixed point for p, that p contracts distances on I, and—icing on the cake—that we won't have to recompute that derivative every time we evaluate p. This sounds promising! Let us call this scheme the "pretty good root finder."

We now reconsider the proof of the implicit function theorem. By looking at the two-variable case, we will be able to connect it to our discussion of quasi-Newton's methods. Suppose we are trying to solve $f(x, y) = 0$ where f is "well behaved."

Step 1. We take a point (a, b) that is known to lie on the $z = \mathbf{0}$-level curve of f and for which

$$\frac{\partial f}{\partial y}(a, b) \neq 0.$$

Figure M.6 shows level curves and a contour diagram for a hypothetical function.

Excursion M ■ The Implicit Function Theorem   405

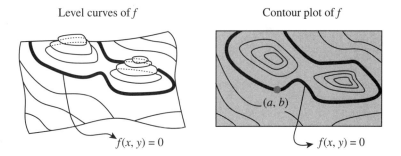

Figure M.6 The level curve/contour $f(x,y) = 0$ is highlighted.

Step 2. Because the partials with respect to y are assumed to be continuous, we narrow our view to a small rectangle R such that for all $(x, y) \in R$,

$$\frac{\partial f}{\partial y}(x, y) \approx \frac{\partial f}{\partial y}(a, b).$$

This is shown in Figure M.7.

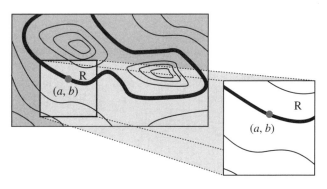

Figure M.7 The partial derivatives with respect to y are all about equal to $\dfrac{\partial f}{\partial y}(a, b)$ in the rectangle R.

Step 3. Then we fix x near a, define

$$\phi_x(y) = y - \frac{f(x,y)}{\frac{\partial f}{\partial y}(a,b)}$$

and iterate ϕ_x. (We may have to restrict the width of the rectangle further to make sure that ϕ_x is a contraction.) This is shown in Figure M.8.

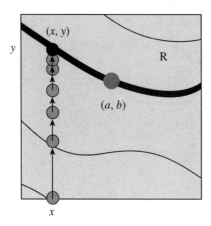

Figure M.8 When we iterate the function ϕ_x, the resulting orbit converges to a point (x, y) on the $f(x, y) = 0$ contour.

Notice that when $x = a$, $\phi_x = \phi_a$ is the "ideal root finder" for the function $g(y) = f(a, y)$. When $x \neq a$, ϕ_x is the "pretty good root finder" for the function $g(y) = f(x, y)$.

The general case of the implicit function theorem has much the same plan as this special case. We are interested in the set where $\mathbf{F}(\mathbf{y}, \mathbf{x}) = \mathbf{0}$, and we wish to solve for $\mathbf{y}$ in terms of $\mathbf{x}$ within a neighborhood of a point $(\mathbf{b}, \mathbf{a})$. Instead of a simple partial derivative, we have the matrix of derivatives $D = d\mathbf{F}(\mathbf{b}, \mathbf{a})$. We cannot divide by a matrix, but we can multiply by a matrix inverse. The function that we iterate is just

$$\phi_{\mathbf{x}}(\mathbf{y}) = \mathbf{y} - D^{-1}(\mathbf{F}(\mathbf{y}, \mathbf{x})).$$

The hypotheses of the theorem and the machinery of the proof are there to establish the existence of this function $\phi_{\mathbf{x}}$ and to guarantee that its iterates converge to the unique fixed point on the **0**-level set of $\mathbf{F}$. For $\mathbf{x} = \mathbf{a}$, this is a multidimensional version of the "ideal root finder"; for other nearby values of $\mathbf{x}$, it is the "pretty good root finder." In either case, the engine of the proof of the implicit function theorem is a quasi-Newton's method.

M.4 The Inverse Function Theorem

The inverse function theorem tells us that a reasonably well-behaved function $\mathbf{f} : \mathbb{R}^n \to \mathbb{R}^n$ is invertible in a neighborhood of a designated point. It is an immediate corollary of the implicit function theorem.

Recall that for sets A and B and a function $f : A \to B$, the following statements are equivalent:

- f is one-to-one and onto.

- There exists a unique function $f^{-1} : B \to A$ such that $f^{-1} \circ f(x) = x$ for all $x \in A$ and $f \circ f^{-1}(y) = y$ for all $y \in B$.

- For each $y \in B$, the equation $y = f(x)$ has a unique solution x in A.

In this case, we say that f is invertible. The function f^{-1} is called the inverse of f.

These three equivalences tell us that the inverse of a function $f : A \to B$ can be "found" by "solving" the equation $y = f(x)$ for x. In other words, the process of inverting a vector field is intimately related to the solving of systems of equations.

If we have a vector field $\mathbf{f} : \mathbb{R}^n \to \mathbb{R}^n$, we can recast the problem as follows: Define $F : \mathbb{R}^{2n} \to \mathbb{R}^n$ by $F(\mathbf{x}, \mathbf{y}) = \mathbf{f}(\mathbf{x}) - \mathbf{y}$, where $\mathbf{x} \in \mathbb{R}^n$ and $\mathbf{y} \in \mathbb{R}^n$. Then, provided it exists, the inverse of the function $\mathbf{f}$ can be obtained by solving the equation $\mathbf{F}(\mathbf{x}, \mathbf{y}) = \mathbf{0}$ for $\mathbf{x}$ in terms of $\mathbf{y}$. In particular, the existence of an inverse is dependent on the existence of a unique solution to the earlier equation, which brings us back to the implicit function theorem.

Theorem M.4.1 [Inverse Function Theorem] Suppose that U is an open subset of $\mathbb{R}^n$. Let $\mathbf{f} : \mathbb{R}^n \to \mathbb{R}^n$ be a function whose partial derivatives all exist and are continuous in U. Suppose that $\mathbf{f}'(\mathbf{a})$ is invertible for some $\mathbf{a} \in U$ and that $\mathbf{b} = \mathbf{f}(\mathbf{a})$. Then there exists $t > 0$ for which $B = B_t(\mathbf{a}) \subseteq U$ and such that the following conditions hold:

1. The restriction of $\mathbf{f}$ to B is one-to-one.

2. The set $V = \mathbf{f}(B)$ is open.

3. The inverse $\mathbf{f}^{-1}$ of $\mathbf{f}|_B$ is continuous on V.

Proof: The proof is Problem 1 at the end of this section. □

Problems M.4

1. Prove the inverse function theorem (Theorem M.4.1).

2. In this excursion, we proved the implicit function theorem and then used it to prove the inverse function theorem. The reverse is also possible. To see that these two theorems are, indeed, equivalent, assume the truth of the inverse function theorem and use it to prove the implicit function theorem.

Excursion N
Spaces of Continuous Functions

> *Required Background*
> Chapters 5, 6, 7, and 12.

Functions are frequently the focus of study in analysis. So far, we have concentrated on single functions or sequences of functions and have proved a large number of theorems about them. This excursion is intended to strengthen our ability to study functions by making a special class of real-valued functions into a metric space in its own right and placing that body of theorems into this larger context. The theorems in this excursion are stated without proof. Think of it as a long series of interconnected exercises.

We need to think about measuring the distance between two functions. Let's start with a loose analogy. Suppose we need to compare prices at two grocery stores, Food Monster and Cheap 'n' Hasty. We get a list of prices from each place. Now how do we make the comparisons?

- *We need to compare prices on many items.* Knowing that Food Monster and Cheap 'n' Hasty both charge $1.49 for a dozen eggs doesn't give us enough information to judge their prices on a global scale.

- *We need to compare prices on like items.* It doesn't make sense to compare the price of a gallon of Food Monster milk to a pound of Cheap 'n' Hasty walnuts. We have to compare Food Monster milk to Cheap 'n' Hasty milk, Food Monster walnuts to Cheap 'n' Hasty walnuts, Food Monster frozen lasagna to Cheap 'n' Hasty frozen lasagna, and so forth.

Once we have these comparisons, how do we make a final determination? Well, there are several things we might do. We could find the *smallest* difference between corresponding prices, the *largest* difference between corresponding prices, or the *average* difference between corresponding prices.[1] If we think of the inputs as the products and the outputs as the prices, we can see how to begin making comparisons between two functions.

It is easy to see that trying to compare the grocery stores by looking at the *smallest difference between corresponding prices* is not a good criterion. To return to the earlier example, if Food Monster and Cheap 'n' Hasty both charge the same amount for a dozen eggs, the smallest price difference would be zero. Such information does nothing to advance our knowledge of price differences between the two grocery stores.

The other two ideas appear more promising. The greatest difference measure asks, "How bad can things get?" Alternatively, if the greatest price difference between items is small, then all price differences will be small. The average difference measure is also promising. If we choose an item at random, how much will the price difference be on average?

With these insights, we return to the question at hand. To help you picture what is going on, suppose we have two functions f and g from $[0,1]$ to $\mathbb{R}$.

What we want to do is to make point-by-point comparisons between the function values of f and g. In Figure N.1, this corresponds to measuring the lengths of vertical lines joining corresponding function values. To compute the distance from f to g, we will want either to average those values (which will require some sort of integration theory) or to find the maximum value. As it happens, either approach is reasonable and is useful in its own way. In this excursion, we will follow the *maximum difference* plan.

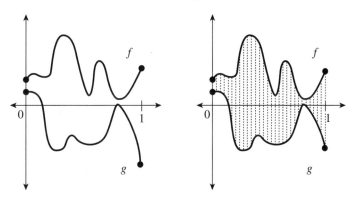

Figure N.1 If we want to compare functions f and g, we need to make comparisons point by point.

---

1. At this point, you can see why the analogy is a "loose" one. Among other things, in our grocery store example we will be profoundly interested in whether the difference between prices is positive or negative. Because distances must be symmetric in analysis, the differences we will end up talking about will be the absolute values of differences. No matter—we can learn a lot from the example, anyhow.

N.1 The Metric Space $C(K)$

Let X be a metric space. Let f and g be real-valued functions on X. We would like to define the distance between the functions by

$$d(f,g) = \max_{x \in X} |f(x) - g(x)|.$$

Under what circumstances can we do this? The only real issue is whether the maximum exists. It may not. (See Figure N.2.)

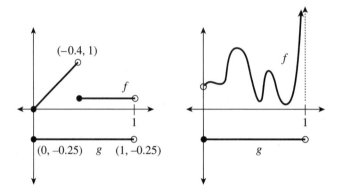

Figure N.2 For these pairs of functions, $\max_{x \in [0,1)} |f(x)-g(x)|$ is not defined. In one case, the differences get arbitrarily close to 1.25 but never reach it. In the second case, the differences get arbitrarily large.

Consider the examples shown in Figure N.2. The difficulty in the first case (left diagram) lies in the fact that f is not continuous. The difficulty in the second case (right diagram) lies in the fact that the domain gives f enough freedom to fly off toward infinity. But we know (by the max–min theorem) that if f and g are continuous and their domain is compact, then the maximum *will* exist.

Lemma N.1.1 Let K be a compact metric space. Let $C(K)$ be the set of all continuous, real-valued functions on K. Then $d : C(K) \times C(K) \to \mathbb{R}$ given by

$$d(f,g) = \max_{k \in K} |f(x) - g(x)|$$

is a metric on $C(K)$.

Using our previous work, we immediately deduce many things about the metric space $(C(K), d)$. Let us consider what we know about the elements of $C(K)$. (Quite a lot, actually!) First, we have at our disposal all the theorems that appeared in Chapter 5 on continuous, real-valued functions. The fact that

K is compact gives us a lot of information, too. If $f : K \to \mathbb{R}$ is a continuous function, then it is uniformly continuous (Theorem 7.2.3). Furthermore, $f(K)$ is compact (Theorem 7.2.1). In particular, f is bounded. Finally, f attains both a maximum and a minimum value (Theorem 7.2.2). The elements of $C(K)$ are extremely well-behaved functions. What can we say about the topology of the metric space $C(K)$?

Exercise N.1.2 [Open balls in $C(K)$] This exercise is intended to help you visualize the open balls of $C(K)$ when K is the real interval $[-1, 1]$ and when K is a compact, connected subset of $\mathbb{R}^2$.

1. Consider the function $f(x) = x^2$ on the compact interval $[-1, 1]$.

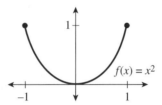

   Describe the open ball of radius $\frac{1}{2}$ centered at f. Both words and a well-labeled diagram are essential parts of your description. Include at least two elements of the ball in your diagram. (*Hint*: Your diagram should be similar to a picture from Chapter 12.)

2. Consider the real-valued function $f(x, y) = x^2 - y^2$ defined on the unit circle in $\mathbb{R}^2$.

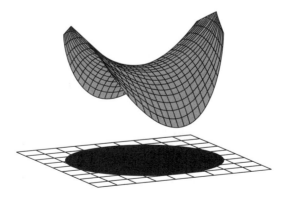

   Describe the open ball of radius $\frac{1}{2}$ centered at f. It is your choice whether to use a diagram or whether to describe the set only in words. Keep in mind, however, that the purpose of this exercise is to *visualize* the set, so the description should evoke an image.

■

What does convergence of sequences "look like" in $C(K)$? A sequence in $C(K)$ is a sequence (f_n) of real-valued, continuous functions on a compact metric space K. We wonder what it would mean to say that this sequence converges to some element f in $C(K)$. But we have considered the convergence of sequences of functions already. In $C(K)$, the limit of (f_n) must be continuous, but we know that the limit of a *pointwise* convergent sequence of continuous functions need not be. Uniform convergence, however, *does* guarantee the continuity of the limit function. It is easy to prove that this is precisely what sequence convergence means in the metric space $C(K)$. (The diagram you drew in the first part of Exercise N.1.2 should help you visualize this situation.)

Theorem N.1.3 Let K be a compact metric space. Let (f_n) be a sequence of real-valued, continuous functions on K. The sequence (f_n) converges to f in $C(K)$ if and only if (f_n) converges uniformly to f on K.

Given this theorem, it is not too hard to conjecture the following fact:

Lemma N.1.4 The sequence (f_n) of elements of $C(K)$ is Cauchy in $C(K)$ if and only if the sequence (f_n) of continuous, real-valued functions on K is uniformly Cauchy on K.

When we couple Lemma N.1.4 with Theorem 12.2.5, Theorem N.1.3 immediately gives us the following important theorem:

Theorem N.1.5 The metric space $(C(K), d)$ is complete.

The propositions in this section may lead you to believe that $C(K)$ is simply a point of view. After all, we have spent most of the section "translating" from statements about the convergence of sequences of functions into statements about the metric space $C(K)$. You are right. It *is* a point of view—but a powerful one.

N.2 Compactness in $C(K)$

We know that if K is a compact subset of a metric space X, it is both closed and bounded in X. In $\mathbb{R}^n$, closed and boundedness are sufficient to guarantee compactness. But recall that even closed balls of radius 1 are not compact in the infinite-dimensional space ℓ_∞, so more than closed and boundedness is generally required for compactness. This is so in $C(K)$.

Let X be a metric space. Let Y be a subset of X. Recall that the following statements are equivalent:

- Y is compact.

- Every sequence in Y has a subsequence converging to a point in Y.

We will use this characterization of compactness in our quest to characterize the compact subsets of the metric space $C(K)$.

Definition N.2.1 Let X be a metric space. Let $\mathcal{F}$ be a family of functions from X to $\mathbb{R}$. The set $\mathcal{F}$ is said to be **pointwise bounded**, if for each $x \in X$, there exists a positive real number M_x such that
$$|f(x)| \leq M_x \text{ for all } f \in \mathcal{F}.$$

$\mathcal{F}$ is said to be **uniformly bounded** if there exists a positive real number M such that
$$|f(x)| \leq M \text{ for all } f \in \mathcal{F} \text{ and for all } x \in X.$$

Exercise N.2.2 Let K be a compact metric space. Let $\mathcal{F}$ be a family of continuous real-valued functions on K. Prove that $\mathcal{F}$ is bounded as a subset of the metric space $C(K)$ if and only if it is uniformly bounded as a family of functions on the metric space K. ∎

Definition N.2.3 Let X and Y be metric spaces. Let $\mathcal{F}$ be a family of functions from X to Y. The set $\mathcal{F}$ is said to be **equicontinuous** if for every $\epsilon > 0$, there exists $\delta > 0$ such that if $f \in \mathcal{F}$ and $x_1, x_2 \in X$ with $d_X(x_1, x_2) < \delta$, then $d_Y(f(x_1), f(x_2)) < \epsilon$.

Theorem N.2.4 Let K be a compact metric space. Let $\mathcal{F}$ be an equicontinuous subset of $C(K)$ that is pointwise bounded on K. Then $\mathcal{F}$ is uniformly bounded on K.

Theorem N.2.5 Let K be a compact metric space. Let (f_n) be a convergent sequence in $C(K)$. Then $\{f_n\}_{n \in \mathbb{N}}$ is equicontinuous.

Lemma N.2.6 Let K be a compact metric space. Let B be a dense subset of K. If (f_n) is a sequence of equicontinuous, real-valued functions on K that converges pointwise on B, then (f_n) converges uniformly on K. (*Hint:* Prove (f_n) is uniformly Cauchy on K.

Lemma N.2.7 Let X be a metric space, let $S = \{s_1, s_2, s_3, \dots\}$ be a countable subset of X, and let (f_n) be a pointwise bounded sequence of real-valued functions on X. Then (f_n) has a subsequence (f_{n_k}) that converges pointwise on S.

Theorem N.2.8 [Arzela–Ascoli Theorem] Let K be a compact metric space, and let (f_n) be a sequence in $C(K)$ that is pointwise bounded and equicontinuous. Then (f_n) has a subsequence that converges in $C(K)$.

Lemma N.2.9 Let K be a compact set. Let T be a subset of $C(K)$ that is not equicontinuous. Then there is a sequence in T that has no convergent subsequence.

We can finally characterize compact subsets of $C(K)$.

Corollary N.2.10 Let K be a compact set. Let $\mathcal{F}$ be a subset of $C(K)$. Then $\mathcal{F}$ is compact in $C(K)$ if and only if it is closed in $C(K)$, pointwise bounded on K, and equicontinuous.

Question to Ponder: If $\mathcal{F} \in \mathcal{C}(\mathcal{K})$ is pointwise bounded and equicontinuous, is the closure of $\mathcal{F}$ in $\mathcal{C}(\mathcal{K})$ compact?

N.3 The Stone–Weierstrass Theorem

Our theme in this section is *approximation*. Given a compact metric space K, the set $C(K)$ includes lots of different functions. Some of these may be easier to deal with than others, so let's suppose that there is an identifiable set $\mathcal{N}$ of "nice" functions. Unfortunately, not every function f in $C(K)$ will be an element of $\mathcal{N}$. But perhaps f can be approximated by something in $\mathcal{N}$. In the best possible situation, every f in $C(K)$ would be arbitrarily close to a nice function in $\mathcal{N}$ (under the metric of $C(K)$). In short, we want $\mathcal{N}$ to be *dense* in $C(K)$.

But what do we mean by a "nice" function? Here is a famous example. The space $C[a,b]$ of continuous real-valued functions on the closed interval $[a,b]$ contains a lot of strange functions. But some of these functions, like the polynomials, are described by relatively simple algebraic formulas. Let $P[a,b]$ be the set of polynomial functions on the interval. These are all in $C[a,b]$. Can we approximate any f in $C[a,b]$ by something in $P[a,b]$? Is every continuous function on $[a,b]$ arbitrarily close to a polynomial?

The answer, first worked out by Karl Weierstrass in 1885, is "yes." Much later (1937), Marshall Stone realized that Weierstrass's theorem is a special case of a much more general principle. His insight allowed him to generalize the approximation theorem to the context of a general compact metric space. Instead of proving the Weierstrass approximation theorem directly, we will eventually deduce it as a special case of the more general Stone-Weierstrass theorem. However, we will need one specific case of the Weierstrass approximation theorem. We must prove explicitly that the absolute value function is the uniform limit of a sequence of polynomials on $[-a, a]$.

Lemma N.3.1 Let $a \in \mathbb{R}$ and fix $\epsilon > 0$. Then there exists a polynomial $P(x) = \sum_{i=1}^{n} c_i x^i$ defined on $[-a, a]$ such that for all $x \in [-a, a]$,

$$|P(x) - |x|| < \epsilon.$$

Proof: The proof is Problem 4 at the end of this section. □

We now return to our discussion of the "nice" set $\mathcal{N}$. What would a "nice" set look like? In the Weierstrass approximation theorem, $\mathcal{N}$ is the set of polynomials. Stone realized that the proof of the approximation theorem depended on certain algebraic and analytic properties of the collection of polynomials and not on the fact that they are polynomials, *per se*.

Definition N.3.2 Let K be a set and let $\mathcal{N}$ be a set of real-valued functions on K.

- $\mathcal{N}$ is an **algebra** provided that if $f, g \in \mathcal{N}$ and $k \in \mathbb{R}$, then $f + g$, fg, and kf are in $\mathcal{N}$.

- $\mathcal{N}$ is said to **separate points** provided that given any two elements x, $y \in K$, there exists some function $f \in \mathcal{N}$ such that $f(x) \neq f(y)$.

Let K be a compact metric space. Stone proved that if $\mathcal{N}$ is an algebra in $C(K)$ that separates points and contains the constant functions, then it is dense in $C(K)$. Before we proceed, we will need to deduce some facts about the set $\mathcal{N}$ and about its closure, which we will initially call $\mathcal{A}$.[2] These results are stated in Lemma N.3.3, with some related details set out in the problems at the end of the Section. The proofs are left to you.

Lemma N.3.3 Let K be a compact metric space. Suppose that $\mathcal{N} \subseteq C(K)$ is an algebra that separates points and contains the constant functions. Let $\mathcal{A}$ be the closure of $\mathcal{N}$ in $C(K)$. The following facts hold:

1. Let $x, y \in K$ with $x \neq y$. Then there exists $f \in \mathcal{N}$ such that $f(x) \neq f(y)$, $f(x) \neq 0$, and $f(y) \neq 0$.

2. Let $x, y \in K$ with $x \neq y$. Let $a, b \in \mathbb{R}$. Then there exists $f \in \mathcal{N}$ such that $f(x) = a$ and $f(y) = b$.

---

2. The goal is eventually to prove that $\mathcal{A} = C(K)$, but at this point, we cannot assert that.

3. $\mathcal{A}$ is an algebra that separates points and contains the constant functions.

4. If $f \in \mathcal{A}$. Then $|f| \in \mathcal{A}$. (*Hint*: If you can approximate $|x|$ in $C[-a,a]$, then you can approximate $|f(x)|$ in $C(K)$. Set $a = \sup|f(x)|$.)

5. If $f_1, f_2, \ldots, f_n$ are in $\mathcal{A}$, then
$$\max(f_1, f_2, \ldots, f_n) \text{ and } \min(f_1, f_2, \ldots, f_n)$$
are both in $\mathcal{A}$. (*Hint*: Problem 1 at the end of this section should be helpful to you.)

We finally come to Stone's generalization of the Weierstrass Approximation Theorem. A proof outline is provided. The details are left to you.

Theorem N.3.4 [Stone–Weierstrass theorem] Let K be a compact metric space. Let $\mathcal{N} \subseteq C(K)$. If $\mathcal{N}$ is an algebra that separates points and contains the constant functions, then $\mathcal{N}$ is dense in $C(K)$.

Proof Outline:
Let $f \in C(K)$. Fix $\epsilon > 0$. We must show that there exists $g \in \mathcal{N}$ that uniformly approximates f to within ϵ.

Step 1. Prove that it is sufficient to show that there exists $g \in \mathcal{A} = \overline{\mathcal{N}}$ such that $d(f,g) < \dfrac{\epsilon}{2}$.

Observe that Step 1 and the third item in Lemma N.3.3 allow us exclusively to consider functions in $\mathcal{A}$.

Step 2. Fix $x \in K$. For each $y \in K$, choose $g_{x,y} \in \mathcal{A}$ such that
$$g_{x,y}(x) = f(x) \text{ and } g_{x,y}(y) = f(y).$$

Step 3. For each $y \in K$, choose $\delta_y > 0$ such that if $t \in B_{\delta_y}(y)$ then
$$g_{x,y}(t) > f(t) - \epsilon.$$
(See Problem 2 at the end of this section.)

Step 4. The balls described in Step 3 form an open cover of K. Extract a finite subcover with centers $y_1, y_2, \ldots, y_n$. Let
$$g_x = \max_{1 \leq i \leq n} (g_{x,y_i}).$$
Prove that $g_x(x) = f(x)$ and that for all $k \in K$, $g_x(k) > f(k) - \epsilon$. (It is probably worth your while to draw a picture to help you see what is going on, even if you can prove the result without doing so!)

Step 5. Steps 2, 3, and 4 tell us how to construct a function for each $x \in K$. Now use the fact that each g_x is continuous at x and that $g_x(x) = f(x)$ to find an open cover $\{B_{\delta_x}(x)\}$ of K such that if $t \in B_{\delta_x}(x)$, then $g_x(t) < f(t) + \epsilon$.

Step 6. Extract a finite subcover of the cover from Step 5. Show that the minimum g of the corresponding functions $g_{x_1}, g_{x_2}, g_{x_3}, \ldots, g_{x_n}$ is the function we need to uniformly approximate f on K.

Be sure to check that all of the functions constructed in the preceding procedure were in $\mathcal{A}$! □

The Stone–Weierstrass theorem is truly a remarkable result! It gives us some very powerful corollaries.

Corollary N.3.5 [Weierstrass Approximation Theorem] The set $P[a, b]$ is dense in $C[a, b]$. That is, every continuous function on $[a, b]$ is the uniform limit of a sequence of polynomials.

Proof: The proof is Problem 5 at the end of this section. □

The discussion of uniform approximation by polynomials on a closed, bounded interval might make you think of Taylor polynomials. What is the relationship between the polynomial approximations in the Weierstrass approximation theorem and Taylor polynomial approximations? Actually, there isn't much of a relationship. Certainly, Taylor polynomials are useful for approximating continuous functions, but remember that analytic functions (sums of Taylor series) are *extremely* well behaved functions. For instance, an analytic function has derivatives of all orders. But the Weierstrass approximation theorem deals with approximations of a much more general class of functions. When you prove Lemma N.3.1 (Problem 4 at the end of this section), you will explicitly construct a sequence of polynomials that converge uniformly to $|x|$ on a closed, bounded interval. These will clearly not be Taylor polynomials, as the absolute value function is not differentiable. Moreover, Weierstrass's theorem tells us that even a really strange function like the everywhere continuous, nowhere differentiable function defined in Excursion K is the uniform limit of a sequence of polynomial functions!

It is worth pointing out a few more differences. Each term of a Taylor series is just the previous term plus a new "higher order" term. Furthermore, once the coefficient of x^n is determined (never later than the $(n+1)^{\text{st}}$ term of the series), it never changes. General polynomial approximations will change drastically from one term to the next. For instance, in Problem 4, you will construct a sequence of polynomials whose uniform limit is $|x|$. It will be easy for you to check that the coefficient of x^2 is different for each term of the sequence.

The Stone–Weierstrass theorem gives us more than just polynomial approximations, however. For instance, it also shows that periodic continuous functions can be uniformly approximated everywhere by relatively simple trig functions.

Corollary N.3.6 Let K be the unit circle in $\mathbb{R}^2$.
$$K = \{(x,y) \in \mathbb{R}^2 : x^2 + y^2 = 1\}.$$
Functions in $C(K)$ can be written in the form $f(\theta)$, where θ is the counterclockwise angle on the circle from the positive x axis. Let $\mathcal{F}$ be the set of all finite linear combinations of the simple trigonometric functions
$$1, \cos(\theta), \cos(2\theta), \cos(3\theta), \ldots,$$
$$\sin(\theta), \sin(2\theta), \sin(3\theta), \ldots.$$
Then $\mathcal{F}$ is dense in $C(K)$. That is, every continuous function on the unit circle can be uniformly approximated by a linear combination of these trigonometric functions.

Proof: The proof is Problem 6 at the end of this section. □

Exercise N.3.7 Before stating the Corollary N.3.6, we talked about approximating "periodic" functions. The corollary itself talked about functions on the unit circle. What is the connection between continuous functions on the circle and periodic functions on $\mathbb{R}$? Was the introduction of the unit circle necessary, or could we have gotten the result by just having K be, say, a closed, bounded interval in $\mathbb{R}$? Explain. ∎

Problems N.3

1. Let K be a set and let f and g be real-valued functions on K. Prove that
$$\max(f,g) = \frac{1}{2}(f + g + |f - g|) \quad \text{and} \quad \min(f,g) = \frac{1}{2}(f + g - |f - g|).$$

2. Let K be a compact metric space and let $y \in K$. Suppose that f and g are real valued, continuous functions on K such that $f(y) = g(y)$. Then for every $\epsilon > 0$, there exists $\delta > 0$ such that if $t \in B_y(\delta)$, then $g(t) > f(t) - \epsilon$.

3. Let K be a compact metric space.

   (a) Suppose that (f_n) is a sequence in $C(K)$ that converges pointwise to the constant function $f(x) = 0$. Also, suppose that, for all $x \in K$ and $n \in \mathbb{N}$, $|f_{n+1}(x)| \leq |f_n(x)|$. Then f_n converges uniformly. (*Hint*: Let $\epsilon > 0$. For each $n \in \mathbb{N}$, let
   $$S_n = \{x \in K : |f_n(x)| \geq \epsilon\}.$$
   Note that it is enough to show that one of these sets is empty. So assume it is not so, and prove there is at least one point $t \in K$ such that $|f_n(t)| \geq \epsilon$ for all $n \in \mathbb{N}$.)

   (b) Use the result from part (a) to prove the following more general result. Suppose that (f_n) is a sequence in $C(K)$ that converges pointwise to a continuous function f and for all $x \in K$ and $n \in \mathbb{N}$, $|f_{n+1}(x) - f(x)| \leq |f_n(x) - f(x)|$. Prove that (f_n) converges uniformly.

4. Let $a \in \mathbb{R}$. In this problem you will prove that the absolute value function is the uniform limit of a sequence on polynomials on $[-a, a]$. (Theorem N.3.1.)

   (a) First show that it is sufficient to approximate the absolute value function by polynomials on $[-1, 1]$. To accomplish this, suppose that (P_n) is a sequence of polynomials on $[-1, 1]$ that converges uniformly to $|x|$. Show that
   $$Q_n(x) = |a| P_n\left(\frac{1}{a} x\right)$$
   converges uniformly to $|x|$ on $[-a, a]$.

   (b) Now we outline one possible approach to approximating $f(x) = |x|$ uniformly by polynomials on $[-1, 1]$.[3] Your job is to fill in the details.

   Step 1. First suppose that (P_n) is a sequence of polynomials containing only *even* powers of x and that $P_n \to x$ uniformly on $[0, 1]$. Show that $(P_n) \to |x|$ uniformly on $[-1, 1]$. In other words, it will suffice to approximate $f(x) = x$ uniformly on $[0, 1]$ by polynomials having only even powers of x.

   Step 2. Let $x \in [0, 1]$. Notice that $f(x) = x$ if and only if $f(x) = f(x) - (f(x))^2 + x^2$. That is, $f(x) = x$ is the unique fixed point for the function $F : C[0, 1] \to C[0, 1]$ given by $F(f) = f - f^2 - q$ where q is the quadratic function $q(x) = x^2$ on $[0, 1]$.

---

3. There are "slicker" ways to prove this theorem, but they usually appear to "pull a rabbit out of a hat." I chose a constructive approach to the proof in which the motivation is clear.

Step 3. We will get our sequence of polynomials by iterating F starting with x^2. Assuming we can show such a sequence converges uniformly on $[0,1]$, show that it must converge to x.

Step 4. Let $P_0(x) = x^2$. Define the sequence of polynomials inductively by
$$P_{n+1}(x) = P_n(x) - (P_n(x))^2 + x^2.$$
Show that this is a sequence of polynomials with only even powers of x. Show also that $P_n(0) = 0$ and $P_n(1) = 1$ for all $n \in \mathbb{N}$. So $P_n(0) \to 0$ and $P_n(1) \to 1$.

Step 5. We need bounds on $P_n(x)$. Prove that for all $n \in \mathbb{N}$, and for all $x \in [0,1]$,
$$0 \leq x^2 \leq P_n(x) \leq 1.$$

Step 6. Show that
$$P_{n+1}(x) - x = (1 - P_n(x) - x)(P_n(x) - x).$$
(This identity will help us with the estimates we need to make.)

Step 7. Now show that for all $n \in \mathbb{N}$ and for all $x \in [0,1]$,
$$|1 - P_n(x) - x| \leq \max(x, 1-x).$$

Step 8. Use the identity from Step 6 and the inequality from Step 7. to show that $(P_n(x))$ converges pointwise on $[0,1]$, and that
$$|P_{n+1}(x) - x| \leq |P_n(x) - x|.$$

Step 9. Now apply the result from Problem 3 to deduce that (P_n) converges uniformly to x on $[0,1]$.

5. Let $a, b \in \mathbb{R}$ with $a < b$. Use the Stone–Weierstrass Theorem to prove Weierstrass's theorem that the polynomials are dense in $C[a,b]$ (Corollary N.3.5).

6. Let K be the unit circle in $\mathbb{R}^2$. Use the Stone–Weierstrass Theorem to prove that linear combinations of simple trigonometric functions are dense in $C(K)$ (Corollary N.3.6). (*Hint*: The only (slightly) tricky part is showing that functions in $\mathcal{F}$ separate points. Suppose two distinct points are labelled by angles θ_1 and θ_2. Without loss of generality, you may assume that $\theta_1 < \theta_2 \leq \theta_1 + \pi$. (*Why?*) The function you need is $f(\theta) = \cos(\theta - \theta_1)$, which of course you must show is in $\mathcal{F}$.)

Excursion O

Solutions to Differential Equations

> *Required Background*
> Chapters 6, 7, 10, 11, and 12 and Excursion N.

This excursion explores some of the fundamental results underpinning the theory of ordinary differential equations. In particular, we will be concerned with existence and uniqueness of solutions.

0.1 Definitions and Motivation

Just so we remain "grounded" as we move ahead, let's review some standard terms and ideas in the context of an easy example.

Example O.1.1 Suppose that your parents are being especially nice to you and loan you $7000 to buy a used car. They propose to charge you an interest rate of 5% per year and ask that you pay them $200 per month ($2400 per year). How long will it take you to finish paying your parents back?

This situation can be modeled by the *differential equation*

$$B' = 0.05B - 2400 \text{ dollars/year}$$

where $B(t)$ is the balance due on your loan at time t. The first term $0.05B$ accounts for the fact that the interest on the loan raises the amount that you owe in proportion to the loan balance. The second term takes into account that your payments reduce the amount you owe at a constant rate over time.[1]

A *solution* to the preceding differential equation is any function B that satisfies
$$B'(t) = 0.05B(t) - 2400 \text{ for all } t \geq 0.$$
A simple separation of variables allows us to deduce that, for any real number C, the function
$$B(t) = Ce^{0.05t} + 48,000$$
solves our differential equation. However, only the value $C = -41,000$ gives a solution for which the initial loan amount is $7000. The solution
$$B(t) = -41,000e^{0.05t} + 48,000$$
to our loan problem really depends on both the differential equation and the *initial condition* $B(0) = 7000$. The pair formed by the differential equation and the initial condition is called an *initial value problem*.

(Now we can easily answer the original question. Solving $B(t) = 0$ tells us that it will take you a little less than 3 years and 2 months to pay your parents back.)

In the loan problem, the rate of change of B depended only on B. In general, however, the derivative will depend on both the dependent and the independent variables. For instance, if the interest rate or the payments in the loan problem had varied with time rather than remaining fixed, then B' would have been expressed in terms of both B and t. Differential equations of the form
$$x' = f(t, x)$$
are the focus of this excursion, although we will discuss generalizations to equations involving higher-order derivatives and to systems of first-order equations as well.

Remark: When we write
$$x' = f(t, x)$$
we are saying that the derivative of x depends on t and on x. This dependency relation may not be defined for all values of t and x. Even if $f(t, x)$ is defined on all of $\mathbb{R}^2$, there are situations where the initial value problem
$$x' = f(t, x) \quad \text{and} \quad x(t_0) = x_0$$

---

1. Modeling with differential equations almost always involves simplifying assumptions. Loans are paid back in discrete monthly chunks, not in a continuous stream. Here we are assuming that your parents will be paid back by a continuous "leaking" of money from your account to theirs. This approximation is not exact, but it's not bad and is frequently used in real loan calculations.

may have a solution only on some finite interval containing t_0 and not on the whole real line. The domain restrictions given in Definition O.1.2 account for these more general situations.

Definition O.1.2 Let U be a subset of $\mathbb{R}^2$ and $f : U \to \mathbb{R}$ be a function. Suppose that I and J are real intervals with $I \times J \subseteq U$.

A differentiable function $x : I \to J$ is said to **solve** the **differential equation** $x' = f(t, x)$ on I provided that $x'(t) = f(t, x(t))$ for all $t \in I$.

If t_0 and x_0 are real numbers with $t_0 \in I$ and $x_0 \in J$, the function x is a solution to the **initial value problem (IVP)**

$$x' = f(t, x) \quad \text{and} \quad x(t_0) = x_0$$

on I provided that it solves the differential equation $x' = f(t, x)$ on I and that it also satisfies the **initial condition** $x(t_0) = x_0$.

Why Existence? Why Uniqueness?

One of my favorite books, *Differential Equations: A Dynamical Systems Approach* by J. H. Hubbard and B. H. West, says

> Chapters on "existence and uniqueness" are usually viewed as absolutely essential and central by mathematicians, and absolutely useless or meaningless by everyone else. [H&W, p.157]

Hubbard and West discuss in detail why anyone studying differential equations needs to be concerned about existence and uniqueness theorems. Given the limited goals of this excursion, I can say much less here. But I do want to give you some idea why existence and uniqueness theorems should be of interest even to the most applied of mathematicians.

As an anti-example, consider the loan problem discussed in Example O.1.1. The circumstances of the loan are such that we know (1) that there will always be a loan balance and (2) that, at any given instant in time, there can be only *one possible* loan balance. That is, our number sense tells us that there *is* a solution to that differential equation and that there is *only* one solution. So, if a mathematician proudly displays some high-powered, theoretical result that guarantees the existence and uniqueness of a solution to the loan problem, people have every reason to be unimpressed.

But suppose we are modeling a more complicated situation for which we don't really understand the underlying dynamics. As with many differential equations, it may be difficult or impossible to solve this problem analytically. Then we turn to numerical approximation techniques. If we set a numerical procedure to work on a problem, we need to know that a solution exists for it to find; that is, we need an existence theorem.

But what about uniqueness of solutions? The study of differential equations is often a way of predicting the future. If, for instance, we want to test a proposed forest management plan, we can devise a differential equation or (more likely) a set of linked differential equations that will model the interactions of replanting, harvesting, natural death and birth of trees, and so forth. Then we set a computer loose on the problem to determine whether our plan will sustain the forest in the long run. Suppose that our numerical technique cranks away and gives us a solution that tells us that the forest will always stay within acceptable parameters. If we have a theorem that says the solution is unique, then we know that our plan is viable. But if more than one solution were possible, our numerical technique might predict an "acceptable" outcome whereas nature might find an alternative, possibly "unacceptable," outcome.

More sophisticated existence and uniqueness theorems than those we will consider here can help us detect situations in which the error in a numerical approximation method is large enough to suggest an erroneous "solution." (I recommend [H&W], with its lively and pragmatic approach, for further study.)

O.2 Picard Iteration Route to Existence and Uniqueness

We are interested in the existence and uniqueness of solutions to the IVP

$$x' = f(t,x) \quad \text{with initial condition} \quad x(t_0) = x_0. \tag{O.1}$$

We will begin by concentrating on the important special case where r is a positive real number, $I = [t_0 - r, t_0 + r]$, and the domain U of f is the "infinite vertical strip"

$$U = \{(t,x) \in \mathbb{R}^2 : t \in I\}.$$

This means that we are considering solutions in a closed neighborhood of t and not restricting the range of the solutions at all. You will consider a more general situation in the problems.

First, we change our point of view from the differential formulation to an equivalent integral formulation of the IVP.

Exercise O.2.1 Let $t_0 \in \mathbb{R}$. Suppose that r is a positive real number, that $I = [t_0 - r, t_0 + r]$, and that $U = \{(t,x) \in \mathbb{R}^2 : t \in I\}$. Let $f : U \to \mathbb{R}$ be a continuous function. Prove that $x : I \to \mathbb{R}$ is a solution to IVP O.1 if and only if it is also a solution to the integral equation

$$x(t) = x_0 + \int_{t_0}^{t} f(u, x(u)) \, du. \tag{O.2}$$

■

By converting IVP O.1 into integral equation O.2, we have managed to do two things at once. First, both the differential equation and the initial condition appear in a single equation. Second, the integral form of the IVP is ideally set up to use fixed point theory.

To see why, we look closely at integral equation O.2. Because f is fixed ahead of time, the left side is $x(t)$ and the right side is an expression depending only on x and t. This equation is of the form $x = F(x)$, where x is a function on I.

We want to define a function $F : C(I) \to C(I)$ by

$$F(y)(t) = x_0 + \int_{t_0}^{t} f(u, y(u)) du.$$

Notice that x is a solution to integral equation O.2 (and thus to IVP O.1) if and only if it is a fixed point of the map F. Since $C(I)$ is a complete space, we might hope to use the contraction mapping theorem to find a unique fixed point. This is the idea of Picard iteration.

Exercise O.2.2 Let $r > 0$, let $t_0 \in \mathbb{R}$, and let $I = [t_0 - r, t_0 + r]$. Let U be the "vertical strip"

$$U = \{(t, x) \in \mathbb{R}^2 : t \in I\}.$$

Suppose that $f : U \to \mathbb{R}$ is a continuous function. Prove that the function $F : C(I) \to C(I)$ given by

$$F(y)(t) = x_0 + \int_{t_0}^{t} f(u, y(u)) \, du$$

is well defined. (*Recall*: F will be well defined provided that given any $y \in C(I)$, $F(y)$ exists, $F(y) \in C(I)$, and only one such function is possible.) ∎

Definition O.2.3 Suppose that $U \subseteq \mathbb{R}^2$ and that $f : U \to \mathbb{R}$ is a function. We say that f satisfies a **uniform Lipschitz condition in the second variable** provided that there exists $L \in \mathbb{R}$ such that

$$|f(t, x) - f(t, y)| \le L|x - y| \quad \text{for all} \quad (t, x) \text{ and } (t, y) \in U.$$

In this case, we say that f satisfies a uniform Lipschitz condition in the second variable **with constant L**.

Remark: Throughout this excursion, we will be interested only in Lipschitz conditions in the second variable that have a uniform constant over the first variable. Therefore, we will use the phrase "Lipschitz condition with constant L" to mean "uniform Lipschitz condition in the second variable with constant L."

Theorem O.2.4 [Picard–Lindelöf Theorem] Let $r > 0$, and let $t_0 \in \mathbb{R}$. Let U be the subset of $\mathbb{R}^2$ given by

$$U = \{(t, x) \in \mathbb{R}^2 : t_0 - r \leq t \leq t_0 + r\}.$$

Suppose that $f : U \to \mathbb{R}$ is a continuous function that satisfies a Lipschitz condition with constant L. Then the initial value problem $x' = f(t, x)$, with initial condition $x(t_0) = x_0$, has a unique solution in the interval $I = [t_0 - r, t_0 + r]$.

Proof: The proof is Problem 2 at the end of this section. □

Corollary O.2.5 Let $r > 0$, and let $t_0 \in \mathbb{R}$. Let U be the subset of $\mathbb{R}^2$ given by

$$U = \{(t, x) \in \mathbb{R}^2 : t_0 - r \leq t \leq t_0 + r\}.$$

Suppose that $f : U \to \mathbb{R}$ is a continuous function and $\frac{\partial f}{\partial x}$ exists and is bounded in U. Then the initial value problem $x' = f(t, x)$, with initial condition $x(t_0) = x_0$, has a unique solution in the interval $I = [t_0 - r, t_0 + r]$.

Proof: The proof is Problem 3 at the end of this section. □

Problems O.2

1. Let $U \subseteq \mathbb{R}^2$, and let $f : U \to \mathbb{R}$ be a function in the variables (t, x). Suppose that $\frac{\partial f}{\partial x}$ exists and is bounded in U. Prove that f satisfies a uniform Lipschitz condition in the second variable.

2. In this problem you will prove the Picard–Lindelöf theorem (Theorem O.2.4). Let $r > 0$, and let $t_0 \in \mathbb{R}$. Let U be the subset of $\mathbb{R}^2$ given by $U = \{(t, x) \in \mathbb{R}^2 : t_0 - r \leq t \leq t_0 + r\}$. Suppose that $f : U \to \mathbb{R}$ is a continuous function that satisfies a Lipschitz condition with constant L. By Exercise O.2.2, we can define a function $F : C(I) \to C(I)$ by

$$F(y)(t) = x_0 + \int_{t_0}^{t} f(u(y), u)\, du.$$

   (a) Prove by induction that for all $n \in \mathbb{N}$, for all $y, z \in C(I)$, and for all $t \in I$,

$$|F^n(y)(t) - F^n(z)(t)| \leq \frac{L^n |t - t_0|^n}{n!} d(y, z).$$

   Note that $d(y, z)$ on the right measures the distance between the functions y and z in $C(I)$.

(b) Use the result from part (a) to prove that some iterate of F is a contraction on $C(I)$.

(c) Use the result from part (b) to finish the proof of the Picard–Lindelöf theorem.

3. Prove the variant of the Picard–Lindelöf theorem in which a Lipschitz condition is replaced by the existence and boundedness of a partial derivative (Corollary O.2.5).

4. Prove the following variant of Theorem O.2.4: Let $r > 0$, $m > 0$, and $(t_0, x_0) \in \mathbb{R}^2$. Suppose that U is the subset of $\mathbb{R}^2$ given by

$$U = \{(t, x) \in \mathbb{R}^2 : t_0 - r \leq t \leq t_0 + r \text{ and } x_0 - m \leq x \leq x_0 + m\}.$$

Suppose further that $f : U \to \mathbb{R}$ is a continuous function that satisfies a Lipschitz condition with constant L. Let $M = \max\{f(t, x) : (t, x) \in U\}$ and set

$$\alpha = \min\left\{r, \frac{m}{M}\right\}.$$

Then the initial value problem $x' = f(t, x)$, with initial condition $x(t_0) = x_0$, has a unique solution in the interval $I = [t_0 - \alpha, t_0 + \alpha]$.

5. In this problem you will generalize the theorem given in Problem 4.

Let U be an open subset of $\mathbb{R}^2$ containing the point (t_0, x_0). Suppose that $f : U \to \mathbb{R}$ is a continuous function that satisfies a Lipschitz condition with constant L. Prove that there exists $\alpha > 0$ such that $x' = f(t, x)$, with initial condition $x(t_0) = x_0$, has a unique solution in the interval $I = [t_0 - \alpha, t_0 + \alpha]$.

O.3 Systems of Equations

Suppose that we have the following third-order differential equation:

$$\frac{d^3 x}{dt^3} + 4t\frac{d^2 x}{dt^2} - 2\frac{dx}{dt} + t^2 x = 7t$$

with initial values

$$x(0) = 1, \quad x'(0) = -2, \quad x''(0) = 2.$$

A simple substitution makes it possible to repace this third-order IVP by an equivalent system of three differential equations.

To see this, define two new variables y and z by

$$y = \frac{dx}{dt} \quad \text{and} \quad z = \frac{dy}{dt} = \frac{d^2x}{dt^2}$$

It follows that

$$\frac{dz}{dt} = \frac{d^3x}{dt^3}$$

So we get a *coupled system* of differential equations.[2]

$$\frac{dx}{dt} = y \qquad\qquad x(0) = 1$$
$$\frac{dy}{dt} = z \qquad\qquad y(0) = -2$$
$$\frac{dz}{dt} = 7t - 4tz + 2y - t^2 x \qquad\qquad z(0) = 2$$

We can rewrite this system more compactly if we use vector notation:

$$\mathbf{v}' = \mathbf{F}(t, \mathbf{v}) \quad \text{with initial condition} \quad \mathbf{v}(0) = (1, -2, 2)$$

where $\mathbf{v} = (x, y, z)$ and $\mathbf{F} = (y, z, 7t - 4tz + 2y - t^2 x)$.

Note that there is nothing special about our third-order example. The same trick can be used on any higher order differential equation of the form

$$\frac{d^n x_1}{dt^n} + g_1(t) \frac{d^{(n-1)} x_1}{dt^{(n-1)}} + g_2(t) \frac{d^{(n-2)} x_1}{dt^{(n-2)}} + \cdots + g_n(t) x_1 = h(t) \qquad (O.3)$$

Exercise O.3.1 Write down the system of differential equations that is equivalent to Equation O.3. ∎

We can immediately see, then, that existence and uniqueness conditions for a system of differential equations will also give us existence and uniqueness conditions for a single higher order equation.

Question to Ponder: We have shown that every differential equation of order n can be rewritten as a system of n coupled differential equations. On the face of it, the study of systems of equations appears to be more general than the study of nth-order differential equations. But is this really true? You might ask

---

2. The fact that the rates of change of one variable depend on the values of one or more of the other variables corresponds to a sort of "feedback loop" in the system. Hence it is impossible to solve one of the equations without solving them all—the differential equations are thus said to be "coupled."

Excursion O ■ Solutions to Differential Equations

yourself whether, given any system of n coupled differential equations, there is *some function* g and some nth-order equation in g that is equivalent to the system. The answer to this question is not as obvious as it might seem at first glance.

We seek an existence and uniqueness theorem for IVP systems of the form

$$\mathbf{x}' = \mathbf{F}(t, \mathbf{x}) \quad \text{with initial condition} \quad \mathbf{x}(0) = (x_1(0), x_2(0), \ldots, x_n(0)).$$

We begin by generalizing the Picard iteration approach that we used in Section O.2. As a solution will be an n-tuple of functions on some interval I, we will use iteration to seek the fixed point of a function on the set of all n-tuples of functions. First, we need a framework for this scheme.

Lemma O.3.2 Let K be a compact metric space, and let $n \in \mathbb{N}$. If $C(K)$ is the set of all continuous, real-valued functions on K, we define a distance function on the set

$$C^n(K) = \underbrace{C(K) \times C(K) \times \cdots \times C(K)}_{n\text{-times}}$$

as follows:

$$d((x_1, x_2, \ldots, x_n), (y_1, y_2, \ldots, y_n)) = \sum_{i=1}^{n} d(x_i, y_i)$$

where each of the summands is a distance measured in $C(K)$. Then $C^n(K)$ is a complete metric space under the metric d.

Proof: The proof is Problem 1 at the end of this section. □

Definition O.3.3 Let $U \subseteq \mathbb{R} \times \mathbb{R}^n$, and let $\mathbf{f} : D \to \mathbb{R}^n$ be a function. The function $\mathbf{f}$ is said to satisfy a **uniform Lipschitz condition with respect to the second variable** provided that there exists a real number L such that for all $(t, \mathbf{x})$ and $(t, \mathbf{y}) \in D$,

$$\|\mathbf{f}(t, \mathbf{x}) - \mathbf{f}(t, \mathbf{y})\| \leq L \|\mathbf{x} - \mathbf{y}\|.$$

As in Section O.2, we will consider only uniform Lipschitz conditions in the second variable. Thus, if $\mathbf{f}$ satisfies such a condition, we will say that it satisfies a Lipschitz condition with constant L.

Lemma O.3.4 Let $a, b \in \mathbb{R}$. Suppose that $g_1 : [a, b] \to \mathbb{R}$, $g_2 : [a, b] \to \mathbb{R}$, ..., $g_n : [a, b] \to \mathbb{R}$ and $h : [a, b] \to \mathbb{R}$ are continuous functions. Let $\mathbf{x}' = \mathbf{f}(t, \mathbf{x})$ be the system of differential equations given by the nth order differential equation

$$\frac{d^n x_1}{dt^n} + g_1(t)\frac{d^{(n-1)}x_1}{dt^{(n-1)}} + g_2(t)\frac{d^{(n-2)}x_1}{dt^{(n-2)}} + \cdots + g_n(t)x_1 = h(t).$$

Then $\mathbf{f}$ satisfies a Lipschitz condition with constant L.

Proof: The proof is Problem 2 at the end of this section. □

Theorem O.3.5 Let $t_0 \in \mathbb{R}$, and let $\mathbf{x}_0 \in \mathbb{R}^n$. Fix a positive real number r. Let U be the set $U = \{(t, \mathbf{x}) : |t - t_0| \leq r\}$. Assume that $\mathbf{f} : U \to \mathbb{R}^n$ is a continuous function, and assume that $\mathbf{f}$ satisfies a Lipschitz condition with constant L. Then the initial value problem

$$\mathbf{x}' = \mathbf{f}(t, \mathbf{x}) \quad \text{and} \quad \mathbf{x}(t_0) = \mathbf{x}_0$$

has a unique solution in $[t_0 - r, t_0 + r]$.

Proof: The proof is Problem 3 at the end of this section. □

Corollary O.3.6 Suppose that $a \leq t_0 \leq b$ and that $g_1 : [a, b] \to \mathbb{R}$, $g_2 : [a, b] \to \mathbb{R}$, ..., $g_n : [a, b] \to \mathbb{R}$ and $h : [a, b] \to \mathbb{R}$ are continuous functions. Then the nth order initial value problem

$$\frac{d^n x_1}{dt^n} + g_1(t)\frac{d^{(n-1)}x_1}{dt^{(n-1)}} + g_2(t)\frac{d^{(n-2)}x_1}{dt^{(n-2)}} + \cdots + g_n(t)x_1 = h(t)$$

with initial conditions

$$x_1(t_0) = \alpha_0, \quad x_1'(t_0) = \alpha_1, \quad \ldots, \quad x_1^{(n)}(t_0) = \alpha_n$$

has a unique solution in $[a, b]$.

Proof: The proof is Problem 4 at the end of this section. □

Problems O.3

1. Prove Lemma O.3.2.

2. Prove Lemma O.3.4.

3. Prove Theorem O.3.5.

4. Prove Corollary O.3.6.

5. Prove the following variant of Theorem O.3.5: Let $t_0 \in \mathbb{R}$, and let $\mathbf{x}_0 \in \mathbb{R}^n$. Fix positive real numbers r and m. Let U be the compact set $U = \{(t, \mathbf{x}) : |t - t_0| \leq r \text{ and } \|\mathbf{x} - \mathbf{x}_0\| < m\}$. Assume that $\mathbf{f} : U \to \mathbb{R}^n$ is a continuous function and assume that $\mathbf{f}$ satisfies a Lipschitz condition with constant L. Let $M = \max\{\|\mathbf{f}(t, \mathbf{x})\| : (t, \mathbf{x}) \in U\}$ and $\alpha = \min\{r, \frac{m}{M}\}$.

   Then the initial value problem $\mathbf{x}' = \mathbf{f}(t, \mathbf{x})$ and $\mathbf{x}(t_0) = \mathbf{x}_0$ has a unique solution in $[t_0 - \alpha, t_0 + \alpha]$.

Bibliography

[AP1] Apostol, Tom M. *Calculus, vol. I: one-variable calculus, with an introduction to linear algebra,* 2nd ed. John Wiley & Sons, New York: 1967.

[AP2] Apostol, Tom M. *Calculus, vol. II: multi-variable calculus and linear algebra, with applications to differential equations and probability,* 2nd ed. John Wiley & Sons, New York: 1969.

[AP3] Apostol, Tom M. *Mathematical analysis,* 2nd ed. Addison-Wesley, Reading, MA: 1974.

[AN] Anton, Howard. *Elementary linear algebra,* 7th ed. John Wiley & Sons, New York, 1994.

[BAR] Bartle, Robert G. *The elements of real analysis,* 2nd ed. John Wiley & Sons, New York: 1976.

[B&S] Bartle, Robert G., and Donald R. Sherbert. *Introduction to real analysis,* 2nd ed. John Wiley & Sons, New York: 1992.

[BRY1] Bryant, Victor. *Metric spaces: iteration and application.* Cambridge University Press, Cambridge, UK: 1985.

[BRY2] Bryant, Victor. *Yet another introduction to real analysis.* Cambridge University Press, Cambridge, UK: 1990.

[BUC] Buck, R. Creighton. *Advanced calculus,* 2nd ed. McGraw-Hill, New York: 1965.

[C&S] Corwin, Lawrence J., and Robert H. Szczarba. *Calculus in vector spaces.* Marcel Dekker, New York: 1979.

[COUR1] Courant, R. *Differential and integral calculus, vol. I,* 2nd ed., translated by E. J. McShane. Wiley Classics Library Edition, Wiley-Interscience, 1988.

[COUR2] Courant, R. *Differential and integral calculus, vol. II,* translated by E. J. McShane. Wiley Classics Library Edition, Wiley-Interscience, 1988.

Bibliography

[GAU] Gaughan, Edward D. *Introduction to analysis,* 4th ed. Brooks/Cole, Pacific Grove, CA: 1993.

[H&J] Hrbacek, Karel, and Thomas Jech. *Introduction to set theory,* 3rd ed. Marcel Dekker, New York, 1999.

[HOL] Holmgren, Richard A. *A first course in discrete dynamical systems,* 2nd ed. Springer-Verlag, New York: 2000.

[H&W] Hubbard, J. H., and B. H. West. *Differential equations: a dynamical systems approach—ordinary differential equations.* Springer-Verlag, New York: 1991.

[WK&AP] Kelley, Walter, and Allan Peterson. *The theory of differential equations classical and qualitative.* Pearson Education, Upper Saddle River, NJ: 2004.

[SK&HP] Kranz, Steven G., and Harold R. Parks. *The implicit function theorem: history, theory, and applications.* Birkhäuser, Boston: 2002.

[McL] McLeod, Robert M. *The generalized Riemann integral.* Carus Mathematical Monographs, No. 20. Mathematical Association of America, 1980.

[ROS] Rosenlicht, Maxwell. *Introduction Analysis.* Scott, Foresman, Glenview, IL: 1968. Reprinted by Dover Publications, Mineola, NY: 1986.

[RUD] Rudin, Walter. *Principles of mathematical analysis,* 3rd ed., McGraw-Hill, New York: 1976.

[SCH] Schumacher, Carol. *Chapter zero: fundamental notions of abstract mathematics,* 2nd ed. Addison-Wesley, Boston, 2001.

[W&M] Wilcox, Howard J., and David L. Myers. *An introduction to Lebesgue integration and Fourier series.* Dover Publications, New York: 1995.

Index

(X, d), 64
$A \times B$, 18
f^{-1}, 22
n^{th} root, 301
n^{th} term test for divergence, 337
p series, 338
δ-separated, 95, 97, 141
$\lim_{n \to \infty} c^{\frac{1}{n}} = 1$, 318
$\lim_{n \to \infty} n^{\frac{1}{n}} = 1$, 317
ℓ_∞, 71, 78, 134, 142
$\emptyset$, 11
$\in$, 9
$\mathbb{Z}$, 9
$\leq$, 44
$\lim_{x \to c^+} f(x)$, 125
$\lim_{x \to c^-} f(x)$, 125
$\mathbb{N}$, 9
$\mathbb{Q}$, 10
$\mathbb{R}$, 10, 37–59
    sequence convergence in, 87–93, 307–318
$\mathbb{R}^n$, 65–67
    cells in, 80
    sequence convergence in, 307–318
    vector space, 262
$\setminus$, 16
$\subseteq$, 11
$\sup(K)$, 52
D-domain, 161

Absolute convergence, 339
Absolute value, 45
Abstraction, 5
Addition, 38
Affine transformation, 262
Alternating harmonic series, 339
Alternating series, 339
Antiderivative, 174
Antisymmetric, 44
Archimedean property, 55
Associative, 28, 38
    composition of functions, 21
Attractors, 195
    in $\mathbb{R}$, 195
Axioms for $\mathbb{R}$, 37–54
    field axioms, 38
    least upper bound axiom, 54
    order axiom, 43

Base, 141
Bernoulli's inequality, 297, 317
Between, 151
Bijection, 20
Binary operation, 28, 38, 39, 41
    associative, 28
    commutative, 28
Boundary, 101
Bounded
    metric space, 77
Bounded function, 116
Bounded sequence, 86
Boundedness
    in $\mathbb{R}$, 51, 77
    in metric spaces, 76–77

Cartesian product, 18

Cauchy criteria
　　Riemann integral, 222–228
　　series, 336
Cauchy Mean Value Theorem, 176
Cauchy Sequences, 131
Cauchy-Schwarz inequality, 67
Cells, 80
Chain rule, 164
　　several variables, 283
Characteristic function, 214
　　rationals, 214, 360
Clopen sets, 155
Closed ball, 73
Closed set, 96
Closure, 99
Codomain, 19
Commutative, 28, 38
Compactness, 135–149
　　and boundedness, 138
　　and closure, 139
　　and completeness, 140
　　and continuity, 143–145
　　and convergence, 139
　　and optimization, 143
　　in $\mathbb{R}^n$, 145–149
　　in $\mathbb{R}$, 139
　　in $C(K)$, 412–414
　　of closed, bounded interval, 137
Comparison test, 342
　　limit version, 342
Completeness, 131–134
　　of $\mathbb{R}$, 133
　　of $\mathbb{R}^n$, 133
　　and compactness, 140
Composition of functions, 20
Conditional convergence, 339
Connectedness, 151–156
Continuity, 105–114
　　of polynomials, 129
　　and compactness, 143–145
　　and inverses, 113, 145, 153
　　essential discontinuity, 125
　　jump discontinuity, 125
　　of rational functions, 129

Continuous Functions, 110–114
Contraction Mapping Theorem, 198–207
Contractions, 199
　　and differentiability, 202
Contracts Distances, 198
Convergence of sequences, 81–86
　　definition, 82
　　in $\mathbb{R}$ and $\mathbb{R}^n$, 307–318
Convex set, 79
Coordinate
　　first and second, 18
Coordinate functions, 259
Countable, 57
Countable base, 141
Countable dense subset, 141
Countably compact, 141

Decreasing function, 178
DeMorgan Laws, 15
Dense, 95, 141
Derivative
　　at an endpoint, 167
　　one variable, 161
　　several variables, 272
Diameter, 76
Difference
　　sets, 16
Difference quotient, 161
Differentiability
　　and Lipschitz condition, 176
　　one variable, 161
　　parametric curve, 285
　　several variables, 272
Differential equations, 421–430
　　higher order, 427–430
　　Picard iteration, 424, 429
　　systems of, 429
Differentiation
　　and algebra, 164, 169
　　and continuity, 163, 169, 273, 278
　　and monotonicity, 177–180
　　and vector algebra, 282

and zooming in, 163
  at an endpoint, 167
  one variable, 159–188
  several variables, 259–287
Discrete metric space, 79
Distance function, 63, 64
Distributive property, 38
Division, 39
Domain, 19
  D-domain, 161
  and differentiation, 159–161
Dot product, 67, 264

Earth
  what on, 395
Element, 9
Element argument, 11
Empty set, 11
Equality of functions, 20
Equality of sets, 12
Essential discontinuity, 125
Euclidean distance or metric, 65
Euclidean metric
  triangle inequality, 66–67
Extreme point, 173

False advertising, 124
Field, 38
Field axioms, 38, 296
Finite diameter, 76
Function, 18
Function value, 19
Fundamental Theorem of Calculus, 235–236

Generalized IVP, 156
Geometric Series, 336
Global extreme point, 173
Global maxima and minima, 173
Gradient, 278, 283
Greatest lower bound, 53, 54

Harmonic series, 337
  alternating, 339
Hausdorff space, 78

Heine-Borel theorem, 145–149

Identity
  additive, 38
  multiplicative, 38
Images, 24
  and set algebra, 24
Immediate predecessors, 29
Immediate successors, 29
Implicit Function, 391
Implicit Function Theorem, 390–402
  and Inverse Function Thm, 406–407
  and Newton's method, 402–406
  statement of, 394
Increasing function, 178
Indexing set, 12
  arbitrary or general, 14
Initial value problem, 423
Integers, 9, 48
Integral
  and arithmetic, 217
  existence of, 229–232
  linearity of, 217
  monotonicity of, 218
Integration, 209–236
Interchanging limits, 248–255
Interior, 100
Intermediate value property, 126, 156, 179
Intermediate value theorem, 151–156
  Generalized, 156
Intersection, 15
Interval notation, 11
Inverse
  additive, 38
    uniqueness of, 39
  multiplicative, 38
    uniqueness of, 39
Inverse function, 22
Inverse Function Theorem, 406–407
Inverse images, 22
  and set algebra, 23
Invertible, 22

Irrational numbers, 10, 48, 57, 296
Isolated point, 79
Iteration, 189–198
    and Contraction Mapping Thm, 198–207
    and fixed points, 189–195
IVT, 151

Jacobian matrix, 277
Jump discontinuity, 125

Least upper bound, 52
Least upper bound axiom, 54
Least upper bound property, 54
Least upper bounds, 52–54
Left-handed derivative, 167
Limit comparison test, 342
Limit point, 94
Linear Algebra, 262–267
Linear transformation
    and one to one, 265
Lipschitz condition, 425
Lipschitz function, 115, 116, 129, 198
    and the derivative, 176
Local extrema, 173
Local extreme point, 173
Local Extreme Theorem, 173
Local maxima and minima, 173
Lower sum, 220

Maclaurin Polynomial, 186
Maps, 19
$\max\{x, y\}$, 45
Max-Min theorem, 143
Maximum, 143, 173
Mean Value Theorem, 171–174
    Cauchy form, 176
    and antidifferentiation, 174
    and monotonicity, 178
    several variables, 287–290
Mesh (of a partition), 211
Method of exhaustion, 210
Metric, 64
    Euclidean metric for $\mathbb{R}^n$, 65
    positive, 64
    Positive Definite, 64
    symmetric, 64
    triangle inequality, 64
Metric space, 64
    bounded subset, 77
    closed ball, 73
    closed set, 96
    closure, 99
    diameter of, 76
    distance function, 64
    open ball, 73
    open set, 73
    subspace of, 64
Minimum, 143, 173
Mixed partials, 286
Monotonic, 126, 178
    strictly, 153
Multiplication, 38
Multiplying series, 354
Multivariable calculus, 257–290

Natural numbers, 9, 295
Negative numbers, 43
Nested interval theorem, 57, 142
Newton function, 383
Newton's method, 380–389
    and Implicit Function Thm, 402–406
Norm, 267
    linear transformation, 268
Normal space, 116

One-to-one, 20
One-to-one correspondence, 20
Onto, 20
Open ball, 73, 74
Open cover, 136
Open sets, 73–81
    intersection of, 75
    relatively open, 79
    union of, 75
Order axiom, 43, 296
Ordered pair, 18

Orderings
  $\leq$ on $\mathbb{R}$, 44–45

Parametric curve, 257
  derivative of, 285
Partial derivatives, 276
Partially Ordered Set, 44
Partition (in integration), 211
Perfect sets, 97, 102
Period-n point, 198
Picard iteration, 424, 429
Point, 64
Pointwise convergence, 237–239
Polynomial approximations, 414–418
Positive (metric), 64
Positive definite, 64
Positive numbers, 42, 43
Power series, 363–373
  continuity of, 368
  convergence of, 364–367
  definition, 364
  differentiation of, 368–369
  integration of, 369
Product rule
  several variables, 283
  one-variable, 164

Quotient rule, 169

Range, 19
Range of a sequence, 31
Ratio test, 345
Rational numbers, 10, 48, 296
Real Analysis, 3
Real numbers, 10, 37–59
  absolute value, 45
  Archimedean property, 55
  field axioms, 38
  infinitude of, 40, 42, 47
  least upper bound axiom, 54
  negative numbers, 43
  open balls in, 78, 81
  order axiom, 43
  ordering $\leq$, 44

positive numbers, 43
sequences in, 307–318
Refinement of a partition, 222
Reflexive, 44
Relation, 12
Relatively open, 79
Removable discontinuity, 126
Repellors, 195
Restriction, 21
Riemann integrable, 211
  Cauchy criteria, 225, 227
  counterexample, 214
  definition, 212, 216, 229
Riemann integral, 209–236
  and arithmetic, 217
  and continuity, 229
  and monotonicity, 229
  examples, 212, 216
  existence of, 229–232
  linearity of, 217
  monotonicity of, 218
  properties of, 217–219
  uniqueness of, 212
Riemann sum, 211
Right-handed derivative, 167
Rolle's theorem, 173
Root test, 346

Sampling points, 211
Scalar field, 257
Scalar multiplication, 262
Second order partials, 286
Sequence rearrangement, 349, 353
Sequences, 30
  n^{th} term, 30
  bounded, 32, 86
  constant, 31
  convergence, 81–86
  convergence and arithmetic, 89–93
  convergence and order, 87–89
  decreasing, 31
  in $\mathbb{R}$, 87–93
  in $\mathbb{R}$ and $\mathbb{R}^n$, 307–318

increasing, 31
monotonic, 31
notation for, 31
of distinct terms, 31
subsequence convergence, 86
subsequences, 33
uniqueness of limit, 85
Sequences of functions, 237–246
and differentiation, 251–253
and integration, 249–251
pointwise convergence, 237
uniform convergence, 240
Series, 334–357
n^{th} term test for divergence, 337
p-series, 338
absolute convergence, 339
alternating, 339
alternating harmonic, 339
Cauchy criterion, 336
comparison test, 342
conditional convergence, 339
geometric series, 336
harmonic, 337
limit comparison test, 342
product of, 354
ratio test, 345
root test, 346
Set difference, 16
Set function, 210
Sets, 9
notation, 11
Square roots, 40, 42, 49, 52, 55
Standard basis, 264
Stone–Weierstrass theorem, 414–418
Strictly monotonic, 153
Subcover, 136
Subinterval (in integration) , 211
Subsequences, 33

Subset, 11
proper, 11
Subspace, 64
open sets in, 79
Subtraction, 39
Super attracting fixed point, 197
Symmetric (metric), 64

Taxicab metric, 69
Taylor Polynomials, 185
Taylor series, 370–373
Taylor's Theorem, 187
Total Derivative, 272
Totally ordered set, 12
Transitive, 44
Triangle inequality, 64
generalized, 68
Trichotomy, law of, 44
Trivial metric, 64

Uncountable, 58
Uniform boundedness, 244
Uniform continuity, 115–117, 129
Uniform convergence, 240–241
Uniformly Cauchy, 242
Union, 14
Unit vector, 267
Unit vector basis, 264
Upper sum, 220

Vector addition, 262
Vector field, 257
coordinate functions, 259

Weierstrass approximation theorem, 414–418
Well definedness, 39

Zooming in, 162